# GIS
# A Computing Perspective
## Second Edition

# GIS
## A Computing Perspective
### Second Edition

**Michael Worboys**
University of Maine
USA

**Matt Duckham**
University of Melbourne
Australia

**CRC PRESS**

Boca Raton   London   New York   Washington, D.C.

### Library of Congress Cataloging-in-Publication Data

Worboys, Michael.
    GIS : a computing perspective / Michael F. Worboys and Matt Duckham.—2nd ed.
        p. cm.
    Includes bibliographical references and index.
    ISBN 0-415-28375-2 (alk. paper)
        1. Geographic information systems. I. Duckham, Matt, 1972- II. Title.

G70.2.W659 2004
910'.285—dc22                                                               2004045490

### Visit the CRC Press Web site at www.crcpress.com

© 2004 by CRC Press LLC

No claim to original U.S. Government works
International Standard Book Number 0-415-28375-2
Library of Congress Card Number 2004045490
Printed in the United States of America  2  3  4  5  6  7  8  9  0
Printed on acid-free paper

To Helen,
and to my parents, Rita and John

MD

To Moira
and to Bethan, Colin, Ruth, Sophie, Tom, and William

MW

# Preface

Geographic information systems (GISs) are computer-based information systems that are used to capture, model, store, retrieve, share, manipulate, analyze, and present geographically referenced data. This book is about the technology, theories, models, and representations that surround geographic information and GISs. This study (itself often referred to as GIS or *geographic information science*) has emerged in the last two decades as an exciting multi-disciplinary endeavor, spanning such areas as geography, cartography, remote sensing, image processing, environmental sciences, and computing science. The treatment in this text is unashamedly biased toward the computational aspects of GIS. Within computing science, GIS is a special interest of fields such as databases, graphics, systems engineering, and computational geometry, being not only a challenging application area, but also providing foundational questions for these disciplines.

The underlying question facing this multidisciplinary topic is "What is special about spatial information?" In this book, we attempt to provide answers at several different levels: the conceptual and formal models needed to understand spatial information; the representations and data structures needed to support adequate performance in GISs; the special-purpose interfaces and architectures required to interact with and share spatial information; and the importance of uncertainty and time in spatial information.

The task of computing practitioners in the field of GIS is to provide the application experts, whether geographers, planners, utility engineers, or environmental scientists, with a set of tools, based around digital computer technology, that will aid them in solving problems in their domains. These tools will include modeling constructs, data structures that will allow efficient storage and retrieval of data, and generic interfaces that may be customized for particular application domains.

The book inevitably reflects the interests and biases of its authors, in particular emphasizing spatial information modeling and representation, as well as developing some of the more formal themes useful in understanding GIS. We have tried to avoid detailed discussion of particular currently fashionable systems, and concentrate instead upon the foundations and general principles of the subject area. We have also tried to give an overview of the field from the perspective of computing science.

Not every topic can be covered and we have deliberately neglected two areas, leaving these to people expert in those domains. The first is the historical background. The development of GIS has an interesting history, stretching back to the 1950s. Readers who wish to pursue this topic will find an excellent introduction in Coppock and Rhind (1991), and more in-depth perspectives from many of the pioneers of GIS in Foresman (1998). The other area that is given scant treatment is spatial analysis, which requires specialized statistical techniques and is judged to be specifically the province of the domain experts. Introductions to spatial analysis include Unwin (1981), Fotheringham et al. (2002), and O'Sullivan and Unwin (2002). The bibliographic notes in Chapter 1 provide further references to texts on specific aspects of spatial analysis.

## WHO SHOULD READ THIS BOOK

This book is intended for readers from any background who wish to learn something about the issues that GIS engenders for computing technology. The reader does not have to be a specialist computing scientist: the text develops the necessary background in specialist areas, such as databases, as it progresses. However, some knowledge of the basic components and functionality of a digital computer is essential for understanding the importance of certain key issues in GIS. Where some aspect of general computing bears a direct relevance to our development, the background is given in the text. This book can be used as a teaching text, taking readers through the main concepts by means of definitions, explications, and examples. However, the more advanced researcher is not neglected, and the book includes an extensive bibliography that readers can use to follow up particular topics.

## CHANGES TO THE SECOND EDITION

The second edition of this book was written with the aim of making the book more accessible to a wider audience, at the same time as retaining the core of tried and tested material. Chapters 1–6 have been extensively revised, updated, and reformatted from the first edition, although in a fast-moving high-technology area like GIS it was encouraging to find that these fundamental aspects of GIS have remained largely unchanged. Chapters 7–10 present almost entirely new material, covering GIS architectures, GIS interfaces, uncertainty in geospatial information, and spatiotemporal information systems. The bibliography, index, and all the diagrams have also been completely revised.

In addition to the changes in content, we have tried to produce a more attractive and readable format for the book. The following section contains more details on the formatting conventions used in this book and on the structure of the book. The spelling, grammar, and usage in second edition has also changed, from British to American English. We hope that this change will further improve the accessibility of this book to an international audience.

## FORMATTING USED IN THIS BOOK

Several formatting conventions, new to the second edition, have been used in this book. Material that is relevant to the main themes in the text, but not essential to the reader, is included in gray inset boxes at the top of a page. Typically insets contain more challenging material, and provide some background to each topic, as well as references and links, which readers may wish to pursue. A list of insets is included in the book's front matter. Every chapter begins with a brief summary, outlining the major ideas in that chapter and highlighting some important terms introduced in the chapter. At its close every chapter ends with itemized bibliographic notes, providing some key references that readers can follow up. The section numbers alongside the bibliographic notes refer to the relevant sections in the main text.

Throughout this book, we have used margin text to allow rapid reference to important terms. When an important term is first defined or introduced, that term will appear in the margin. A corresponding entry can be found in the index, with the page reference in bold typeface. This enables the reader to use the index rather like an extensive glossary of terms used in this book. Each index term has at most one bold typeface page reference, and a term can be rapidly located within a page by finding the corresponding margin entry. In addition to normal- and bold-typeface index entries, those index entries that appear in italics refer to terms that appear within a gray inset box.

## STRUCTURE OF THIS BOOK

Figure 0.1 indicates the overall structure of interdependencies between chapters. Readers may find it helpful to refer to Figure 0.1 to tailor their use of this book to their own particular interests.

*Chapter 1*: Motivation and introduction to GIS; preparatory material on general computing.

*Chapters 2–3*: Background material on general databases and formalisms for spatial concepts.

*Chapters 4–6*: Exposition of the core material, forming a progression from high-level conceptual models, through representations and algorithms, to indexes and access methods that allow acceptable performance.

*Chapters 7–8*: Discussion of the types of system architectures and user interfaces needed for GIS.

*Chapter 9*: Introduction to spatial reasoning theory and techniques, with particular focus on reasoning under uncertainty.

*Chapter 10*: Introduction to temporal and spatiotemporal information systems.

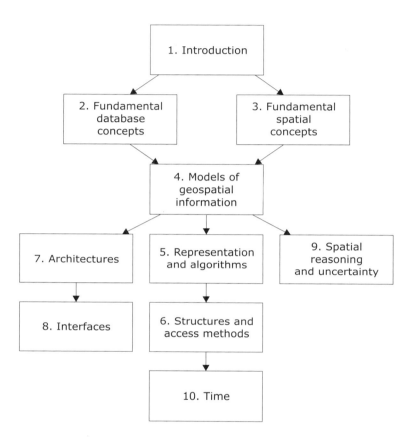

**Figure 0.1:** Relationships between chapters

## ONLINE RESOURCES

The website that accompanies this book can be found at:

http://worboys.duckham.org

The resources at this site are constantly under development, but include resources such as sample exercises, lecture slides and notes, open-source computer code, sample material, useful links, errata, and contact information. We, the authors, welcome suggestions from readers as to resources that we should include on the website, or indeed any feedback or comments on the book itself. We can be contacted on email at gisacp@worboys.duckham.org; other up-to-date contact information can be found on the website.

# Acknowledgments

The second edition of this book could never have been written without those people who helped in completing the first edition. Consequently, the acknowledgments to the first edition are reproduced in their entirety below. We are indebted to many old and new faces who have contributed to the second edition. Lars Kulik provided invaluable assistance with both the format and content of the book, reading and commenting on several early drafts. University of Maine graduate students in the Department of Spatial Information Science and Engineering provided very helpful comments on Chapters 8 and 9. The staff at CRC and Taylor & Francis, Randi Cohen, Matthew Gibbons, Samar Haddad, Sarah Kramer, and Tony Moore, have been most patient and professional. Other help and comments came from many quarters, including Jane Drummond, Max Egenhofer, Jim Farrugia, Mike Goodchild, Lars Harrie, Chris Jones, David Mark, Jörg-Rüdiger Sack, and Per Svensson. Helen Duckham commented on and corrected several early drafts. Moira Worboys greatly assisted with the final draft.

Matt Duckham and Mike Worboys, 2003

## ACKNOWLEDGMENTS TO THE FIRST EDITION

Many people have contributed to this book, either explicitly by performing tasks such as reading drafts or contributing a diagram, or implicitly by being part of the lively GIS community and so encouraging the author to develop this material. Waldo Tobler, of Santa Barbara, California, contributed a diagram of an interesting travel-time structure for Chapter 3 from his doctoral thesis, and cheerful encouragement. Also, from the National Center for Geographic Information and Analysis (NCGIA) at the University of California Santa Barbara, Mike Goodchild, Helen Couclelis, and Terry Smith provided a congenial environment for developing some of the ideas leading to this book. Max Egenhofer, from another NCGIA site at the University of Maine, provided hospitality and a useful discussion about the structure of the book. At Keele University, Keith Mason and Andrew Lawrence provided invaluable help with some of the example applications introduced in Chapter 1. Dick Newell of Smallworld has generously provided Smallworld GIS as a research and teaching tool, enabling me to develop and implement many ideas using a state-of-the-art system. Tigran Andjelic and Sonal Davda checked some of the technical material. Richard Steele from Taylor & Francis has been encouraging and patient with my failure to meet deadlines. Mostly, I am indebted to my wife Moira for her invaluable assistance with the final draft, and for her moral and practical support throughout this work.

Mike Worboys, 1995

# Authors

Mike Worboys is a professor in the Department of Spatial Information Science and Engineering, and at the National Center for Geographic Information and Analysis (NCGIA), University of Maine, USA. He holds a PhD in pure mathematics, and has worked for many years on computational aspects of geographic information. He is the author of the first edition of this book, editor of several collections, and serves on the editorial boards of various international journals.

Matt Duckham is a lecturer in the Department of Geomatics at the University of Melbourne, Australia. From 1999 to 2004, Matt worked as a postdoctoral researcher with Mike, first at the Department of Computer Science, University of Keele, UK, where Mike was head of department, and subsequently at the NCGIA, Department of Spatial Information Science and Engineering, University of Maine, USA. Matt is an editor of the book *Foundations of Geographic Information Science*.

# List of Insets

# List of Algorithms

# Contents

# Introduction

1

**Summary**

A **geographic information system** (GIS) is a special type of computer-based **information system** tailored to store, process, and manipulate **geospatial** data. This chapter sets the scene, describing what a GIS is and giving examples of what it can do. At the heart of any GIS is the **database**, which organizes data in a form that is easy to store and retrieve. The chapter also describes some of the **hardware** technology surrounding GIS, including computer processors, storage devices, user input/output devices, and computer networks.

"What makes GIS special?" Most people who work with geographic information systems have asked themselves this question at one time or another. This chapter starts to answer the question by describing the field of GIS against the general background of computing and identifying what distinguishes geographic information systems from other information systems. First, we define the terms "information system" and "GIS," and outline the main components of a GIS (section 1.1). Then, in section 1.2, we look at some example applications that illustrate what a GIS can do, and provide a motivation for studying this topic. The most important technology underlying GIS is the database system. Databases are introduced in section 1.3 along with an outline of the key features that are distinctive in geospatial databases. The chapter concludes with an overview of the basic computing hardware common to all GISs (section 1.4).

## 1.1   WHAT IS A GIS?

A good starting point for defining a geographic information system, is to look at a general definition of an *information system*. An information system is an association of people, machines, data, and procedures working together to collect, manage, and distribute information of importance

information system

1

to individuals or organizations. The term "organization" is meant here in a wide sense that includes corporations and governments as well as more diffuse groupings, such as global colleges of scientists with common interests or a collection of people looking at the environmental impact of a proposed new rail-link. The World Wide Web (WWW) is an example of an information system. The WWW comprises data (web pages) and machines (web servers and web browsers), but also the many people across the world who use the WWW and the procedures for maintaining information on the WWW.

A GIS is a special type of information system concerned with *geographically* referenced data. Specifically:

geographic information system

> A geographic information system is a computer-based information system that enables capture, modeling, storage, retrieval, sharing, manipulation, analysis, and presentation of geographically referenced data.

geospatial

In this book we use the term *geospatial* to mean "geographically referenced." Thus geospatial data is a special type of spatial data that relates to the surface of the Earth. The key components of a GIS are shown schematically in Figure 1.1. Several other terms and types of information system are closely related to GIS (see "GIS terminology" inset, page 3).

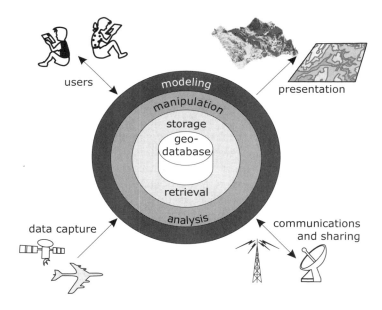

**Figure 1.1:**
Schematic of a GIS

### 1.1.1 The shape of GIS

The world around us is both spatial and temporal, so we have a need for information that has spatial and temporal dimensions. Future decisions that affect us all, for example, in planning new roads or cities, formulating

**GIS terminology** *There are several terms in common use that are more or less synonymous with GIS. A spatial information system (SIS) has the same functional components as a GIS, but may handle data that is referenced at a wider range of scales than simply geographic (for example, data on molecular configurations). A spatial database provides the database functionality for a spatial information system. A geographic database (or geodatabase) provides the database functionality for a GIS. An image database is fundamentally different from a spatial database or geodatabase in that the images have no structural interrelationships or topological features. For example, a database of brain-scan images may be termed an image database. Computer-aided design (CAD) has some elements in common with GIS, and historically some GIS software packages have developed from CAD software. Unlike CAD, GIS software is used to manipulate data sets that are geographically referenced. As a consequence, GIS software is usually used with larger data sets and more complex data models than CAD software. Conventionally, and in this book, the singular form of GIS may be used to refer to "the field of GIS," where no confusion is likely. While not synonymous with GIS, the terms geographic information science (GIScience) and geoinformatics are relevant to GIS as they are used to describe the systematic study of geographic information and geographic information systems.*

agricultural strategies, and locating mineral extraction sites, rely upon properly collected, managed, distributed, analyzed, and presented spatial and temporal information. A GIS may be thought of as a tool that is able to assist us with these tasks. This tool relies on several underlying elements. In the remainder of this section we look in more detail at these different elements, so describing the "shape" of GIS.

*Database element*    At the heart of any GIS is the *database*. A database is a collection of data organized in such a way that a computer can efficiently store and retrieve the data. Databases are introduced in Chapter 2. An important element of a database is the *data model*. Some applications require relatively simple data models. For example, in a library system data about books, users, reservations, and loans is structured in a clear way. Many applications, including most GIS applications, demand more complex data models (consider, for example, a model of global climate). One of the main issues addressed in this book is the provision of facilities for handling these complex data models. Chapter 2 introduces some fundamental data modeling concepts, while Chapters 3 and 4 examine what makes geospatial data and geospatial data models special.

In this book we discuss how complex geospatial data models can be represented in an information system. However, we do not discuss the suitability of particular data models for particular application areas: this is a topic for application domain experts. For example, a transportation geography textbook should offer insights into whether a particular model of transportation flow is appropriate for a particular application. We are only concerned with providing the general facilities to represent geospatial data models, which might include models of transportation flow.

*database*

*Data processing element*   Data models are stepping-stones to the manipulation and processing of data. GIS software needs to have sufficiently
complete functionality to provide the higher-level analysis and decision
support required within an application domain. For example, a GIS that
is used for utility applications will require network processing operations
(optimal routing between nodes, connectivity checks, and so forth). A
GIS for geological applications will require three-dimensional operations.
Identifying and specifying a general set of primitive operations required
by a generic GIS is a major concern of the book, covered primarily in
Chapter 5. Again, the discussion in this book does not address whether
specific approaches to geospatial data analysis are suitable for particular
applications: those questions we leave to application domain experts.

*Data storage and retrieval element*   Efficient storage and retrieval of
data depend on not only properly structured data in the database to
provide satisfactory performance, but also optimized structures, representations, and algorithms for operating on data. Retrieval, operations,
and performance raise many interesting questions for geospatial data and
will be a further main theme of this text (Chapter 6).

*Data sharing element*   A GIS may comprise many different software and
hardware components, such as geographic data servers, web browsers,
and mobile computing devices. For example, in order to access real-time
information about the best route to drive home from work, allowing for
traffic jams, road works, and changing road conditions, several different
information system components must work together. An important characteristic of any GIS is the capability to *share* data between different
information systems, or between different components within a single
information system. Understanding and achieving data sharing are key
topics in this book, discussed under the general heading of *system architecture* (Chapter 7).

*Data presentation element*   Data within a GIS, and the operations performed upon that data, are not useful unless they can be communicated to
the people using that GIS. Many traditional information systems require
only limited presentational forms, usually based around tables of data,
numerical computation, and textual commentary. While these forms are
also required to support decision-making using a GIS, the geospatial nature of data in a GIS allows a whole new range of possibilities, including
*cartographic* (map-based) presentation as well as more exploratory forms
of presentation. Presentation of the results of retrievals and analyses using
a GIS therefore takes us beyond the scope of most traditional databases
and is a further concern of this text (Chapter 8).

*Spatial reasoning element*   To get the most from geospatial data, it is
important that we are aware of and able to reason about its limitations.

*Accuracy*, *precision*, and *reliability* are examples of terms we often use when talking about the limitations of data. Understanding such terms, and how they affect the way we use a GIS to *reason* about geospatial data, is another important topic in GIS, covered in Chapter 9.

*Spatiotemporal element*   Finally, as stated above, the world is both spatial *and* temporal. Extending GIS to allow the storage and processing of dynamic geospatial phenomena that change over time is a major topic for current research. In the final chapter, we introduce the key models of *spatiotemporal* phenomena, and show how the next generation of GIS is beginning to move beyond static representations of the world around us.

### 1.1.2   Data and information

The structure of interrelationships between data and how data is collected, processed, used, and understood within an application forms the *context* for data. An understanding of the data model and of the limitations of data, discussed above, are elements of the context for data. Data only becomes useful, taking on value as information, within this context.

*context*

*data*

For example, data about atmospheric conditions is recorded by meteorological stations across the world: there is likely to be one near you recording right now. However, on its own the raw data from your nearest recording station is unlikely to be useful to you. Useful information, such as a weather forecast that helps you decide if you need to carry an umbrella with you, is produced within the context of careful modeling and analysis of raw data from multiple sources (such as satellite imagery, data from other recording stations, and historical data). Accordingly, information can be defined as "data plus context":

*information*

information = data + context

We refine the concepts of data and information in later chapters (particularly Chapter 9), but this basic distinction between data and information is needed throughout the book.

## 1.2   GIS FUNCTIONALITY

As highlighted above, a GIS is an information system that has some special characteristics. To demonstrate the range of capabilities and functionality of a GIS, seven example applications are described in this section. These applications have been chosen for a region of England, familiarly called "The Potteries" due to its dominant eponymous industry. The Potteries comprise the six pottery towns of Burslem, Fenton, Hanley, Longton, Stoke, and Tunstall, along with the neighboring town of Newcastle-under-Lyme (see Figure 1.2). The Potteries region developed rapidly during the English industrial revolution, local communities producing ware of the highest standard (for example, from the potteries of Wedgwood and Spode) from conditions of poverty and cramp. The region's landscape is scarred by the extraction of coal, ironstone, and clay. In the 20th century, the Potteries declined in prosperity although the area is now in a phase of regeneration.

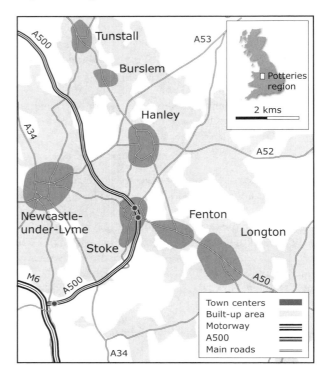

**Figure 1.2:**
The Potteries region

We emphasize that the example applications are here merely to show functionality, rather than serious descriptions of application areas. For application-driven texts, see the bibliography at the end of the chapter. The final part of this section summarizes typical GIS functionality.

### 1.2.1   Resources inventory: A tourist information system

The Potteries, because of its past, has a locally important tourist industry based upon the industrial heritage of the area. A GIS may be used to support this, by drawing together data on cultural and recreational facilities within the region, and combining this data with details of local transport infrastructure and hotel accommodation. Such an application is an example of a simple *resource inventory*. The power of almost any information system lies in its ability to relate and combine data gathered from disparate sources. This power is increased dramatically when the data is geographically referenced, provided that the different sources of geospatial data are made compatible through some common spatial unit or transformation process. Figure 1.3 shows the beginnings of such a system, including some of the local tourist attractions, the major road network, and built-up areas in the region.

resource
inventory

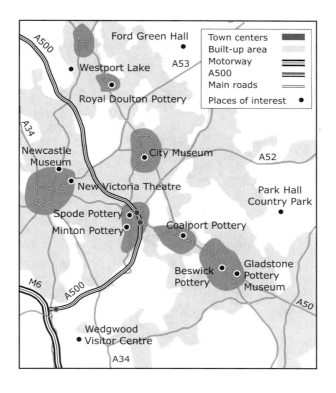

**Figure 1.3:**
Places of
interest in the
Potteries region

### 1.2.2 Network analysis: A tour of the Potteries

*Network analysis* is one of the cornerstones of GIS functionality. Applications of network analysis can be found in many areas, from transportation networks to the utilities. To provide a single simple example, we stay with the tourist information example introduced above. The major potteries in the Potteries area are famous worldwide. Many of these potteries offer factory tours and have factory outlet stores. The problem is to provide a route using the major road network, visiting each pottery (and the City Museum) only once, while minimizing the traveling time. The data set required is a travel-time network between the potteries: a partial example is given in Figure 1.4a.

network analysis

The travel-time network in Figure 1.4a was derived from average times on the main roads shown on the map. The network may then be used as a basis for generating the required optimal route between Royal Doulton and the Wedgwood Visitor Centre, shown in Figure 1.4b. The route visits the potteries in sequence: Royal Doulton, City Museum, Spode, Minton, Coalport, Gladstone, Beswick, Wedgwood. The specific network analysis technique required here is the *traveling salesperson algorithm*, which constructs a minimal weight route through a network that visits each node at least once. The analysis could be dynamic, assigning weights to the edges of the network and calculating optimal routes depending upon changeable road conditions.

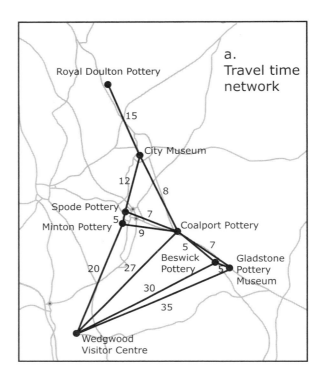

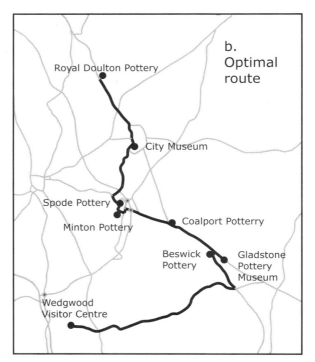

**Figure 1.4:**
Travel network
based upon
travel times (in
minutes) and
optimal route

### 1.2.3   Distributed data: Navigating around the potteries

Much of the data used in a GIS is located in different formats at physically remote locations. A GIS needs to be able to overcome these barriers to data sharing. Continuing the tourist information example, the resources inventory and network analysis described above demand the sharing and analysis of disparate geospatial data about the locations of places of interest, the transport infrastructure, and cultural, hospitality, and recreational resources. These resources are commonly held by different organizations at different locations. Base map data may be held in one place, such as by a national mapping agency. Data about transport infrastructure might be compiled by the local government or held by individual bus or train companies. The tourist information bureau will hold some data about the local amenities, although more will often reside with the individual amenities themselves, such as museums or hotels.

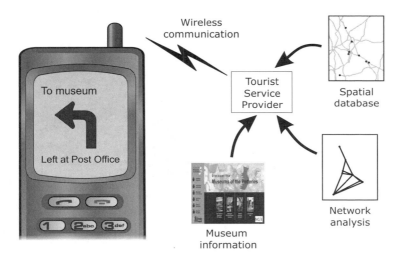

**Figure 1.5:** Schematic view of a distributed tourist information system

Figure 1.5 shows a schematic version of a distributed information system that might be used as the basis of a Potteries tourist application. Before a tourist visiting the Potteries can receive navigation directions and information about local attractions (for example, on their cell phone or handheld computer), data from all these different sources must be integrated, processed, and transmitted to the tourist. Since the tourist will not usually be a GIS expert, these complex tasks would normally be done on behalf of the tourist by some tourist service provider. The service provider might gather all the information needed, either in advance or dynamically when requested, and perform the network analysis necessary to find the best route for a particular user at a particular time and location. Although the information actually presented to a tourist at any moment in time may be simple (such as "turn left" indicators, Figure 1.5), the task of integrating data from different sources may be complex.

### 1.2.4   Terrain analysis: Siting an opencast coal mine

Terrain analysis is usually based upon data sets of topographic elevations at point locations. Basic information about degree of slope and direction of slope (termed *aspect*) can be derived from such data sets. A more complex type of analysis, termed *visibility analysis*, concerns the visibility between locations and the generation of a *viewshed* (a map of all the points visible from some location).

aspect<br>visibility analysis

The applications of terrain analysis are diverse. For our example, in the Potteries conurbation the search for new areas of opencast coal mining has in the past resulted in much interest from local communities, which might be concerned about the effects of such operations. One factor in this complex question is the visual impact of proposed opencast sites. Visibility analysis can be used to evaluate visual impact, for example, by measuring the size of the local population within a given viewshed. Sites that minimize this population may be considered desirable. The terrain surface of the area around Biddulph moor may be represented using a contour map, as in Figure 1.6a.

**Figure 1.6:**
Contour map
and perspective
projection of
terrain surface

a. Contour map                    b. Perspective projection

Figure 1.6b shows an perspective projection of the same surface shown in Figure 1.6a. Such projections provide a powerful depiction of the terrain. Figure 1.7 shows the same surface as before, this time draped by the viewshed. The darker shaded regions give the area from which the marked point would not be visible. If we assume the point represents the location of the opencast mine, then the lighter areas provide a first approximation to the visual impact of the mine. Of course, a real case would take into account much more than just the visibility of a single point, but the principle remains.

### 1.2.5   Layer-based analysis: The potential of extraction sites for mineral ore deposits

The Potteries area is rich in occurrences of superficial and bedrock sand and gravel, although few such sites have been worked in recent times. To determine the potential of different locations for sand and gravel extraction demands the drawing together and analysis of data from a variety of sources. Geological data describing the location of appropriate

**Figure 1.7:**
Perspective
projection of
terrain surface
draped with
viewshed for
the marked
point

deposits is, of course, needed. Other important considerations are local
urban structure (e.g., urban overbuilding), water table level, transportation
network, land prices, and land zoning restrictions. Figure 1.8 shows a
sample of the available data overlaid on a single sheet, including data
on built-up areas, known sand and gravel deposits, and the major road
network.

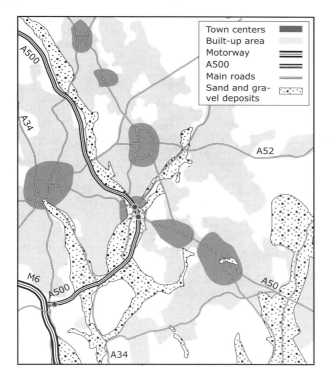

**Figure 1.8:**
Locations of
sand and gravel
deposits in the
Potteries region

Layer-based analysis results from posing a query such as: "Find all
locations that are within 0.5 km of a major road, not in a built-up area,
and on a sand/gravel deposit." Figure 1.9 illustrates the construction of
an answer to this question. The gray areas in Figure 1.9a show the region
within 0.5 km of a major road (not including the motorway), termed a
*buffer*. The gray areas in Figure 1.9b indicate known sand and gravel

deposits, and in Figure 1.9c the shaded areas indicate locations that are not built up. Figure 1.9d shows the overlay of the three other layers, and thus the areas that satisfy our query. The analysis here is simplistic. A more realistic exercise would take into account other factors, like the grading of the deposit, land prices, and regional legislation. However, the example does show some of the main functionality required of a GIS engaged in layer-based analysis, including:

- buffering

  - The formation of areas containing locations within a given range of a given set of features, termed *buffering*. Buffers are commonly circular or rectangular around points, and corridors of constant width about lines and areas.

- Boolean overlay

  - The combination of one or more layers into a single layer that is the union, intersection, difference, or other Boolean operation applied to the input layers, termed *Boolean overlay*.

Layer-based functionality is explored further in the context of field-based models and structures later in the book.

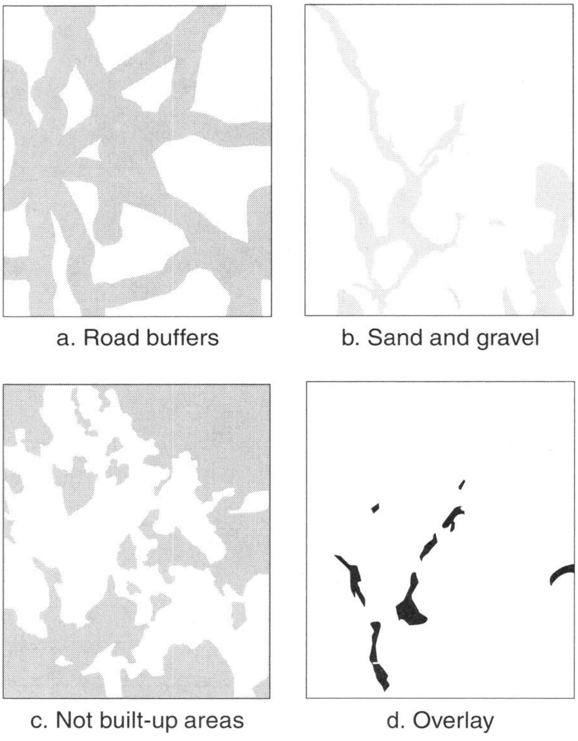

**Figure 1.9:** Layer-based analysis to site a mineral ore extraction facility

a. Road buffers          b. Sand and gravel

c. Not built-up areas          d. Overlay

### 1.2.6    Location analysis: Locating a clinic in the Potteries

Location problems have been solved in previous examples using terrain models (opencast mine example) and layer-based analysis (estimating the potential of sites for extracting sand and gravel). Our next example is the location of clinics in the Potteries area. A critical factor in the decision to use a particular clinic is the time it takes to travel to it. To assess this we may construct the "neighborhood" of a clinic, based upon positions of nearby clinics and travel times to the clinic. With this evidence, we can then support decisions to relocate, close, or create a new clinic.

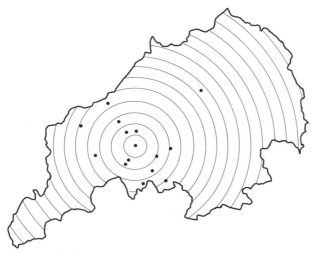

a. Clinics, with isochrones around one clinic

**Figure 1.10:**
Potteries
clinics and
their proximal
polygons

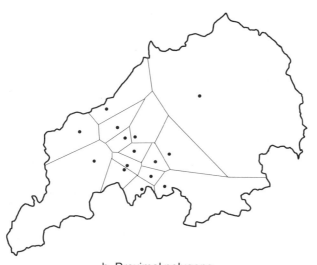

b. Proximal polygons

Figure 1.10 shows the idealized positions of clinics in the Potteries region. Assuming an "as the crow flies" travel time between points (i.e., the time is directly related to the Euclidean distance between points), Figure 1.10a shows lines connecting locations that are equally far from the clinic in terms of travel time, termed *isochrones*. It is then possible    isochrone to partition the region into areas, each containing a single clinic, such that each area contains all the points that are nearest (in travel time) to its clinic, termed *proximal polygons* (see Figure 1.10b). Of course, we are making a simplistic assumption about travel time. If the road network is accounted for in the travel-time analysis, then the isochrones will no longer be circular and the areal partition will no longer be polygonal. This more general situation is discussed later in the book.

### 1.2.7   Spatiotemporal information: Thirty years in the Potteries

Geospatial data sometimes becomes equated with purely static data, thus neglecting the importance of change and time. In a temporal GIS, data is referenced to three kinds of dimensions: space, time, and attribute. Our last example suggests possibilities for a future dynamic GIS functionality, just beyond the reach of present systems. We return to the problem of adding temporal dimensions in the final chapter.

The main period of industrial activity in the Potteries is long since past. The history of the region in the latter half of the 20th century has been one of industrial decline. Figure 1.11 shows the Cobridge area of the Potteries, recorded in snapshot at two times: 1878 and 1924. It is clear that as time has passed many changes have occurred, such as the extension of residential areas in the northwest and southeast of the map. Examples of questions that we may wish to ask of our spatiotemporal system include:

- Which streets have changed name in the period 1878–1924?

- Which streets have changed spatial reference in the period 1878–1924?

- In what year is the existence of the Cobridge Brick Works last recorded in the system?

- What is the spatial pattern of change in this region between 1878 and 1924?

### 1.2.8   Summary of analysis and processing requirements

The examples of applications reviewed in this section demonstrate some of the specialized processing functionality that a GIS needs to provide, usually termed *spatial analysis*. Geospatial database functionality was illustrated with the resources inventory in section 1.2.1. Geospatial data is commonly presented using computer graphics, as shown in most of the figures in this section. However, we have seen that a GIS is more than just a graphics database. A graphical system is concerned with the manipulation and presentation of screen-based graphical objects, whereas a GIS handles phenomena embedded in the geographic world and having not only geospatial dimensions but also structural placement in multidimensional geographic models. Some of the analytical processing requirements that give a GIS its special flavor include:

*spatial analysis*

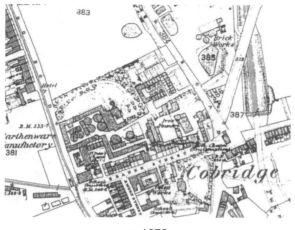

a. 1878

**Figure 1.11:**
History of the
Cobridge area,
recorded in
snapshots at
times 1878 and
1924 (Source:
Ordnance
Survey)

b. 1924

*Geometric, topological, and set-oriented analyses*: Most if not all geographically referenced phenomena have geometric, topological, or set-oriented properties. Set-oriented properties include membership conditions, Boolean relationships between collections of elements, and handling of hierarchies (e.g., administrative areas). Topological operations include adjacency and connectivity relationships. A geometric analysis of the clinics (section 1.2.6) produced their proximal polygons. All these properties are key to a GIS and form a main theme of this book.

*Field-based analysis*: Many applications involve spatial *fields*, that is, variations of attributes over a region. The terrain in section 1.2.4 is a variation of topographic elevation over an area. The gravel and sand deposits and built up areas of section 1.2.5 are variations of

other attributes. Fields may be discrete (a location is either built up
or not) or continuous (e.g., topographic elevation). Fields may be
*scalar* (variations of a scalar quantity and represented as a surface)
or *vector* (variations of a vector quantity such as wind velocity).
Field operations include overlay (section 1.2.5), slope and aspect
analysis, path finding, flow analysis, and viewshed analysis (section
1.2.4). Fields are discussed further throughout the text.

*Network analysis*: A network is a configuration of connections between
nodes. The maps of most metro systems are in network form, nodes
representing stations and edges being direct connections between
stations. Networks may be directed, where edges are assigned
directions (e.g., in a representation of a one-way street system)
or labeled, where edges are assigned numeric or non-numeric
attributes (e.g., travel-time along a rail link). Network operations
include connectivity analysis, path finding (trace-out from a single
node and shortest path between two nodes), flow analysis, and
proximity tracing (the network equivalent of proximal polygons).
The Potteries tourist information system application (section 1.2.2)
gives an example of a network traversal. Networks are considered
further in Chapter 3.

### 1.3   DATA AND DATABASES

Any information system can only be as good as its data. This section
introduces some of the main themes associated with databases and data
management. To support the database, there is a need for appropriate capabilities for data capture, modeling, retrieval and analysis, presentation,
and dissemination. An issue that arises throughout the book is the extra
functionality that a geodatabase needs over and above the functionality
of a general-purpose database. This section provides an overview, but the
topic is taken up further in Chapter 2 and throughout the book.

#### 1.3.1   Spatial data

First, a reminder of some data basics. Data stored in a computer system is
measured in *bits*. Each bit records one of two possible states: 0 (off, false)
and 1 (on, true). Bits are amalgamated into *bytes*, each byte representing
a single character. A character may be encoded using 7 bits with an extra
bit used as an error check; thus each byte allows for $2^7 = 128$ possible
character combinations. This is enough for all the characters normally
used in the English language (this encoding is the basis of ASCII code).
To represent all the characters found in other languages requires more
bits. Two-byte characters allow for $2^{16} = 65536$ different character
combinations (the basis for Unicode).

Geospatial data is traditionally divided into two great classes, *raster*
and *vector*. Traditionally, systems have tended to specialize in one or other
of these classes. Translation between them (*rasterization* for vector-to-

**Margin notes:** scalar, vector, bit, byte, rasterization

raster and *vectorization* for raster-to-vector) has been a thorny problem, to be taken up later in the book. Figure 1.12 shows raster and vector data representing the same situation of a house, outbuilding, and pond next to a road.

vectorization

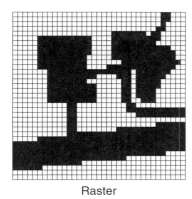
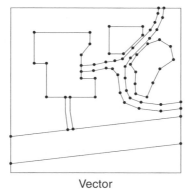

Raster                                        Vector

**Figure 1.12:** Raster and vector data

*Raster* data is structured as an array or grid of cells, referred to as *pixels*. The three-dimensional equivalent is a three-dimensional array of cubic cells, called *voxels*. Each cell in a raster is addressed by its position in the array (row and column number). Rasters are able to represent a large range of computable spatial objects. Thus, a point may be represented by a single cell, an arc by a sequence of neighboring cells and a connected area by a collection of contiguous cells. Rasters are natural structures to use in computers, because programming languages commonly support array handling and operations. However, a raster when stored in a raw state with no compression can be extremely inefficient in terms of usage of computer storage. For example, a large uniform area with no special characteristics may be stored as a large collection of cells, each holding the same value. We will consider efficient computational methods for raster handling in later chapters.

raster
pixels
voxels

The other common paradigm for spatial data is the vector format. (This usage of the term "vector" is similar but not identical to "vector" in vector fields). A *vector* is a finite straight-line segment defined by its end-points. The locations of end-points are given with respect to some coordinatization of the plane or higher-dimensional space. The discretization of space into a grid of cells is not explicit as it is with the raster structure. However, it must exist implicitly in some form, because of the discrete nature of computer arithmetic. Vectors are an appropriate representation for a wide range of spatial data. Thus, a point is just given as a coordinate. An arc is discretized as a sequence of straight-line segments, each represented by a vector, and an area is defined in terms of its boundary, represented as a collection of vectors. The vector data representation is inherently more efficient in its use of computer storage than raster, because only points of interest need be stored. A disadvantage is that, at least in their crude form, vectors assume a hard-edged boundary

vector

model of the world that does not always accord with our observations. We
return to these issues in later chapters.

### 1.3.2   Database as data store

A database is a repository of data that is logically related, but possibly
physically distributed over several sites, and required to be accessed by
many applications and users. A database is created and maintained using a
general-purpose piece of software called a *database management system*
DBMS    (DBMS). For a database to be useful it must be:

*Reliable*:  A database must be able to offer a continual uninterrupted service when required by users, even if unexpected events occur, such
as power failures. For example, a database must be able to contend
with the situation where we deposit money into an automatic teller
machine (ATM) and the power fails before our balance has been
updated.

*Correct and consistent*:  Data items in the database should be correct and
consistent with each other. This has been a problem with older file
systems, when data held by one department contradicts data from
another. While it is not possible to screen out all incorrect data, it
is possible to control the problem to some extent, using integrity
checking at input. Clearly, the more we know about the kinds of
data that we expect, the more errors can be detected.

*Technology proof*:  A database should evolve predictably, gracefully, and
incrementally with each new technological development. Both
hardware and software continue to develop rapidly, as new processors, storage devices, software, and modeling techniques are in-
vented. Database users should be insulated from the inner workings
of the database system. For example, when we check our balance at
the ATM, we do not really want to know that on the previous night
new storage devices were introduced to support the database.

*Secure*:  A database must allow different levels of authorized access and
prevent unauthorized access. For example, we may be allowed to
read our own, but no one else's, bank balance (read access). At
the same time we may not be allowed to change our bank balance
(write access).

### 1.3.3   Data capture

The process of collecting data from observations of the physical envi-
data capture    ronment is termed *data capture*. The primary source of data for a GIS is
from *sensors* that measure some feature of the geographic environment.
Satellite imagery, for example, uses sensors to measure the electromagnetic radiation reflected or emitted from the Earth's surface at different

wavelengths. Surveyors use sensors to measure distances and angles between features located on the Earth's surface.

Sometimes humans are needed to convert the data produced by a sensor into a form that an information system can use, such as when a human enters data using a keyboard. In most circumstances, however, data capture is automated, so that digital data from a sensor is fed directly to an information system and, after processing, stored within a database. The variety of common digital sensors is steadily increasing, while their size and cost is steadily decreasing. As a result, GISs often need to store and process sensor-based information from many different sources. The different types of sensor that exist for determining location are discussed in Chapter 7.

A secondary data capture stream for GISs is from a *legacy data* source, such as paper maps. Maps combine the functions of data presentation and data storage, functions that GISs keep separate. Converting geospatial data stored in a paper map into a form that can be stored in a GIS can be difficult and costly. Automatic conversions, such as scanning a map, cannot easily capture the complex structure of the map, and the results are more similar to an image database than a GIS. Manual conversions, for example, where humans trace the features of a map using a digitizer (see section 1.4.4 and Figure 1.17), can capture more structure, but are time-consuming and laborious.

*legacy data*

### 1.3.4  Data modeling

The choice of appropriate *data model* can be the critical factor for the success or failure of an information system. In fact, the data model is the key to the database idea. Data models function at all levels of the information system. The process of developing a database, or indeed any information system, is essentially a process of model building. At the highest level is the *application domain model*, which describes the core requirements of users in a particular application domain, based on an initial study. At the next level, the *conceptual computational model* provides a means of communication between the user and the system that is independent of the details of the implementation. The process of developing a conceptual computational model focuses on questions of *what* it is the system will do, termed *system analysis*.

*data model*

*system analysis*

The *logical computational model* is tailored to a particular type of implementation. The process of developing a logical computational model focuses on questions of *how* the system will implement the conceptual computational model, termed *system design*. Finally, the low-level *physical computational model* is the result of a process of programming and *system implementation*. It describes the actual software and hardware application, including how data is processed and organized on a particular type of machine.

*system design*

*system implementation*

The different stages and processes involved in system development are shown in Figure 1.13. The stages form a logical progression from

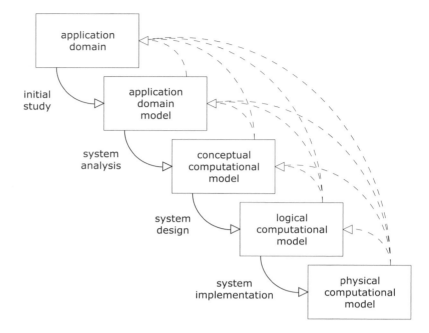

**Figure 1.13:**
Modeling
processes and
the stages of
system
development

application domain to actual implementation. This logical progression
is often referred to as the *waterfall model* of system development. In
practice, system development is more cyclic (indicated in Figure 1.13
with the dashed arrows), for example, with a new round of analysis and
design following from a pilot implementation. The term *system life-cycle*
is therefore preferred to the waterfall model, because it emphasizes the
evolutionary and iterative nature of system development.

waterfall model

system life-cycle

The different stages of the system life-cycle form a bridge between
the human conception of the application domain and the computational
processes used in the information system. Without this structure, the
system development process can become unmanageable. Typically, dis-
regarding the system life-cycle leads to systems that are either difficult to
use, because they neglect the human aspects of the information system,
or inefficient, because they neglect the computational aspects.

Two important elements of the system life-cycle are not depicted
in Figure 1.13, because they are not primarily modeling tasks. *System maintenance* follows implementation, making ongoing changes and
improvements to a system in response to actual system usage. *System
documentation* occurs throughout every stage of the system life-cycle,
ensuring that adequate documentation accompanies the system.

### 1.3.5   Data retrieval and analysis

A key function of a database is the capability to rapidly and efficiently
retrieve and analyze information. To retrieve data from the database, we
may apply a filter, usually in the form of a logical expression, or *query*.

query

For example:

1. Retrieve names and addresses of all opencast coal mines in Staffordshire.

2. Retrieve names and addresses of all employees of Wedgwood Pottery who earn more than half the sum earned by the managing director.

3. Retrieve the mean population of administrative districts in the Potteries area.

4. Retrieve the names of all patients at Stoke City General Hospital who are over the age of 60 and have been admitted on a previous occasion.

Assuming that the first query accesses a national database and that the county in which a mine is located is given as part of its address, then the data may be retrieved by means of a simple look-up and match. For the second query, each employee's salary would be retrieved along with the salary of the managing director and a simple numerical comparison made. For the third, populations are retrieved and then a numerical calculation is required. For the last, a more complex filter is required, including a check for multiple records for the same person.

There are spatial operators in the above queries. Thus, in the first query, whether a mine is *in* Staffordshire is a spatial question. However, it is assumed that the processing for this query needs no special spatial component. In our example, there is no requirement to check whether a mine is within the spatial boundary of the county of Staffordshire, because this information is given explicitly by textual means in the database. A GIS allows real geospatial processing to take place. For example, consider the following two queries:

1. *What locations in the Potteries satisfy the following set of conditions*:

   - less than the average price for land in the area;
   - within 15 minutes drive of the motorway M6;
   - having suitable geology for gravel extraction; and
   - not subject to planning restrictions?

2. *Is there any correlation between*:

   - the location of vehicle accidents (as recorded on a hospital database); and
   - designated "accident black spots" for the area?

Satisfying such queries often requires the integration of both spatial and non-spatial information. The first of these examples is rather more prescriptive than the second, specifying the type of relationship required.

The second example requires a more open-ended application of spatial analysis techniques.

All of the above functionality is attractive and desirable, but will be useless if not matched by commensurate performance. Computational performance is usually measured in terms of storage required and processing time. Storage and processing technology are continually advancing. However, not all computational performance problems can be solved by using technology that is more advanced. Some problems, such as the traveling salesperson problem mentioned in section 1.2.2, require so many calculations that in general it is not possible to compute an optimal solution in a reasonable time.

Performance is an even bigger issue for a geodatabase than a general-purpose database. Spatial data is notoriously voluminous. In addition, spatial data is often hierarchically structured (a point being part of an arc, which is part of a polygon). We shall see that these structures present problems for traditional database technology. Although DBMSs are designed to handle multidimensional data, the dimensions are assumed independent. However, geospatial data is often embedded in the Euclidean plane: special storage structures and access methods are required. We shall return to the question of spatial data storage, access, and processing later in the book.

### 1.3.6   Data presentation

Traditional general-purpose databases provide output in the form of text and numbers, usually in tabular form. A *report generator* is a standard feature of a DBMS that allows data from a database to be laid out in a clear human-readable format. Many databases also have the capacity to offer enhanced presentations, like charts and graphical displays, termed *business graphics*.

In addition to these general presentation mechanisms, a GIS needs to output maps and map-based material. Map-based presentation is a highly distinctive feature of a GIS compared with a general-purpose database. Indeed, the ability to produce hard copy maps automatically, termed

digital
cartography

*automated* or *digital cartography*, was one of the primary reasons for the initial development of GIS and remains an important function of some GIS software. A few DBMSs and GISs go further than business graphics and maps, and provide tools to help users explore a data set and discover

data mining

relationships and patterns embedded in their data, termed *data mining* (see "Spatial data mining" inset, on the facing page). Questions of the kind "Here is a lot of data: is there anything significant or interesting about it?" are prototypical for data mining. Systems that support data mining have usually highly flexible presentation capabilities, so that users can interactively combine and re-express multidimensional data in many different forms, including graphs, trees, and animations.

**Spatial data mining**  *Data mining refers to the process of discovering valuable information and meaningful patterns within large data sets. Data mining is particularly important for geospatial data because this type of data is typically voluminous and a rich source of patterns. Data mining differs from basic database querying, in that users may not know what information or patterns they are seeking in advance. The sorts of information and patterns that can be discovered using data mining include association rules and clustering patterns. One common application of spatial data mining is customer relationship management (CRM). A CRM company might use data mining to discover if there is an association between different types of retail outlet, perhaps based on data about credit or debit card use. For example, if shortly after seeing a movie customers tend to eat out at restaurants close to the cinema, data mining should help reveal this pattern as a basis for specialized promotions (such as "See a movie at the Regal cinema, get a free dessert at Friendly Joe's Pizza Restaurant!"). Data mining can also be used to discover clusters as a basis for classification. As an example, an environmental agency might use data mining to discover geospatially clustered patterns in the location of industrial pollution.*

### 1.3.7  Data distribution

A centralized database system has the property that the data, the DBMS, and the hardware are all gathered together in a single computer system. Although the database may be located at a single site, that does not preclude remote access to the database. Indeed, this is one of the guiding principles of a database: facilitating shared access to a data.

If all applications could use a centralized database, then the study of databases would be a much simpler topic. However, as illustrated by the examples of the tourist navigation and mineral extraction systems (sections 1.2.3 and 1.2.4 above), many applications require access to data from multiple databases connected by a digital communication network, termed a *distributed* database. There are natural reasons for data to be distributed in this way. Data may be more appropriately associated with one site rather than another, allowing a greater degree of autonomy and easier update and maintenance. For example, details of local weather conditions may be more usefully held at a local site where local control and integrity checks may be maintained. Another advantage of a distributed database is increased reliability; failure at one site will not mean failure of the entire system. Distributed databases may also offer improved performance. In the example of the local weather conditions database, commonly occurring accesses to the local site from local users will be more efficient. Performance will be weaker for those accesses to remote sites, but such accesses may be fewer.

The downside is that distributed databases have a more intricate structure to support. For example, distributed databases must handle queries where the data is fragmented across sites, and maintain the consistency of data that is replicated across sites. Distributed geodatabases are one topic covered in more detail in Chapter 7.

## 1.4   HARDWARE SUPPORT

hardware
software

The term *hardware* is used to refer to the physical components of a computer system, like computer chips and keyboards, while *software* refers to instructions or programs executed by a computer system. While the software needed for a GIS is highly specialized, the hardware used in a GIS is broadly the same as that used within general-purpose computer systems. This section discusses the structure and function of computer hardware, to the degree necessary for understanding of the role that it plays in supporting GIS. We assume readers are already somewhat familiar with the architecture of a computer, so this section contains only a brief overview of those components directly relevant to GIS. More detailed texts on hardware are referenced in the bibliographic notes.

### 1.4.1   Overall computer architecture

von Neumann
architecture

Despite the wide range of tasks required of computers, almost all contemporary computers conform to the *von Neumann architecture*, developed by the Hungarian-born mathematician John von Neumann at Princeton University during the 1940s and 50s. According to the von Neumann architecture, a computer system can be thought of as comprising four major subsystems, which are illustrated in Figure 1.14 and described below.

*Processing*: Data processing consists of operations performed to combine and transform data. Complex data processing functions may be reduced to a small set of primitive operations.

*Storage*: Data is held in storage so that it may be processed. This storage may range from short term (held only long enough for the processing to take place) to long term (held in case of future processing needs).

*Control*: The storage and processing functions must be controlled by the computer, which must manage and allocate resources for the processing, storage, and movement of data.

*Input/output*: Computers must be able to accept data input and to output the results of processing operations. Two important classes of input/output are discussed in more detail in later chapters: input/output between humans and computers and input/output between different computer systems, especially when mediated by a digital communication network.

In terms of components rather than functionality, the key computer system components are the CPU (responsible for processing and control), memory devices (responsible for data storage), and input/output devices (such as human input/output devices and computer networks). Each of these components is considered in turn in the following sections.

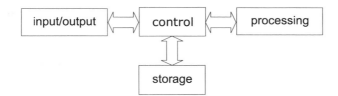

**Figure 1.14:**
The four major functional components of a computer

### 1.4.2 Processing and control

Processing of data in the computer hardware is handled by the *central processing unit* (CPU). The CPU's main function is to execute machine instructions, each of which performs a primitive computational operation. The CPU executes machine instructions by fetching data into special registers and then performing computer arithmetic upon them.

CPU

The CPU is itself made up of several subcomponents, most important of which are the *arithmetic/logic unit* (ALU) and the *control unit*. The control unit is responsible for the control function, managing and allocating resources. The ALU is responsible for the actual processing functions.

ALU
control unit

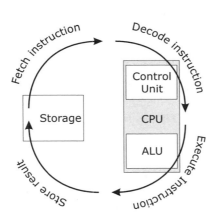

**Figure 1.15:**
The instruction cycle

Operations are performed upon data sequentially, by retrieving stored data, executing the appropriate operation, and then returning the results to storage. The process of execution is known as the *instruction cycle* (also termed the *machine cycle* or the *fetch-execute cycle*). The instruction cycle involves four steps, shown in Figure 1.15. The first step is for the control unit to retrieve an instruction from storage. Next, the control unit decodes the stored instruction to determine what operation must be performed. The control unit then passes this instruction to the ALU, which completes the actual task of executing the operation. Finally, the results of the execution are returned to storage, ready to be retrieved for a subsequent instruction cycle. Connectivity between the CPU and other components in the computer is provided by dedicated communication wires, each called a *bus*. While almost all computers rely on the

instruction cycle

bus

**CISC and RISC** *We usually think of hardware as completely separate from software: hardware is real computing objects you can touch, software is digital instructions and data. However, the distinction is not always so clear-cut. The set of instructions supported by a CPU (hardware) is a determining factor in the complexity of programs (software) running on that CPU. Most personal computers use a complex instruction set computer (CISC) architecture, such as that found in the Intel Pentium family of CPUs. In a CISC architecture, the CPU supports tens or even hundreds of instructions, so making software simpler and cheaper to program. However, complex instructions are generally slower to execute than simple instructions. A reduced instruction set computer (RISC) architecture supports fewer instructions than a CISC architecture. RISC has the advantage of leading to generally faster and cheaper CPUs with lower power consumption. Most small and mobile computing devices use a low-power RISC architecture, such as the ARM family of CPUs. By making the hardware simpler, RISC architectures need more complex software to achieve the same tasks. CISC and RISC are good illustrations of why the distinction between hardware and software can be blurred. In fact, the distinction between CISC and RISC is also becoming blurred, as the latest CISC CPUs incorporate many of the ideas developed for RISC architectures, and new RISC CPUs use as many instructions as older CISC architectures.*

instruction cycle, computer processors differ in the types and range of instructions they implement (see "CISC and RISC" inset, on this page).

### 1.4.3   Storage devices

Digital data must be physically kept somewhere in the computer system. Storage devices differ in their capacity (how much data can be stored), performance (how quickly the data can be accessed), volatility (whether stored data persists after power to the system is turned off), and price. Storage devices can be divided into two categories:

- Storage that can be directly manipulated by the CPU is termed *primary storage*, and is usually based on semiconductor technology. Primary storage is relatively expensive per stored bit, compared to secondary storage, and is generally *volatile* (stored data is lost when the power to the storage device is turned off).

primary storage

volatile

- Storage that can be accessed only indirectly by the CPU (via input/output controllers) is termed *secondary storage*. Secondary storage is usually based upon magnetic or optical technology. Secondary storage is relatively cheaper than primary storage and is normally *non-volatile* (stored data persists after the power to the storage device is turned off).

secondary storage

Figure 1.16 shows some examples of common primary and secondary storage devices. Primary storage generally provides faster access to data than secondary storage. However, primary storage normally offers lower data storage capacity than secondary storage. The balance between primary and secondary storage is critical to the day-to-day performance of

**Primary storage**

ROM and CPU cache memory
(on board CPU)

RAM

**Figure 1.16:**
Common
storage devices

**Secondary storage**

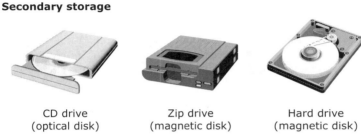

CD drive
(optical disk)

Zip drive
(magnetic disk)

Hard drive
(magnetic disk)

a GIS. In general, because of the large sizes of data sets, it is impractical to handle all the data required by a GIS in primary storage. Therefore, efficient structuring of data files on secondary storage devices is an important factor in GIS performance. Much effort and ingenuity has been spent by computer scientists devising suitable data structures to ensure good performance with spatial data. To understand the issues that arise with such data structures, it is worth examining in more detail some of the properties of the typical primary and secondary storage technology.

Primary storage is directly accessible to the CPU. The CPU requires its own local memory in the form of *registers* so that data may be operated on by the ALU, and the control unit also requires some internal memory. Such *register memory* is the fastest and most expensive of all types, the speed of which must be comparable with instruction cycle times.

register memory

Other types of primary storage are referred to as *main memory*. The most common main memory is *random access memory* (RAM), a misnomer because all main memory is randomly accessible (i.e., directly addressable). RAM is volatile, thus requiring a continuous power supply, and capable of both read and write access. RAM is used for holding data that is coming and going between the CPU and peripheral devices. *Read-only memory* (ROM) is non-volatile and contains data that is "hard-wired," in the sense that it is written during the ROM chip manufacture and cannot subsequently be changed. ROM is therefore only useful for storing permanent data, such as systems routines for often-used functions (for example, graphics functions in a GIS). Other types of ROM allow data to be stored after manufacture. PROM (Programmable ROM) allows the writing of data to be done, once, after chip manufacture. EPROM (Erasable PROM) and EEPROM (Electrically EPROM) allow multiple

main memory
RAM

ROM

non-volatile writing and rewriting of data, although usually at the cost of
flash memory    lower data access speeds than RAM. *Flash memory* is a type of EEPROM
with relatively high speed data access, often used in small storage devices
and configurable mobile computing devices, like cell phones.

The most important form of secondary storage is the magnetic disk.
A magnetic disk is coated with a thin layer of magnetic material. The
polarization of minute regions of the disk can be accessed or changed
by an electrical read/write head. A combination of disk rotation and
movement of the read/write head provides access to the entire surface
of the disk. The disk itself rotates at several thousand RPM, so the time
seek time    taken to move the head, termed *seek time*, is the overriding factor in access
to data on a disk. Thus, there is an advantage in physically structuring
the data on the disk so as to minimize seek time as far as possible.
The physical structure of geospatial data on a disk is a question that has
received attention from the computer science community. Optical storage,
like CDs and DVDs, has a similar structure (rotating disk and read/write
heads), but uses laser technology to read and write from a photosensitive
(rather than magnetic) disk.

Magnetic and optical disks read and write data by finding a specific
*block* of data on the disk based on a unique physical address, a process
direct access    known as *direct access*. After reaching the desired block, the data from the
block is scanned in sequence until the precise read/write location is found,
sequential access    a process known as *sequential access*. By analogy, when locating a bar of
music on a CD we may go directly to the desired track (direct access) and
then scan through the music track for the specific bar (sequential access).
The time taken to access data varies for direct access, while the time taken
for sequential access is dependent on how far through the sequence we
need to search. Continuing the musical analogy, the time taken to find the
bar of music within a track will depend on how far the bar is from the
beginning of the music.

### 1.4.4   Human input/output devices

An array of different devices exists to enable humans to input data into
computers and receive a computer's output. One of the most common
input devices is the keyboard (or keypad), used for entering numerical and
textual data. Pointing devices are a second common class of input devices.
Unlike keyboards, which use fairly standard configurations of buttons
(for example, the keypad buttons on a telephone), pointing devices come
in many shapes and sizes. The mouse, touchpad, and stick pointer are
examples of pointing devices that can be found on many desktop or
portable PCs, while joysticks and touch screens are also familiar pointing
devices.

One specialized type of pointing device that is of particular relevance
digitizer    to GIS is the *digitizer*. A digitizer is a combination of a large flat *tablet*
and a *puck* or *stylus* that can be moved across the tablet (see Figure
1.17). Digitizers are needed for legacy data capture: converting geospatial

information on paper maps into a digital form. A human operator places the map securely on the tablet and then uses the puck to trace out the shape of map features, such as the course of a river or the boundary of a an administrative district. As the puck moves across the tablet its position is recorded by a computer. Other input devices can be used to input data in an audible form, such as a microphone used in combination with a speech recognition system, or in a optical form, such as scanners used for converting maps or pictures into digital images.

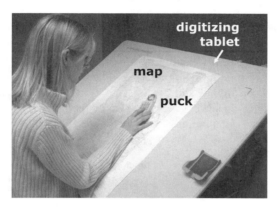

**Figure 1.17:** Digitizing tablet

Output from the computer to a human user can be classified into two categories: *hard copy* and *soft copy*. Hard copy output devices, such as printers and plotters, produce output with some physical permanence. GIS has often been associated with expensive high-quality plotters, needed to produce large format detailed paper maps. However, most GIS users rely largely or entirely on soft copy output devices, primarily the computer screen or other video display unit (VDU). Soft copy output devices produce output that is transient and intangible, like the image on a VDU. Sound from audio devices, such as speakers, is also a type of soft copy output. For GIS applications, the highly visual nature of human perception and of geospatial information means that audio output is of less importance than visual display. Chapter 8 discusses human input/output in more detail, including some examples of GIS applications where audio output is important.

hard copy

soft copy

### 1.4.5   Computer networks

Communication between computing devices connected by a network has become a fundamental part of most of today's information systems, including GISs. Most communication networks are *digital*, meaning a series of bits is transmitted using signal bursts at different intensities corresponding to the binary values 0 and 1 (Figure 1.18a). Older data communication technology sometimes uses *analog* signals, meaning the signal strength can vary continuously like a sine wave (Figure 1.18b).

There are a variety of techniques for encoding digital data within an analog signal, including modulating (varying) the amplitude (intensity)

digital

analog

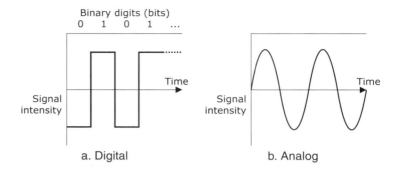

**Figure 1.18:**
Digital and
analog signals

a. Digital                                 b. Analog

of the wave signal. An example of amplitude modulation for encoding
digital data within an analog signal is illustrated in Figure 1.19. Converting between digital and analog signals requires a device called a

*modem*

*modem*. Analog signals degrade more easily than digital signals, so most
remaining analog communication technology is being replaced by digital
technology.

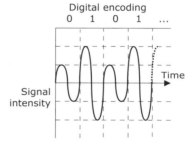

**Figure 1.19:**
Amplitude
modulation
encoding
digital data
within an
analog signal

Irrespective of whether digital or analog signals are being transmitted, all communication networks use electromagnetic (EM) radiation to
propagate signals. EM radiation can be thought of as a wave traveling
through a medium. The *frequency* and *wavelength* of EM radiation affect

frequency
wavelength

its transmission properties. Frequency is the number of cycles a wave
completes per unit time, while the wavelength is the length of each cycle.
For a wave traveling at constant speed, such as EM radiation traveling
through a vacuum, frequency is inversely proportional to wavelength (that
is to say, as wavelength increases frequency decreases).

EM radiation with shorter wavelengths, such as infrared and visible
light, can carry more data than radiation with a longer wavelength, such
as microwaves and radio waves, because the shorter wavelengths allow
the signal bursts to be shorter. However, shorter wavelength EM signals
degrade more quickly than longer wavelengths, and so are harder to use
over longer distances. The range of wavelengths or frequencies available

bandwidth

for data transmission is called the *bandwidth*. There is a clear relationship
between the bandwidth of a signal and the amount of data that can be
carried by the signal: higher bandwidth means greater data transmission
capacity.

Figure 1.20 summarizes the magnetic spectrum and its data-transmission capabilities. Radio waves, microwaves, infrared, and visible light can all be used for computer networks. High-frequency ultraviolet, X-ray, and gamma ray EM radiation are not used for data transmission, because their high energies can be hazardous to the environment and human health.

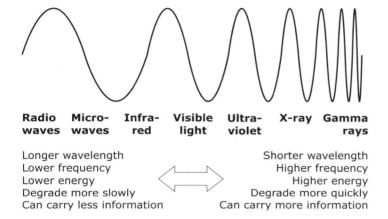

**Radio waves  Micro-waves  Infra-red  Visible light  Ultra-violet  X-ray  Gamma rays**

Longer wavelength          Shorter wavelength
Lower frequency            Higher frequency
Lower energy               Higher energy
Degrade more slowly        Degrade more quickly
Can carry less information  Can carry more information

**Figure 1.20:** Electromagnetic spectrum

Signals carried by EM radiation can be transmitted using different media. The most common media used for data transmission are metal, normally copper, wires like those used in conventional phone lines and coaxial computer network cables. Conventional telephone networks often have low data transmission capacities, because they make inefficient use of the bandwidth available in the copper wires. Normal voice communications require only a small frequency range for transmission, leaving most of the available bandwidth unused. Digital communications using the telephone network often have to squeeze into the small bandwidth available for voice communications. Many telephone networks have now been upgraded to make better use of bandwidth, and are able to transmit digital data alongside voice communications, using frequencies that are not needed for voice communications (termed digital subscriber line, DSL).

Increasingly, fiber-optic cables are replacing copper wire-based cables. Fiber-optic cables use visible light to transmit data within fine glass fibers. Visible light offers greater bandwidth for data transmission than radio waves, so fiber-optic cables can carry more data than copper cables. Additionally, fiber-optic cables are more reliable and less susceptible to interference than copper wires.

Data transmission can also take place unguided through the Earth's atmosphere, termed *wireless* communication. Wireless communications normally use radio wave or microwave EM radiation to transmit data. To combat signal degradation and interference from environmental radiation, wireless signals must be relayed via antennas, either on the ground or on satellites. The shorter wavelengths of visible light and infrared make

wireless

these types of EM radiation less suitable for wireless communications. Using radio and microwave signals for wireless networks carries the advantage that the low-frequency radiation can penetrate non-metallic obstacles, such as walls and floors, while visible light and infrared cannot. As a result, networks based on visible light and infrared signals are line-of-sight often termed *line-of-sight* technology, although some diffuse optical and infrared wireless networks have been developed that do not require a direct line of sight.

Most computers today are able to communicate with other computers via a digital communication network, either using a *connection-oriented* or *connectionless* service. A connection-oriented service requires that a user's computer must explicitly connect with a remote computer before any communication can occur. For example, to access a network via a telephone line, it is normally necessary to first dial up a remote computer connectionless to initiate the connection. A connectionless service allows computers to continually communicate with each other without the need to establish an explicit connection before communication begins. Most dedicated data communication networks rely on connectionless services, and many telephone networks are now beginning to offer connectionless services to replace their older connection-oriented services.

Communication networks are distinguished according to the size of geographic area they cover. A *local-area network* (LAN) connects LAN groups of computers located within a small geographic area. A *wide-area network* (WAN) connects groups of LANs together. The Internet is the WAN largest WAN in existence, estimated in 2001 to connect more than 100 million hosts.

From a GIS perspective, digital communications are radically altering the way people use geospatial information. Digital communications promote sharing of geospatial information. Many companies and mapping agencies now offer a variety of geospatial information that can be downloaded from the Internet. This in turn reduces the costs of data capture and analysis for other organizations that use this data. Digital communications are also *rapid*. The speed of digital communications makes possible certain modes of computing, that would not otherwise be possible. For example, mobile GIS applications, like the tourist navigation system described in section 1.2.3, rely heavily on rapid digital communications. The topic of digital communications in GIS is examined in more detail in Chapter 7.

## BIBLIOGRAPHIC NOTES

1.1 There are now a great many general GIS textbooks, of which only a few examples are mentioned here. Longley et al. (2001) is an excellent introductory GIS textbook, which includes a thorough review of basic GIS functionality. Clarke (2002) is another popular introductory-level GIS textbook. Jones (1997) emphasizes the cartographic applications in GIS, while Burrough and

McDonnell (1998) is aimed toward environmental applications. Affectionately known as the "Big Book," Longley et al. (1999) is a more advanced two volume collection of chapters covering many aspects of GIS, both theory and applications. The first edition of this book is still relevant and at the time of writing some chapters are freely available online (Maguire et al., 1991).

1.2  Volume two of Longley et al. (1999) provides many examples of specific applications of GIS. Examples of other application-oriented GIS texts include Clarke et al. (2001) (environmental modeling applications of GIS) and Miller and Shaw (2002) (transportation applications in GIS). The topic of spatial analysis is introduced in Fotheringham et al. (2002) and O'Sullivan and Unwin (2002). Openshaw (1991) summarizes the methods available for spatial analysis and suggests spatial analysis functionality relevant for GIS. Burrough and McDonnell (1998) and Weibel and Heller (1991) give overviews of visibility analysis. The inspiration for some of the example Potteries applications was provided by Phillips (1993).

1.3  Databases are the primary topic of the next chapter. Rigaux et al. (2001) and Shekhar and Chawla (2002) are two books that address GIS from a spatial databases perspective. Elmasri and Navathe (2003) is a more general databases textbook that includes a short section on GIS.

1.4  O'Leary and O'Leary (2003) and Williams et al. (2001) are examples of very basic introductory texts on information systems, which contain chapters on hardware and computer networks. Thompson and Thompson (2002) is a slightly more advanced book on PC hardware. Tannenbaum (2002) and Stallings (1999) cover both basic and more advanced networking topics.

# Fundamental database concepts

2

**Summary**

*This chapter is about **databases**, and the role they play in a GIS. Many databases rely on the **relational model**, which structures data into a set of tables called **relations**. **Query languages** allow users to interact with the database. A common query language for relational databases is **SQL**. Relational databases are often developed using a modeling technique called **entity-relationship** (E-R) modeling. **Object-oriented** modeling is another approach, based on **objects** with **state** and **behavior** as instances of **classes**.*

The database is the foundation of a GIS. A knowledge of the fundamental principles of databases is necessary to understand GIS technology. Many existing GISs are built upon general-purpose relational databases; certainly all GISs will connect with such systems in a distributed environment. This chapter introduces the reader to the main principles of databases. The general database approach is introduced in section 2.1, along with the high-level architecture of a database system. The most common database model is the *relational model*, described in detail in section 2.2. Based on an understanding of the relational model, the principles of database development are introduced in section 2.3. The object-oriented model extends the relational model in a way that offers several particularly useful advantages for GIS. The object-oriented model and object-oriented database systems are described in section 2.4.

## 2.1 INTRODUCTION TO DATABASES

To introduce the database approach, it is useful to contrast databases with more traditional computing paradigms. Before databases, computers were almost universally used to convert one data set into another by means of a large and complex transformation process. For example, we might wish to use a transportation model to predict average annual traffic flows. The input to this system is data about city populations and about the road network. The transformation is encoded as a procedural program based upon the transportation model. The output is the traffic flow prediction data.

This approach of thinking of the computer as a "giant calculator" is illustrated in Figure 2.1. The approach is sometimes called *file processing*, because the input and output data sets are commonly held in individual computer files.

file processing

**Figure 2.1:**
The "computer as a giant calculator" paradigm

input                                                    output

process

Treating the computer as a giant calculator has several disadvantages, discussed in more detail below. In short, such an approach tends to lead to significant duplication of both data and processing. The alternative offered by the database approach is shown in Figure 2.2. In this case, the computer acts as a useful repository of data, allowing the deposit, storage, and retrieval of data. Stored data can be accessed, modified, and analyzed in a standard way while it is in the store, ensuring that these and other basic functions are never duplicated.

**Figure 2.2:**
The "computer as data repository" paradigm

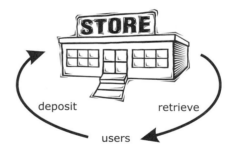

deposit                                        retrieve

users

The remainder of this section provides more detail on the database approach. A summary of the key features of the database approach can be found in the "Databases in a nutshell" inset, on the facing page.

### 2.1.1   The database approach

The "computer as giant calculator" and "computer as data repository" are extreme positions. Most applications require a balance of calculation (processing) and a repository of data upon which the processes are to act. We illustrate this balance with a fictitious example.

**Databases in a nutshell**  *In order to act effectively as a data store, a computer system must have the confidence of its users. A database must be dependable and continue to operate correctly even in the case of unexpected events, such as power or hardware failure (reliability). As far as possible data in the database should be correct and consistent (integrity). A database must be able to prevent data being used in unauthorized ways (security), but offer sufficient flexibility to give different classes of users different types of access to the store (user views). Ideally, the database interface should be easy to use for casual or first-time users as well as offering more powerful functions for regular users (user interface). Most users will not be concerned with how the database works and should not be exposed to low-level database mechanisms (data independence). Users should be able to find out what is in the database (self-describing). Many users may wish to use the store, perhaps at the same time or accessing the same data (concurrency). Databases should be able communicate with each other in order to access remote data (distributed database). Finally, a database should be able to retrieve data rapidly (performance). All these functions are managed by a dedicated software application (database management system).*

"Nutty Nuggets" is a vegetarian fast-food restaurant, established in the 1970s. The owner-manager, a computer hobbyist, decided to apply computer technology to help with some aspects of the management of the restaurant. She began by writing a suite of programs to handle the menu. Menu items were stored as records in a menu file. Programs were written to allow the file to be modified (items deleted, inserted, and updated) and for the menu to be printed each day. Figure 2.3 shows on the left the menu file being operated on by the two programs and on the right the constitution of a menu item record of the menu file. The menu file is held in the operating system and accessed when required by the programs.

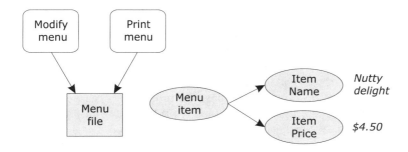

**Figure 2.3:**
Nutty Nuggets
stage one:
Menu system

Time passed, the menu system was successful, and the owner gained the confidence to extend the system to stock management. A similarly structured stock system was set up, consisting of a stock file, and programs to modify the stock file and periodically print a stock report. Once stock and menu details were in the system, it became apparent that changes in stock costs influenced menu prices. A program was written to use stock costs to price the menu. Stage 2 of the Nutty Nuggets system is shown in Figure 2.4.

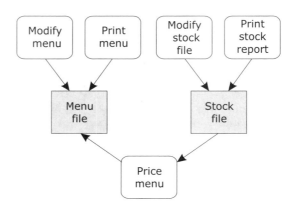

**Figure 2.4:**
Nutty Nuggets
stage two:
Menu, stock,
and menu
pricing system

The system continued to grow, with new files for supplier and customer details added. However, as the system became enlarged, some
problems began to emerge, including:

*Loss of integrity*:  Linkages between programs and files became complex.
    The programs encoded the relationships between the data items in
    the files. If the relationships changed, then the programs had to be
    changed. The development of software was becoming complex and
    costly, leading to errors. For example, the only supplier of a crucial
    raw material went out of business and was deleted from the supplier
    file, but the material supplied was not deleted from the stock file.

*Loss of independence*: The close linkage between program and data
    caused high software maintenance costs. For example, a change in
    secondary storage medium required a partial rewriting of many of
    the programs.

*Loss of security*:  A personnel file was added without regard to security,
    and it was discovered that a member of staff had been accessing
    personal information about colleagues.

The database philosophy is an attempt to solve these and other
problems that occur in a traditional file processing system. Figure 2.5
shows a reorganization of the system, so that all data files are isolated
from the rest of the system and accessible to the processes only through
a controlled channel. The idea is to place as much of the structure of
the information into the database as possible. For example, if there is
a relationship between suppliers and the goods they supply, then this
self-describing          relationship should be stored with the data. Databases are *self-describing*,
as they encode both data and the structure of that data. The means of
expressing the structure and relationships in the data is provided by the
data model. The data model also allows the user to enter into the database
any properties of the data that are expected always to be true (*integrity
integrity constraint    constraints*). Integrity constraints are an aid to maintaining correctness
of the data in the database, because they only allow modifications to the

database that conform to these constraints. An example of an integrity constraint would be that the price of any menu item must be greater than zero but less than $10.

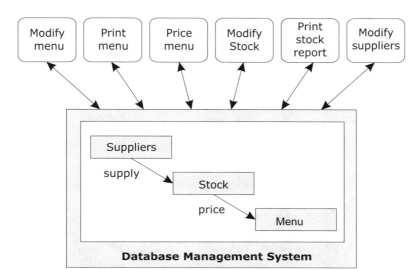

**Figure 2.5:** Nutty Nuggets stage three: Database approach

The data is collected in one logical, centralized location. A DBMS (database management system) manages the database by insulating the data from uncontrolled access. A DBMS allows the definition of the data model, supports the manipulation of the data, and provides controlled two-way access channels between the exterior and the database. A database should allow different users to customize their *view* of the data (see "ANSI/SPARC architecture" inset, on the next page). A database also enables a designer to define the structure of the data in the database, providing levels of authorization that permit different groups of users secure access to data. Different users may access the database at the same time, termed *concurrency*. A DBMS also allows users to access the data in the database without precise knowledge of implementation details, termed *data independence*.

view

concurrency

data independence

A database can now be more precisely defined as a unified computer-based collection of data, shared by authorized users, with the capability for controlled definition, access, retrieval, manipulation, and presentation of data. Examples of common types of database application include:

*Home/office database*: Home/office database systems often run on general-purpose hardware and do not require concurrent multi-user access. As a result they are relatively inexpensive. An example is the Nutty Nuggets system.

*Commercial database*: Databases are widely used to manage the information of businesses. Commercial databases must be secure and reliable, run on many platforms, offer high performance, and allow

**ANSI/SPARC architecture** *A key element of database philosophy is the ability to provide multi-user access to the data in a database. Not all users have the same requirements of the database: each user group may require a particular window onto the data. This is the concept of a* view. *Views provide users with their own customizable data model, which is a subset of the entire data model, and the authorization to access the sectors of the database that fall within their domain.*

*There is a distinction between the data model for the entire database (*global conceptual schema*) and a data model for a subclass of users (*local conceptual schema *or* external view*). Data independence is provided by separating the implementation details (handled in the* internal schema*) from the higher-level, user-oriented considerations provided by the* conceptual schema*. This layered model of a generic database architecture is embodied by the so-called ANSI/SPARC database architecture shown right. The DBMS provides mappings between each layer. Users may only interact with the database by means of an external view.*

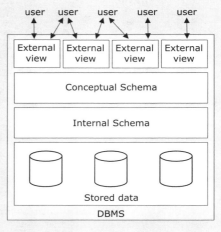

concurrent access by different groups of users. An example might be the database used by a bank to manage customer accounts.

*Engineering database*: Engineering databases are often used for design tasks, and are characterized by graphics-based data. The management of multiple versions of designs is usually important, with long and complex transactions with the database (when modifying a design). An example would be a computer-aided design (CAD) system for developing a new engine or turbine.

*Image and multimedia database*: Multimedia databases allow the storage and retrieval of a wide range of complex data types, which often require much storage space, such as image, audio, and video. Examples of searchable multimedia databases can be found on many search engines on the WWW.

*Geodatabase*: Geodatabases store a combination of spatial and non-spatial data and require complex data structures and analyses. They are discussed throughout this book.

### 2.1.2   Elements of a database management system

This section describes the main components of a DBMS that allow it to perform the complex functions described above, summarized in Figure

2.6. To illustrate, imagine that someone at Nutty Nuggets wants to know the quantity of rice in the stock room. This task will require retrieval of data from the database, in particular from the stock file. A retrieval request must be communicated to the database, either using a database interaction language (usually called a *query language*) or through a user interface. Let us suppose that a query language expression is used.

query language

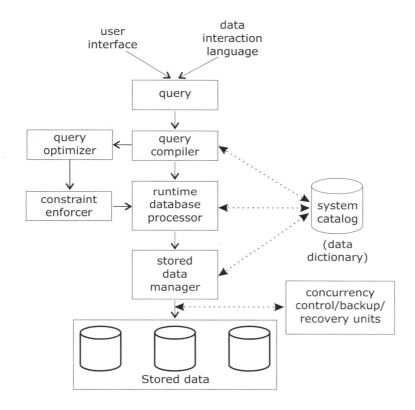

**Figure 2.6:** DBMS components used to process user queries

The DBMS has a *query compiler* that will parse and analyze the query and, if all is correct, generate the execution code that is passed to the *runtime database processor*. Along the way, the compiler may call the *query optimizer* to optimize the code, so that performance on the retrieval is as good as possible (usually there are several possible execution strategies). The compiler may also use the *constraint enforcer* to check that any modifications to the database (none in this case) satisfy any integrity constraints in force. Access to DBMS data is handled by the *stored data manager*, which calls the operating system for control of the physical access to storage devices. Auxiliary units may be used to handle transaction management, concurrency control, back-up, and recovery in the case of system failure. Further DBMS elements, not considered here, include utilities for loading large files into the database and for monitoring performance so that database tuning can take place.

query compiler

constraint enforcer

stored data manager

To retrieve the results of the query, the data that is physically located on the storage device must be mapped to the high-level objects in the query language statement. These mappings are made using the *system catalog*, also termed the *data dictionary*, which stores the information about the data model (such as the internal schema, conceptual schema, and external views described in the "ANSI/SPARC architecture" inset, on page 40).

system catalog

### 2.1.3 Transaction management

transaction

A *transaction* is an atomic unit of interaction between user and database. Transactions are primarily of the following types:

- Insertion of data into the database
- Modification of data in the database
- Deletion of data in the database
- Retrieval of data from the database

The different transaction types depend on two primitive operations: read an item from store (usually a disk) into main memory, and write an item from main memory into store. Two key issues are addressed by transaction management: support for *concurrency* and *recovery control*.

**Table 2.1:**
Lost update for non-atomic interleaved transactions, $T_1$ and $T_2$, with variables $X$ and $Y$ and bank balance $B$

| $T_1$ | $T_2$ | $B$ | $X$ | $Y$ |
|---|---|---|---|---|
| | | $1000 | | |
| $X \leftarrow B$ | | $1000 | $1000 | |
| $X \leftarrow X + \$300$ | | $1000 | $1300 | |
| | $Y \leftarrow B$ | $1000 | $1300 | $1000 |
| | $Y \leftarrow Y - \$400$ | $1000 | $1300 | $600 |
| $B \leftarrow X$ | | $1300 | $1300 | $600 |
| | $B \leftarrow Y$ | $1300 | $600 | |

Concurrent access confronts us with this problem: if the same data item is involved in more than one concurrent transaction, the result may be a loss of database integrity. For example, suppose that my bank balance is \$1000. Two transactions are in progress: $T_1$ to credit my account with \$300, and $T_2$ to debit \$400 from my account. Table 2.1 shows a particular sequence of the constituent operations of each transaction, termed *interleaving*. Transaction $T_1$ begins by reading my balance $B$ from the database into a program variable $X$ and increasing $X$ to \$1300. Transaction $T_2$ then starts by reading the same balance from the database into $Y$ and decreasing $Y$ to \$600. $T_1$ then concludes by writing $X$ to the database as the new balance of \$1300, and $T_2$ writes $Y$ to the database as the new balance of \$600. It is as if transaction $T_1$ never occurred, a problem known as *lost update*. Interleaving can improve database performance, because shorter operations may be executed while more lengthy

interleaving

lost update

operations are still in progress. However, interleaving must be controlled to avoid problems such as lost update.

*Recovery management* concerns the retrieval of a valid database state following a system failure. Imagine that you are able to update the balance of your bank account by inserting a check at an ATM. You insert the check. The system logs the check as accepted and then crashes due to an unforeseen failure. Has the database been updated with your new balance or not? The DBMS must ensure that either the entire transaction has completed or, if not, that it can recover its state before the transaction started.

*recovery management*

To avoid lost update and to aid recovery management transactions should aim to ensure that:

- The constituent operations of a transaction must either all have their effect on the database or all make no change to the database, termed *transaction atomicity*.

*transaction atomicity*

- Each transaction should have an effect on the database independently of the interleaving of constituent operations of other transactions submitted concurrently, termed *transaction independence*.

*transaction independence*

These aims are achieved by two DBMS operations, commit and rollback transactions.

*Commit transaction*: signals a permanent change to the database when all constituent operations have been successfully completed.

*Rollback transaction*: recovers the state of the database immediately prior to the transaction if there is any problem with the transaction.

## 2.2  RELATIONAL DATABASES

There are two main models of the overall structure of a database system (but see "Alternative database architectures" inset, on the next page). The most widely used model, found in most of today's database systems, is the *relational* model. The other important model is the *object-oriented* model, mainly used in particular application areas, like GIS. The focus of this section is the relational model, introduced in a classic 1970 paper by Ted Codd. The object-oriented model is tackled in section 2.4.

### 2.2.1  The relational model

In the Nutty Nuggets example, we have seen that a database holds not only primary data files but also connections between the data items in the files. In the traditional file-processing approach, these connections would be maintained by software programs, but it proves more efficient to hold the connections within the database. These connections are at the heart of the relational model. No single piece of data provides information on its own: it is the relationships between data items that provide much of the context for the data.

**Alternative database architectures** *Aside from relational and object-oriented models, three other models are less commonly encountered in connection with databases. Many early database systems were based on the hierarchical model. Hierarchies are efficient storage structures, but are not expressive enough for modeling many everyday phenomena. For example, a library where books may have multiple authors and authors may have written multiple books cannot be represented in hierarchical database. The network model overcame the restrictions of the hierarchical model, by allowing such many-to-many relationships. However, the flexible structure of the network model is too expressive, leading to highly complex database systems, particularly compared with the relational model. Both hierarchical and network models are now relatively infrequently used in GIS or database systems generally. The third model is the deductive database model. The relational model can be viewed as allowing the expression of a set of facts, such as "Screen 1 at the Majestic is showing Malcolm X." In addition to facts, deductive databases allow the storage of rules, such as "Screen 1 at the Majestic and Screen 2 at the Regal will always show the same films." A deductive database has the ability to make inferences based on the stored facts and rules. For example, the fact and rule given above would allow a deductive database to infer "Screen 2 at the Regal is showing Malcolm X." In this way, storing general propositions can obviate the storage of much data. Deductive databases are an active research area, although not yet widely used in practice. Further information on the use of deductive databases with geospatial information can be found in Abdelmoty et al. (1993) and Paton et al. (1996).*

The structure of a relational database is very simple (this is what makes it so powerful). A relational database is a collection of tabular *relations*, often just called *tables*. Table 2.2 shows part of a relation called FILM (from the CINEMA database given in full in Appendix A) containing data about some popular films, their names, directors, country of origin, year of release, and length in minutes. A relation has associated with it a set of *attributes*: in Table 2.2 the attribute names are TITLE, DIRECTOR, CNTRY, YEAR, and LNGTH, labeling the columns of FILM. The data in a relation is structured as a set of rows. A row, or *tuple*, consists of a list of values, one for each attribute. Each cell contains a single attribute occurrence, or *value*. The tuples of the relation are not assumed to have any particular order.

attribute

tuple

**Table 2.2:**
Part of the
relation FILM

| TITLE | DIRECTOR | CNTRY | YEAR | LNGTH |
|-------|----------|-------|------|-------|
| A Bug's Life | Lasseter | USA | 1998 | 96 |
| Traffic | Soderbergh | USA | 2000 | 147 |
| Die Another Day | Tamahori | UK | 2002 | 132 |
| Malcolm X | Lee | USA | 1992 | 194 |
| American Beauty | Mendes | USA | 1999 | 122 |
| Eyes Wide Shut | Kubrick | USA | 1999 | 159 |
| ... | ... | ... | ... | ... |

We make a distinction between *relation scheme*, which does not include the data but gives the structure of the relation, and *relation*, which includes the data. Data items in a relation are taken from *domains* (like data types in programming). Each attribute of the relation scheme is associated with a particular domain. In basic database systems the possible domains are often quite limited, comprising character strings, integers, floats, dates, etc. In our example, the attribute DIRECTOR might be associated with character strings of length up to 20. We may now give some definitions.

- A *relation scheme* is a set of attribute names and a mapping from    relation scheme
  each attribute name to a domain.

- A *relation* is a finite set of tuples associated with a relation scheme    relation
  in a relational database such that:

  - Each tuple is a labeled list containing as many data items as
    there are attribute names in the relation scheme.
  - Each data item is drawn from the domain with which its
    attribute type is associated.

- A *database scheme* is a set of relation schemes and a *relational*    database scheme
  *database* is a set of relations.

- The database software that manages a relational database model is
  termed a *relational database management system* (RDBMS).    RDBMS

Relations have the following properties:

- The ordering of tuples in the relation is not significant.

- Tuples in a relation are all distinct from one another.

- Columns are ordered so that data items correspond to the attribute
  in the relation scheme with which they are labeled.

Most relational systems also require that the data items are themselves *atomic*; i.e., they cannot be decomposed as lists of further data items. Thus a single cell cannot contain a set, list, or array. Such a relation, which contains only atomic attributes, is said to be in *first normal form* (1NF).    1NF
In the example of the FILM relation, first normal form means that films are not allowed to have multiple directors or alternative titles. The *degree* of    degree
the table is the number of its columns. The *cardinality* of the table is the    cardinality
number of its tuples. As tuples come, go, and are modified the relation will change, but the relation scheme is relatively stable. The relation scheme is usually declared when the database is set up and then left unchanged through the lifetime of the system, although there are operations that will allow the addition, subtraction, and modification of attributes.

The theory of a relational database so far described has concerned the structuring of the data into relations. The other aspects of the relational model are the operations that may be performed on the relations (database manipulation) and the integrity constraints that the relations must satisfy. The manipulative aspects will be considered next, after we have described our working example.

*Relational database example:* CINEMA

We will work with a hypothetical relational database called the CINEMA database. The full CINEMA database can be found in Appendix A. The idea is that data on the cinemas in the Potteries region is to be kept in a single relational database. The database holds data on cinemas and the films they are showing at a given time. The database is not *historical* because it only stores the current showings of films, and does not keep records of past showings nor projections of future showings. The CINEMA database scheme is as follows:

CINEMA (<u>CIN ID</u>, NAME, MANAGER, TELNO, TOWN, GRID_REF)
SCREEN (<u>CINEMA ID</u>, <u>SCREEN NO</u>, CAPACITY)
FILM (<u>TITLE</u>, DIRECTOR, CNTRY, YEAR, LNGTH)
SHOW (<u>CINEMA ID</u>, <u>SCREEN NO</u>, <u>FILM NAME</u>, STANDARD,
   LUXURY)
STAR (<u>NAME</u>, BIRTH_YEAR, GENDER, NTY)
CAST (<u>FILM STAR</u>, <u>FILM TITLE</u>, ROLE)

Each relation scheme is given as its name followed by the list of its attributes. Thus, each cinema in the database is to have an identifier, name, manager name, telephone number, town location, and geospatial grid reference. Cinemas have screens, given by number, each with an audience capacity. Films have titles, directors, country of origin, year of release, and length in minutes. A film may be showing on a screen of a cinema. STANDARD and LUXURY refer to the two-tier ticket prices. Each film star has a name, year of birth, gender, and nationality (NTY), and has roles in particular films. For brevity, the corresponding domains have been omitted. Note that the same attribute may have different names in different relations (e.g., CIN_ID and CINEMA_ID).

candidate key     A *candidate key* is an attribute or minimal set of attributes that will serve to uniquely identify each tuple of the relation. On its own, the screen number is not a candidate key (because different cinemas may each have a "Screen One"). The combination of the two attributes CINEMA_ID and SCREEN_NO form a candidate key for relation SCREEN. There may be several such candidate keys for a relation. One candidate key is usually

primary key     chosen as the *primary key*. There is a convention that the set of attributes constituting the primary key of the relation is underlined.

## 2.2.2   Operations on relations

A relation is nothing more than a structured table of data items. Each column of a relation is named by an attribute, and has all its data items taken from the same domain. The basic operations supported by a relational database are therefore simple. There are five fundamental

relational operator     relational operators: *union, difference, product, project,* and *restrict*. The first three of these are traditional set-based operators, introduced in the next chapter, while the project and restrict operators are described below. Three further relational operators *intersection, divide,* and *join,* termed *derived* relational operators, can be expressed using different combinations of the fundamental five operators. Of these, intersection is another set-based operator introduced in the next chapter, join is described below, and divide is a less commonly used operator not discussed further here. The structure of these operations and the way that they can be combined is called *relational algebra*.

relational algebra

The relational model is *closed* with respect to all the above relational operations, because they each take one or more relations as input and return a relation as result. The set operations union, intersection, product, and difference work on the relations as sets of tuples. Thus, if we have relations holding female and male film stars, then their union will hold stars of both genders and their intersection will be empty. For all the set operations except product, the relations must be *compatible*, in that they must have the same attributes; otherwise the new relation will not be well formed.

The *project* operation is unary, applying to a single relation. It returns a new relation that has a subset of attributes of the original. The relation is then modified so that any duplicate tuples formed are coalesced. The project operator $\pi$ has the following syntax:

project operator

$$\pi_{<attribute\ list>}(relation)$$

For example, $\pi_{\text{NAME,TOWN}}(\text{CINEMA})$ returns the relation shown in Table 2.3a, and $\pi_{\text{NAME}}(\text{CINEMA})$ returns the relation shown in Table 2.3b. Note that in the second case the two identical tuples containing the value "Regal" have been coalesced into a single tuple.

(a)

| NAME | TOWN |
|------|------|
| Majestic | Stoke |
| Regal | Hanley |
| Regal | Newcastle |

(b)

| NAME |
|------|
| Majestic |
| Regal |

(c)

| TITLE | DIRECTOR | CNTRY | YEAR | LNGTH |
|-------|----------|-------|------|-------|
| The Hunted | Friedkin | USA | 2003 | 94 |
| The Hours | Daldry | USA | 2002 | 114 |
| Die Another Day | Tamahori | UK | 2002 | 132 |
| X2 | Singer | USA | 2003 | 133 |

**Table 2.3:** Results of relational projections and restrictions

(d)

| DIRECTOR |
|----------|
| Friedkin |
| Daldry |
| Tamahori |
| Singer |

restrict operator

The *restrict* operation is also unary. The restrict operator works on the tuples of the table rather than the columns, and returns a new relation that has a subset of tuples of the original. A condition specifies those tuples required. The restrict operator is often referred to as the *select* operator, and consequently is denoted with the Greek symbol $\sigma$ (sigma). The syntax used here is:

$$\sigma_{<condition>}(relation)$$

For example, films released after 2001 can be retrieved from the database using the expression $\sigma_{\text{YEAR}>2001}(\text{FILM})$. This will return the relation shown in Table 2.3c. Operations can be combined, for example:

$$\pi_{\text{DIRECTOR}}(\sigma_{\text{YEAR}>2001}(\text{FILM}))$$

returns the directors of films released after 2001, as shown in Table 2.3d.

join operator

With the *join* operation, the relational database begins to merit the term "relational." Join is a binary operator that takes two relations as input and returns a single relation. The join operation allows connections to be made between relations. There are several different kinds of relational

natural join

join but we describe only the *natural join* of two relations, defined as the relation formed from all combinations of their tuples that agree on a specified common attribute or attributes. The join operator $\bowtie$ has the following syntax:

$$\bowtie_{att_1=att_2}(rel_1, rel_2)$$

to indicate that relations $rel_1$ and $rel_2$ are joined on attribute combinations $att_1$ of $rel_1$ and $att_2$ of $rel_2$. For example, to relate details of films to the screens on which they are showing, relations SHOW and FILM are joined on the film title attribute in each relation. The expression is:

$$\bowtie_{\text{FILM\_NAME}=\text{TITLE}}(\text{SHOW}, \text{FILM})$$

The resulting relation, shown in Table 2.4a, combines tuples of SHOW with tuples of FILM, provided that the tuples have the same film name. Notice that the join has not repeated the duplicate attribute. If we only require the directors of films showing at cinema 1, we may restrict the relation to tuples whose CINEMA_ID has value '1'; then project the relation in Table 2.4a to have only attributes CINEMA_ID, SCREEN_NO, FILM_NAME, and DIRECTOR. The result of this compound operation is shown in Table 2.4b, and can be represented symbolically as:

$$\pi_{\text{CINEMA\_ID},...}(\sigma_{\text{CINEMA\_ID}=1}(\bowtie_{\text{FILM\_NAME}=\text{TITLE}}(\text{SHOW}, \text{FILM})))$$

The last example may be used to demonstrate an important property of relation operations: the order in which operations are performed will affect performance. The join operation is the most time-consuming of all relational operations, because it needs to compare every tuple of one relation with every tuple of another. To extract data for Table 2.4b we performed operations join, project, and restrict. In fact, it would have been

(a)

| TITLE | DIRECTOR | CNTRY | YEAR | LNGTH | CINEMA_ID | SCREEN_NO | STANDARD | LUXURY |
|---|---|---|---|---|---|---|---|---|
| X2 | Singer | USA | 2003 | 133 | 1 | 1 | £5.50 | £7.00 |
| American Beauty | Mendes | USA | 1999 | 122 | 1 | 2 | £5.50 | £6.50 |
| The Hours | Daldry | USA | 2002 | 114 | 1 | 3 | £5.00 | |
| Training Day | Fuqua | USA | 2001 | 120 | 2 | 1 | £5.00 | £6.00 |
| Traffic | Soderbergh | USA | 2000 | 147 | 2 | 2 | £4.50 | £6.00 |
| X2 | Singer | USA | 2003 | 133 | 3 | 1 | £6.00 | |

(b)

| TITLE | DIRECTOR | CINEMA_ID | SCREEN_NO |
|---|---|---|---|
| X2 | Singer | 1 | 1 |
| American Beauty | Mendes | 1 | 2 |
| The Hours | Daldry | 1 | 3 |

**Table 2.4:** Results of relational joins, projections, and restrictions

more efficient to have first done a restrict operation on the SHOW table, then joined the resulting smaller table to FILM, and then projected. The result would be the same but the retrieval would perform better, because the join involves smaller tables.

In general, reordering the elements of a relational algebra expressions may not lead to an equivalent expression. For example, the relational algebra expression $\pi_{\text{DIRECTOR}}(\sigma_{\text{YEAR}>2001}(\text{FILM}))$ given above is not equivalent to $\sigma_{\text{YEAR}>2001}(\pi_{\text{DIRECTOR}}(\text{FILM}))$, because YEAR is not a valid attribute of the relation $\pi_{\text{DIRECTOR}}(\text{FILM})$. The topic of *query optimization* is a critical study for high-performance databases, concerned with processing queries as efficiently as possible. An important component of query optimization involves performing transformations (such as reordering) upon queries, to produce equivalent queries that can be processed more efficiently. The bibliographic notes provide starting points for readers wishing to research the literature on query optimization.

query optimization

### 2.2.3   Structured query language

The *structured query language* (SQL) provides users of relational databases with the facility to define the database scheme (data definition), and then insert, modify, and retrieve data from the database (data manipulation). The language may either be used on its own, as a means of direct interaction with the database, or embedded in a general-purpose programming language. SQL originally arose out of the relational database language SEQUEL (structured English query language). The aim of this section is to provide an introduction to SQL, without making any attempt to be a complete SQL reference (see the bibliographic notes for further references on SQL).

*Data definition using SQL*

The data definition language (DDL) component of SQL allows the creation, alteration, and deletion of relation schemes. Normally, a relation scheme is altered only rarely once the database is operational. A relation scheme provides a set of attributes, each with an associated data domain. SQL allows the definition of a domain by means of the expression below (square brackets indicate an optional part of the expression).

> **CREATE DOMAIN** *domain-name data-type*
> [*default definition*]
> [*domain constraint definition list*]

The user specifies the name of the domain and associates that name with a predefined data type, like character string, integer, float, date, or time. The default definition allows the user to specify a default value for a tuple: a usual default value is NULL. The domain constraint definition list acts as an integrity constraint by restricting the domain to a set of specified values. An example of the definition of a domain for the attribute GENDER is as follows:

**CREATE DOMAIN** GENDER CHARACTER(1)
    **CHECK VALUE IN** {'M', 'F'};

A relation scheme is created as a set of attributes, each associated with a domain, with additional properties relating to keys and integrity constraints. For example, the relation scheme CAST can be created by the command:

**CREATE TABLE** CAST
    (FILM_STAR STAR,
    FILM_TITLE FILM_TITLE,
    ROLE ROLE,
    **PRIMARY KEY** (FILM_STAR, FILM_TITLE),
    **FOREIGN KEY** (FILM_STAR)
      **REFERENCES** STAR (NAME),
    **FOREIGN KEY** (FILM_TITLE)
      **REFERENCES** FILM (TITLE),
    **CHECK** (FILM_STAR **IS NOT_NULL**),
    **CHECK** (FILM_TITLE **IS NOT_NULL**));

This statement begins by naming the relation scheme (called a table in SQL) as CAST. The attributes are then defined by giving their name and associated domain (assume that we have already created domains STAR, FILM_TITLE, and ROLE). The primary key is next, given as the attribute combination FILM_STAR, FILM_TITLE. A *foreign key* is a primary key of another relation contained in the given relation. In the CAST table, there are two foreign keys: FILM_STAR and FILM_TITLE. For example, FILM_TITLE occurs as the primary key (TITLE) of the FILM relation. Referential integrity can be maintained by the database by ensuring that if a film is deleted (or updated) from the FILM relation, then any reference to it is also deleted (or updated) in CAST. Finally, two further integrity checks are added to limit the insertion of NULL values on data entry to the ROLE attribute only: any attempt to insert a row with no entry for FILM_STAR or FILM_TITLE will be disallowed.

*foreign key*

### Data manipulation using SQL

Having defined the relation schemes, the next step is to insert data into the relations. These SQL commands are quite straightforward, allowing insertion of single or multiple tuples, update of tuples in tables, and deletion of tuples, and will not be covered here. Data retrieval forms the most complex aspect of SQL: a large book could be written on this topic alone. Our treatment is highly selective, giving the reader a feel for SQL in this respect. The general form of the retrieval command is:

**SELECT** *select-item-list*
    **FROM** *table-reference-list*
    [**WHERE** *condition*]
    [**GROUP BY** *attribute-list*]
    [**HAVING** *condition*]

A simple example of data retrieval, already considered in the relational algebra section, is to find the names of all directors of films that were released after 2001. The corresponding SQL expression is:

**SELECT** DIRECTOR
    **FROM** FILM
    **WHERE** YEAR>2001;

The **SELECT** clause serves to project (how confusing!) on the required DIRECTOR attribute. The **FROM** clause tells us from which table the data is coming, in this case FILM. The **WHERE** clause provides the restrict condition.

Relational joins are effected by allowing more than one relation (or even the same relation called twice with different names) in the **FROM** clause. For example, to find details of films and where they are showing, we would give the following SQL command:

**SELECT** CINEMA_ID, SCREEN_NO, FILM_NAME, DIRECTOR
    **FROM** SHOW, FILM
    **WHERE** SHOW.FILM_NAME=FILM.TITLE;

In this case, the **WHERE** clause provides the *join condition* by specifying that tuples from the two tables are to be combined only when the values of the attributes FILM_NAME in SHOW (indicated by SHOW.FILM_NAME) and TITLE in FILM (FILM.TITLE) are equal. A more complex case, using all the clauses of the **SELECT** expression, is the following expression, which retrieves the average lengths of films for actors that have at least three USA films in the database.

**SELECT** FILM_STAR, **AVG**(LNGTH)
    **FROM** FILM, CAST
    **WHERE** CNTRY = 'USA' **AND** CAST.FILM_NAME=FILM.TITLE
    **GROUP BY** FILM_STAR
    **HAVING COUNT**(*) > 2;

**AVG** and **COUNT** are built-in SQL functions. With the exception of **AVG**(LNGTH), the first three lines of code act to retrieve the cast of USA films, using the join of FILM and CAST. The **GROUP BY** clause serves to logically construct a table where the tuples are in groups, one for each film star. This table (Table 2.5) is not a legal first normal form relation because values in some cells are not atomic. But we are concerned with the average film length for each film star, and will eventually project out titles. The **HAVING** clause comes into play to operate as a condition on the groups in the grouped relation. It selects only groups that have a tuple count of at least three. The final stage of the retrieval, after projecting out the unwanted attributes, is shown in Table 2.6.

### 2.2.4  Relational databases for spatial data handling

In an unmodified state relational databases are unsuitable for spatial data management. Although relational databases have all the characteristics

| FILM_STAR | FILM_TITLE | ROLE | DIRECTOR | CNTRY | YEAR | LNGTH |
|---|---|---|---|---|---|---|
| | The Usual Suspects | Fred Fenster | Singer | USA | 1995 | 106 |
| Benicio Del Toro | Traffic | Javier Rodriguez Rodriguez | Soderbergh | USA | 2000 | 147 |
| | The Hunted | Aaron Hallam | Friedkin | USA | 2003 | 94 |
| | Malcolm X | Malcolm X | Lee | USA | 1992 | 194 |
| Denzel Washington | Philadelphia | Joe Miller | Demme | USA | 1993 | 125 |
| | Training Day | Alonzo Harris | Fuqua | USA | 2001 | 120 |
| Halle Berry | X2 | Storm | Singer | USA | 2003 | 133 |
| | Monster's Ball | Leticia Musgrove | Forster | USA | 2001 | 111 |
| | The Usual Suspects | Roger "Verbal" Kint | Singer | USA | 1995 | 106 |
| | American Beauty | Lester Burnham | Mendes | USA | 1999 | 122 |
| Kevin Spacey | A Bug's Life | Hopper | Lasseter | USA | 1998 | 96 |
| | Midnight in the Garden of Good and Evil | James Williams | Eastwood | USA | 1997 | 155 |
| | Moulin Rouge | Satine | Luhrmann | USA | 2001 | 127 |
| Nicole Kidman | The Hours | Virginia Woolf | Daldry | USA | 2002 | 114 |
| | Eyes Wide Shut | Alice Harford | Kubrick | USA | 1999 | 159 |

**Table 2.5:** Evaluation of an SQL query to the CINEMA database: intermediate stage

**Table 2.6:**
Evaluation of
an SQL query
to the CINEMA
database: final
result

| FILM_STAR | AVG(LNGTH) |
|---|---|
| Benicio Del Toro | 116 |
| Denzel Washington | 146 |
| Kevin Spacey | 120 |
| Nicole Kidman | 133 |

required for general corporate data management, there are difficulties when the technology is applied to spatial data. The main issues are:

*Structure of spatial data*: Spatial data has a structure that does not naturally fit with tabular structures. Vector areal data is typically structured as boundaries composed of sequences of line segments, each of which is a sequence of points. Such sequences of arbitrary length violate first normal form. Later chapters construct models of spatial data, where the problems become apparent.

*Performance*: In general, to reconstruct a spatial object requires the joining of several tables. As discussed above, joins impose performance overheads. Typically, many such spatial objects are required quickly, for example, to show a map on the screen. Consequently, it is difficult to achieve good performance with standard relational database technology.

*Indexes*: Indexing questions will be considered in Chapters 5 and 6. An index increases the speed of access to data. Relational databases provide indexes that perform well with standard tabular data (e.g., a file of stock items). Spatial data requires specialized indexes, which are not always supported by proprietary RDBMSs, although many do now offer some such indexes.

The key issue facing spatial data handling in relational databases is that more complex structures than pure relations are required. This, in turn, leads to a need for specific operations and appropriate indexing techniques for these structures. An *extensible RDBMS* is designed to provide specialist users with the facilities to add extra functionality, specific to their domains. Such facilities may include:

extensible
RDBMS

- user-defined data types;
- user-defined operations on the data;
- user-defined indexes and access methods; and
- active database functions (e.g., triggers).

Many of these aims are shared by object-oriented databases, to be discussed later in this chapter. The characteristic of extensible RDBMSs is that the extra functionality is built upon the relational model with as little change as possible.

## 2.3   DATABASE DEVELOPMENT

A DBMS is a general-purpose information system that must be customized to meet the requirements of particular applications. In order to
do this, we need to have a precise idea of the way that information is
structured in the system and the kinds of algorithms that will act upon
the data. Here, we are not concerned with the actual data in the database,
but with the kinds of data that we expect. For example, in a cartographic
application, we are not so much concerned with individual data items,
"London," "New York," "France," "Ben Nevis," as with data types, **city**,
**country**, and **mountain**. We abstract from information system content to
information structure.

Development of a GIS, like any information system, should follow
the system development process introduced in section 1.3.4. This section
focuses particularly on system analysis and design, with an eye to implementation in a relational database. The main task for the GIS analyst is
the construction of the conceptual computational model for the database,
termed a *conceptual data model*. A GIS designer will then tailor the
conceptual data model to the particular kind of DBMS on which the
system will be implemented, called a *logical data model*. For example, if
the DBMS is relational, then part of the design stage will be the creation
of relation schemes. Thus, the logical model may be mapped directly to an
implementation, while itself being independent of the details of physical
implementation.

### 2.3.1   Conceptual data modeling

A *conceptual data model* is a model of the proposed database system that
is independent of any implementation details. A conceptual model must
express the structure of the information in the system: that is, the types
of data, and their interrelationships. Such structural properties are often
termed *static*, but the system will also have a *dynamic* component related
to its behavior in operation. For example, the allowable transactions with
the system shape its dynamic properties.

The correctness (integrity) of the information in a system is often a
critical factor in its success. Correctness is maintained as much as possible
by the specification of integrity constraints that impose conditions on the
static and dynamic structures of the system. A data model should allow
the specification of a range of integrity constraints.

A good conceptual data model can act as an efficient means of
communication between the analyst, designer, and potential users. This
will aid the design and implementation of the system. Further, when the
system is eventually implemented, the conceptual data model provides a
basic reference for users who need to understand the structure of the data
in the system (see Figure 2.7). In summary, conceptual models:

conceptual data
model

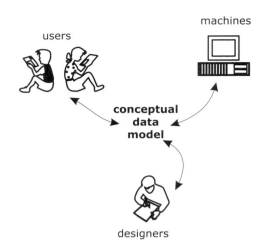

machines

users

**conceptual data model**

designers

**Figure 2.7:**
The conceptual
data model as
mediator
between users,
designers, and
machines

1. Provide a framework that allows the expression of the structure of
   the system in a way that is clear and easy to communicate to non-
   specialists.

2. Contain sufficient modeling constructs so that the complexity of the
   system may be captured as completely as possible.

3. Have the capability for translation into implementation-dependent
   models (i.e., logical and physical models), so that the system may
   be designed and built.

*Entity-relationship model*

Imagine that we are designing a database that contains spatially referenced information about a specific location, for example, the Potteries
region. This hypothetical system should contain data on administrative
units (e.g., towns, districts, wards); transportation networks (e.g., road,
rail, canals); physical features (e.g., lakes, rivers); and spatially referenced
attributes (e.g., areas and heights of physical features, populations of
towns and cities, traffic loads on stretches of roads). How would we
make a start? Well, we have already started in that we have elicited some
requirements of the system, and expressed these requirements in the form
of collections of entities and their relevant properties. This is the simple
and powerful idea behind one of the most compelling and widely used
approaches to forming a conceptual model of an information system: the
*entity-relationship model* (E-R model).

E-R model
entity type

An *entity type* is an abstraction that represents a collection of similar
objects, about which the system is going to contain information. In our
example, some of the entity types might be **town**, **district**, **road**, **canal**,
and **river**. We make a distinction between the *type* of an entity and an
*occurrence* or *instance* of an entity type. For example, we have entity
type **town** and occurrences such as 'Newcastle-under-Lyme', 'Hanley',
and 'Stoke'. By convention for conceptual modeling, types are rendered

entity instance

in bold, and values in single quotes. Constituents of a relational database (relation and attribute names) are rendered in uppercase.

When we write about the town of Newcastle-under-Lyme, we are not being absolutely precise, since, of course, 'Newcastle-under-Lyme' is not a town but a data item that serves to name the town. The data item 'Newcastle-under-Lyme' is associated with a particular occurrence of the entity **town** as its name. Thus, entity types have properties, called *attribute types*, that serve to describe them. For example, entity type **town** has attribute types **name**, **population**, **centroid** (the geometric center of the town, discussed further in the next chapter). A particular occurrence of **town** would have associated with it occurrences of these attributes, such as the value 'Newcastle-under-Lyme' assigned to the attribute **name**. The attachment of attribute types to an entity type may be represented diagrammatically, as in Figure 2.8. Entity types are shown in rectangular boxes. The E-R diagrammatic notation presented here is common but not standard; there are many variations in use.

*attribute type*

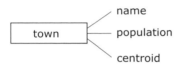

**Figure 2.8:**
An entity type
and its attribute
types

An important characteristic of an entity is that any occurrence of it should be capable of unique identification. The means of unique identification is through the value of a subset of its attribute types. The name of a place, for example, is not always sufficient to identify it uniquely. In the case of 'Newcastle-under-Lyme' the town's name is sufficient, as there is only one 'Newcastle-under-Lyme'. However, there are many places named simply 'Newcastle' (for example, Newcastle near Sacramento, California, and Newcastle in southeast Australia). An attribute type or combination of attribute types that serves to identify an entity type uniquely is termed an *identifier*. By definition, an entity type must have at least one identifier. As in the notation for a relation scheme, the attributes comprising the chosen identifier are often underlined. In our example, attribute type **centroid**, which gives the grid reference of the centroid of a town, is an identifier of **town**.

*identifier*

So far, the model comprises a number of independent entity types, each with an associated set of attribute types, one or more of which serve to identify uniquely each occurrence of the type. The real power in this model comes with the next stage, which provides a means of describing connections between entity types. A diagrammatic language called an *entity-relationship diagram (E-R diagram)* is used for expressing the features and properties of an E-R model. For example, suppose that we have defined the two entity types:

- **town** with attributes <u>**centroid**</u>, **name**, **population**
- **road** with attributes <u>**road id**</u>, **class**, **start_point**, **end_point**

A question that we might want to ask of our finished system is "Which towns lie on which roads?" This question can only be answered by forming a link between towns and the roads upon which the towns

relationship

lie. This connection is called a *relationship*. It is unfortunate that the terms "relationship" (connection between entities) and "relation" (table

relationship type

in a relational database) are so similar. A *relationship type* connects one or more entity types. In the example, the relationship type will be called

relationship occurrence

**lies_on**. A *relationship occurrence* is a particular instance of a relationship type. Thus, the incidence of the town Hinckley with the A5 road from London to Anglesey is an occurrence of the relationship **lies_on**. The relationship **lies_on** between entities, towns, and roads may be shown in an E-R diagram as shown in Figure 2.9. Relationships are shown in a diamond-shaped box.

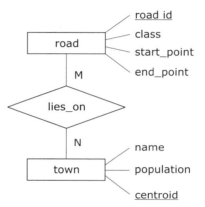

**Figure 2.9:**
Two entities
and a
many-to-many
relationship

Relationships may have their own attributes, which are independent of any of the attributes of the participant entities. In the above example, the relationship **lies_on** might have the attribute **length**, which gives the length of the road within the town boundary. A further attribute might indicate whether the road goes through the town center or bypasses the town.

The entity-relationship model allows the expression of a limited range of integrity constraints. Relationship types are subdivided into many-to-many, many-to-one, and one-to-one relationships. The relationship

many-to-many

**lies_on** is an example of a *many-to-many* relationship, because each town may have several (potentially more than one) roads passing through it and each road may pass through several (potentially more than one) towns. This constraint on the relationship is shown diagrammatically by the $M$ and $N$ on each side of the relationship.

Not all relationships are many-to-many. For example, consider the relationship **located** between types **cinema** and **town** shown as an E-R diagram in Figure 2.10 (attributes omitted). Each cinema is located in at most one town, but a town may contain several (potentially more than

many-to-one

one) cinemas. Such a relationship is called a *many-to-one* relationship. The positioning of the symbols $N$ and 1 indicates the nature and direction of the relationship. Note that the relative positioning of the symbols $N$ and 1 is significant. If these symbols were reversed, the diagram would have a different interpretation (i.e., a rather strange world in which a cinema could be located in several towns, but each town could have at most one cinema).

The third and final relationship category is exemplified by the relationship **manages** between **staff** and **cinema**. In this relationship, a member of staff may be the manager of at most one cinema and a cinema may have at most one manager. This is a *one-to-one* relationship. The form of the E-R diagram is shown in Figure 2.10. The positioning of the two 1's indicates the nature of the relationship.

one-to-one

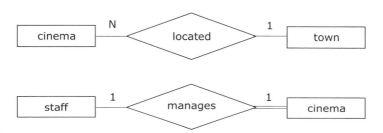

**Figure 2.10:** Many-to-one and one-to-one relationships

In allowing the modeler to express information about the category of a relationship, the E-R modeling tool provides facilities for the expression of *cardinality conditions* upon entity occurrences entering into relationships. The entity-relationship model also allows the expression of another class of integrity constraint, namely, *participation conditions*. Entity types are involved in relationships with other (or the same) entity types. Consider a single entity type $E$ and a single relationship $R$ in which it is involved. There is an important distinction to be made regarding the way that $E$ participates in $R$. It may be that every occurrence of $E$ participates in the relationship. In this case, we say that the *membership class* or *participation* of $E$ in $R$ is *mandatory*. On the other hand, if not every occurrence of $E$ need participate in the relationship $R$, we say that the participation is *optional*.

mandatory participation

optional participation

For example, the participation of entity **staff** in **manages** is optional, because not every member of staff in an imagined database system will be a manager (there may also be office staff, projectionists, ushers, and so forth in the full system). However, the participation of entity **cinema** in **manages** is mandatory, because we will assume that every cinema must have a manager. These conditions are displayed on the E-R diagram by using double lines for mandatory and single lines for optional participation (see Figure 2.10).

All the relationships given as examples so far have been binary, in that they have connected together precisely two entity types. It sometimes happens that these two types are the same. In this case, the relationship is one that relates an entity type to itself. Such a relationship is call *involutory*. For example, the information that roads intersect other roads

involutory

may be represented by the relationship **intersect** between entity type **road** and itself.

*ternary relationship*

A relationship that connects three entity types is called a *ternary* relationship. Imagine that data showing major road usage by different bus companies in the region is to be stored in our Potteries system. The daily totals of vehicle-miles along major roads for each bus company have been recorded for each district in the region. We can model this using a ternary relationship **usage** connecting entities **bus_company**, **road**, and **district**, with attributes **date** and **vehicle_miles**. This is shown in Figure 2.11. Participation conditions apply as with the binary case, but with extra complexity. In this example, bus companies may have routes along roads in several districts; districts may have many roads through them that are used by several bus companies; and roads may pass through several districts and be used by several bus companies. This relationship is many-to-many-to-many.

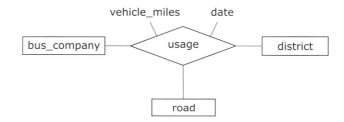

**Figure 2.11:**
A ternary
relationship

*dependent entity*

A *dependent* or *weak* entity is an entity type that cannot be identified uniquely by the values of its own attributes. The complete identification

*identifying relationship*

occurs through an *identifying relationship* between the dependent and weak entities. A good example of this phenomenon is the entity type screen in the CINEMA database. The attribute **screen_no** is insufficient to identify a particular screen; after all, "Screen One" tells us very little.

*partial identifier*

The attribute **screen_no** is a *partial identifier* for the dependent entity. We need the **cinema_id** to complete the identification. The notation for this is shown in Figure 2.12. Weak entities and identifying relationships are distinguished by double-lined boxes and diamonds. A partial identifier is underlined with a dotted line. A dependent entity always has mandatory participation in the identifying relationship; otherwise some occurrences could not be identified. For the same reason, the relationship is one-to-one or many-to-one from dependent to parent.

*Example E-R model*   The E-R model is now applied to the example of the CINEMA database. Imagine that we are at the analysis stage of system development. There is a need to hold information about cinemas and the films that they show. Choosing entities and attributes for a model is a matter of judgment, often with more than one acceptable solution. Sometimes it is difficult to decide whether to characterize something as an attribute or an entity. In the modeling of the CINEMA database, there

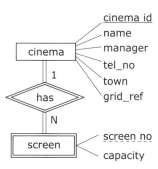

are several such choices. For example, should we make the town in which
a cinema is located an entity or attribute? Some guidelines include:

- If the data type is relatively independent and identifiable, with its
  own attributes, then it is probably an entity; if it is just a property
  of something, then it is an attribute.

- If the data type enters into relationships with other entities (apart
  from being a property of something), then it is probably an entity.

In our system the **town** is just a name and a property of the cinema,
so we choose to make it an attribute of **cinema**. If **town** had possessed
its own attributes, such as **population**, then we would probably have
made it an entity and constructed a relationship between entities **town**
and **cinema**. The same reasoning applies to **manager**, which we choose to
make an attribute of **cinema** because only the manager's name is involved
(it may be less confusing to call these attributes **manager_name** and
**town_name**). Initial investigation reveals that the following entity types
and their attributes are needed:

- **cinema(cinema_id, name, manager, tel_no, town, grid_ref)**
- **screen(screen_no, capacity)**
- **film(title, director, cntry, year, length)**
- **star(name, birth_year, gender, nty)**

Most of the attribute names are self-explanatory, except that attribute
**year** of **film** denotes the year that the film was released and attribute
**nty** of **star** denotes the nationality of the star. It is taken that **cinema_id**
is by definition an identifier of the entity **cinema**. Identifiers for the
other entities are less clear. For the purposes of this system, a film is
to be identified by its title, and a star identified by his or her name (the
assumption is that there cannot then be more than one film with the same
title or more than one star with the same name in the database). Screens
are not identified by their number alone and must also take the identifier
of the cinema for complete identification.

Each screen may be showing one film and the cinemas have a two-tier
ticket pricing scheme of standard and luxury seats; the prices depend upon

the film and on which screen it is shown. Film stars are cast in particular roles in films. We can therefore construct the following relationships:

- **shows** between **screen** and **film**, with attributes **standard**, **luxury**
- **cast** between **star** and **film**, with attribute **role**

There are also some other system rules that we can glean from common knowledge about the film industry.

1. Each cinema may have one or more screens.

2. Each screen is associated with just one cinema.

3. Each screen (not cinema) may be showing zero or one films (the database holds only current information, not past or future information).

4. A film may be showing on zero, one, or several screens.

5. A film star may be cast in one or more films.

6. A film may contain zero, one, or more stars.

All this information can be modeled and represented by the E-R diagram shown in Figure 2.13. The dependent entity and identifying relationship between **screen** and **cinema** have been discussed earlier. The relationship **shows** between **screen** and **film** is constrained, by system rules 3 and 4, to be many-to-one with optional participation from **screen** and **film** (rule 3). The relationship **cast** between **star** and **film** is constrained, by system rules 5 and 6, to be many-to-many with mandatory participation from **star** and optional from **film**.

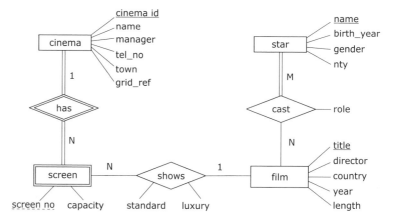

**Figure 2.13:**
E-R diagram
for the CINEMA
database

The E-R model is still the most widely used modeling tool for database design. Measured against the three criteria of a modeling approach given at the beginning of section 2.3.1, it scores highly. First, the method is based upon the intuitive notions of entity, attribute, and relationship, so is easily grasped by non-specialists. E-R modeling therefore provides an excellent means of communication between systems analysts and users during the specification of the system. On the third criterion, we shall see later in the chapter how readily a conceptual data model framed using E-R may be translated into a relational logical data model.

It is the second criterion, regarding the existence of sufficiently powerful modeling constructs, where the shortcomings of the basic E-R model are sometimes said to lie. This is particularly the case when the application domain to be modeled does not fit into the standard pattern and requires complex data types and relationships. Since it is precisely such systems that are the concern of this book, we move on to look at extensions of the E-R model that set out to remedy this deficiency. The price to be paid for any increase in modeling power that such extensions can provide is the slight loss of the natural feel and simple diagrams that the basic E-R model possesses.

### The extended entity-relationship (EER) model

The *extended* or *enhanced* entity-relationship (EER) model has features additional to those provided by the standard E-R model. These include the constructs of *subtype*, *supertype*, and *category*, which are closely related to *generalization* and *specialization* and the mechanism of *attribute inheritance*. This method allows the expression of more meaning in the description of the database. EER leads toward object-oriented modeling. However, the object-oriented approach models not only the structure but also the dynamic behavior of the information system.

*EER model*

An entity type $E_1$ is a *subtype* of an entity type $E_2$ if every occurrence of type $E_1$ is also an occurrence of type $E_2$. We also say that $E_2$ is a *supertype* of $E_1$. The operation of forming subtype from types is called *specialization*, while the converse operation of forming supertypes from types is called *generalization*. Specialization and generalization are inverse to each other.

*subtype*

*supertype*
*specialization*
*generalization*

Specialization is useful if we wish to distinguish some occurrences of a type by allowing them to have their own specialized attributes or relationships with other entities. For example, the entity **travel_mode** may have generic attributes and relationships, such as a relationship with type **traveler**. Entities **road** and **railway** might be modeled as subtypes of **travel_mode** with their own specific properties. Entity **road** might have attributes such as **max_speed** in addition to the attributes of **travel_mode**. Similarly, entity **railway** might have different attributes in addition to those of **travel_mode**.

The E-R diagram can be extended to show specialization. Figure 2.14 shows the example in diagrammatic form. The subtype-supertype relationship is indicated by the subset symbol ⊂. The figure also shows that **canal** and **railway** are *disjoint* subtypes of **travel_mode**, indicated by the letter "d" in the circle. Types are *disjoint* if no occurrence of one is an occurrence of the other, and vice versa. In our example, no arc can be both a canal and a railway. On the other hand, Figure 2.14 also shows two

*disjoint type*

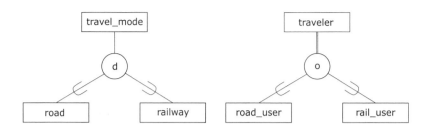

**Figure 2.14:**
Specialization
in the EER
model

overlapping subtypes. Entities **road_user** and **rail_user** are overlapping
subtypes of entity **traveler**, indicated by the letter "o" in the circle. In
our example, a traveler may use both road and rail. As for E-R diagrams,
the double lines indicate mandatory participation (i.e., that every traveler
must use at least one of road or rail).

Specialization may be summarized as the creation of new types such
that:

- each occurrence of the subtype is also an occurrence of the supertype;
- the subtype has the same identifying attribute(s) as the supertype;
- the subtype has all the attributes of the supertype, and possibly
  some more; and
- the subtype enters into all the relationships in which the supertype
  is involved, and possibly some more.

The entire collection of subtype-supertype relationships in a particular
EER model is called an *inheritance hierarchy*.

Generalization is the reverse modeling process to specialization.
Given a set of types, we sometimes want to identify their common properties by generalizing them into a supertype. It might be that our initial
modeling led us to entities **road_user** and **rail_user**. We then realized
that these types had many common features, for example, having similar
identifiers and other attributes, such as **name**, **age**, **gender**, **address**, and
also a common relationship to entity **town** in which they lived. The
commonalities between **road_user** and **rail_user** may be "pushed up"
into a new generalized entity **traveler**. The resulting model is the same
in the case of either specialization or generalization, but the process of
arriving at the model was different.

*inheritance
hierarchy*

### EER for spatial information

E-R or EER modeling may be used to model configurations of spatial
entities. The structure that we use as an example is a simplification of the
model underlying most vector-based GISs (discussed in more detail in
Chapter 5). Figure 2.15 shows a finite region of the plane partitioned into
sub-regions. Each sub-region, which we shall call an **area**, is bounded by
a collection of **directed arcs**, which have **nodes** at their ends. Figure 2.16
shows an EER diagram for these types of configurations.

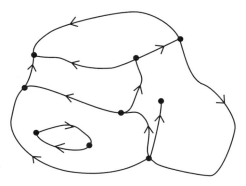

**Figure 2.15:**
Planar
configuration
of nodes,
directed arcs,
and areas

In order to relate a node to a directed arc, we specify the nodes that begin and end the directed arc. To relate an area to a directed arc, we specify the areas to the left and right of it. Subtypes of **area** and **node** are created so that these relationships can be easily defined. Note that both specializations are overlapping: a node may begin one directed arc and end another; an area may be the left area of one directed arc and the right area of another. Also, the mandatory nature of the participation of nodes and areas in the specialization relationships indicates that each node must either begin or end at least one arc (this disallows isolated nodes), and each area must be to the left or right of at least one arc (this disallows areas with no bounding arcs). All relationships are many-to-one between **directed arc** and the related entities. Thus, a directed arc must have exactly one area on its left and one area on its right, and it must have exactly one begin and one end node. These constraints imply, for example, that a directed arc cannot exist without begin and end nodes and that there must be a exterior area defined for the finite planar partition. Further details of these kinds of spatial data models are discussed in Chapter 5.

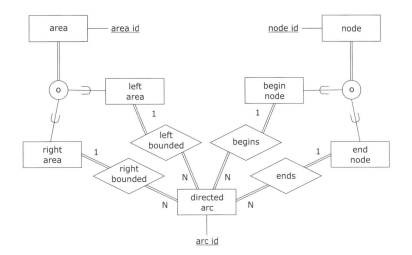

**Figure 2.16:**
EER model of
planar
configurations
of nodes,
directed arcs,
and areas

### 2.3.2    Relational database design

Database design, in the case of a relational database, concerns the construction of a relational database scheme. This section will consider some of the principles upon which the design is based and show how an E-R model can be transformed into a relational database scheme. The central question is "What characterizes a good set of relations for the target application?" Two advantageous features are:

redundancy

- lack of *redundant* data in relations (redundant data wastes space in the database and causes integrity problems); and

- fast access to data from relations.

These two features essentially trade off space (and integrity) against time. The relational join is the most expensive of relational database operations, so fewer joins mean faster data access, which in turn implies fewer relations. Consequently, it is not usually efficient to have many small relations that must be joined in order to respond to common queries. On the other hand, while using fewer relations leads to fewer joins, it leads to other problems, shown by the following example from the CINEMA database.

Suppose that we decide to have all the information about stars and the films in which they are cast in a single relation in the database. The scheme would look something like:

STAR_FILM (FILM_TITLE, DIRECTOR, FILM_CNTRY, RELEASE_YEAR, LNGTH, STAR_NAME, BIRTH_YEAR, STAR_GENDER, STAR_NTY, ROLE)

This relation scheme of large degree might appear suitable if there were likely to be many retrievals requiring star and film details together. The problem is that this scheme results in redundant duplication of data. Table 2.7 shows an example of the problem, with just a few attributes of the relation given. The relation includes redundant data. Thus 'Singer' is shown as the director of 'The Usual Suspects' in two tuples, while Kevin Spacey's and Benicio Del Toro's birth years are both given in triplicate. As mentioned above, such data redundancy wastes space and causes integrity problems (for example, if Kevin Spacey's year of birth is updated in one cell but not in the other two).

The specific problems in the STAR_FILM relation above can be solved by splitting the scheme so that the film data is held in one relation and the star data is held in another. There is also a need for a third relation to hold the castings of stars in films; that is, to make the connections between the two tables of information. The modified relation schemes are:

FILM(TITLE, DIRECTOR, CNTRY, YEAR, LNGTH)
STAR(NAME, BIRTH_YEAR, GENDER, NTY)
ROLE(FILM STAR, FILM TITLE, ROLE)

| TITLE | DIRECTOR | ... | FILM_STAR | BIRTH_YEAR | ... | ROLE |
|---|---|---|---|---|---|---|
| A Bug's Life | Lasseter | | Kevin Spacey | 1959 | | Hopper |
| American Beauty | Mendes | | Kevin Spacey | 1959 | | Lester Burnham |
| The Usual Suspects | Singer | | Kevin Spacey | 1959 | | Roger "Verbal" Kint |
| The Hunted | Friedkin | | Benicio Del Toro | 1967 | | Aaron Hallam |
| The Usual Suspects | Singer | | Benicio Del Toro | 1967 | | Fred Fenster |
| Traffic | Soderbergh | | Benicio Del Toro | 1967 | | Javier Rodriguez Rodriguez |

**Table 2.7:** Some rows and columns of the STAR_FILM relation

(a)

| TITLE | DIRECTOR | CNTRY | YEAR | LNGTH |
|-------|----------|-------|------|-------|
| A Bug's Life | Lasseter | USA | 1998 | 96 |
| American Beauty | Mendes | USA | 1999 | 122 |
| The Hunted | Friedkin | USA | 2003 | 94 |
| The Usual Suspects | Singer | USA | 1995 | 106 |
| Traffic | Soderbergh | USA | 2000 | 147 |

**Table 2.8:**
The
STAR_FILM
relation split
into three
relations FILM,
STAR, and
CAST

(b)

| NAME | BIRTH_YEAR | GENDER | NTY |
|------|-----------|--------|-----|
| Benicio Del Toro | 1967 | M | Puerto Rico |
| Kevin Spacey | 1959 | M | USA |

(c)

| FILM_STAR | FILM_TITLE | ROLE |
|-----------|-----------|------|
| Benicio Del Toro | The Hunted | Aaron Hallam |
| Benicio Del Toro | The Usual Suspects | Fred Fenster |
| Benicio Del Toro | Traffic | Javier Rodriguez Rodriguez |
| Kevin Spacey | A Bug's Life | Hopper |
| Kevin Spacey | American Beauty | Lester Burnham |
| Kevin Spacey | The Usual Suspects | Roger "Verbal" Kint |

The relations with the tuples from Table 2.7 appropriately distributed are shown in Table 2.8a, b, and c. There is now no redundant repetition of data.

Apart from avoiding redundant duplication of information, decomposition of relation schemes has the advantage that smaller relations are conceptually more manageable and allow separate components of information to be stored in separate relations. Of course, relations cannot be split arbitrarily. Relations form connections between data in the database and inappropriate decomposition can destroy these connections. One important guideline for appropriate decomposition has already been introduced in section 2.2.1, with relations in first normal form (1NF) having atomic attributes. In fact, there exists a hierarchy of normal forms for relational databases; higher normal forms require higher levels of decomposition of the constituent database relations. The process of appropriately decomposing relations into normal form is termed *normalization*. Normal forms are useful guidelines for database design. However, in any logical data model the level of normal form (i.e., degree of relation decomposition) must be balanced against the decrease in performance resulting from the need to reconstruct relationships by join operations.

normalization

An E-R model of a system may be straightforwardly transformed into a set of relation schemes that are a good first pass at a suitable database design. The general principle, with modifications in some cases, is that each entity and relationship in the E-R model results in a relation in the database scheme. The following example shows the concepts involved in the transformation from E-R model to logical model. Consider the E-R model of the CINEMA database given in Figure 2.13. The independent entities and their attributes provide the following partial set of relation schemes:

CINEMA(<u>CIN ID</u>, NAME, MANAGER, TEL_NO, TOWN, GRID_REF)
FILM(<u>TITLE</u>, DIRECTOR, CNTRY, YEAR, LNGTH)
STAR(<u>NAME</u>, BIRTH_YEAR, GENDER, NTY)

Special care is required for the dependent entity **screen**. We must add the identifier CINEMA_ID from CINEMA to SCREEN (renamed from CIN_ID) in order to properly identify each tuple in the relation. This relation scheme is then:

SCREEN(<u>CINEMA ID</u>, <u>SCREEN NO</u>, CAPACITY)

It only remains to consider the relationships in the E-R model. Relationships result in relations in two different ways. In our example, each relationship results in a further relation. This relation takes for its identifier the identifiers of the participating entities. Thus, the relationship **cast** leads to the relation scheme CAST with identifier (FILM_STAR, FILM_TITLE), where FILM_STAR is a renaming of NAME from relation STAR and FILM_TITLE renames TITLE from relation FILM. The renaming is just to make the relation schemes more readable and less ambiguous. The relation CAST also picks up the attribute ROLE from the **cast** relationship.

CAST(<u>FILM STAR</u>, <u>FILM TITLE</u>, ROLE)

The other relationship leads in a similar manner to the relation:

SHOW(<u>CINEMA ID</u>, <u>SCREEN NO</u>, <u>FILM NAME</u>, STANDARD, LUXURY)

It may not be necessary to provide a new relation for each relationship in the E-R model. For example, Figure 2.17 shows an E-R diagram of the vector spatial data model, given in EER form earlier in Figure 2.16.

If we adopt the principle that each entity and relationship defines a relation, we have the following database scheme:

AREA(AREA_ID)
NODE(NODE_ID)
ARC(ARC_ID)
LEFT_BOUNDS(ARC_ID, AREA_ID)
RIGHT_BOUNDS(ARC_ID, AREA_ID)
BEGINS(ARC_ID, NODE_ID)
ENDS(ARC_ID, NODE_ID)

This database scheme can be simplified with no loss of connectivity (in the database structure sense). The **begins** relationship between **directed arc** and **node** has the property that each arc has exactly one begin

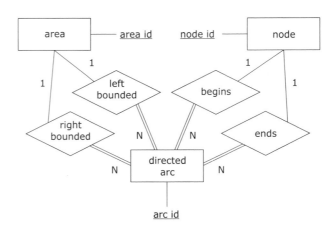

**Figure 2.17:**
E-R model of
planar
configurations
of nodes,
directed arcs,
and areas

node. Thus, rather than have the extra relation BEGINS in the database, it is simpler to add the identifier NODE_ID, in its capacity as a begin node, to the ARC relation. This construction is called *posting the foreign key*. In technical language, we have posted the identifier of NODE as a foreign key into ARC. In fact, because of the nature of the four relationships, each leads to a posting of a foreign key into ARC. The resulting simplified database scheme is:

posting the
foreign key

AREA(<u>AREA ID</u>)
NODE(<u>NODE ID</u>)
ARC(<u>ARC ID</u>, BEGIN_NODE, END_NODE, LEFT_AREA, RIGHT_AREA)

Posting a foreign key, rather than adding another relation, is always an option if the relationship between entities is as shown in Figure 2.18.

**Figure 2.18:**
A relationship
where posting
the foreign key
is an option

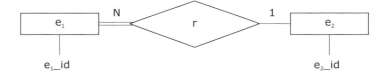

The relationship **r** is many-to-one and mandatory on the side of $\mathbf{e}_1$, so it is the case that each occurrence of $\mathbf{e}_1$ relates to exactly one occurrence of entity $\mathbf{e}_2$. Therefore, if we allow the posting of the foreign key of $\mathbf{e}_2$ into $\mathbf{e}_1$ to make the connection, then for each tuple in the relation for $\mathbf{e}_1$ there will be one and only one value for the foreign key attributes. As a result, there will be no possibilities of either null or repeating values.

### 2.3.3  Summary

Relational databases have a simple and formally convincing underlying model, and are able to implement many aspects of the database philosophy discussed in the first part of the chapter. Despite having its origins more than 30 years ago, the relational model, with the associated RDBMS and interaction language SQL, remains the global leader among database paradigms. However, there are problems with more complex forms of data that are difficult to handle using standard relational technology. Database construction is an art supported by many tools and methodologies. We have spent some time describing the data modeling approach provided by the entity-relationship model. This modeling approach has the advantage of simplicity and ease of transformation to an appropriate database scheme. However, there are other approaches, referenced in the bibliographic notes at the end of the chapter.

## 2.4   OBJECT-ORIENTATION

This section introduces the object-oriented approach as an alternative to the relational model paradigm for information systems. The staged representation of the information system development process (section 1.3.4) leaves us with an important difficulty that motivates the object-oriented approach. The gap between the constructs that are available at different stages of the development process makes the transition from one stage to the next inefficient; information may be lost or a simple concept may get hidden in a complex modeling paradigm. The term *impedance mismatch* is often used to describe this problem of translating information from one modeling stage or level of abstraction to another. An important instance of impedance mismatch occurs if an application domain model, expressed using high-level and domain-specific constructs, must be translated into a low-level computational model in order that it can be implemented. In a nutshell, object-orientation attempts to raise the level of the logical and physical computational modeling environment so that the problem of impedance mismatch is lessened.

*object-oriented*

*impedance mismatch*

### 2.4.1   Foundations of the object-oriented approach

As might be expected, the concept of an *object* is central to the object-oriented approach. Rather like a tuple of a relation in the relational model, an object models the static, data-oriented aspects of information. For example, a **city** object might have **name**, **center**, **population** among its attributes. A particular **city** object might take the value 'Newcastle-under-Lyme' for the **name** attribute. The totality of attribute values for a given object at any time constitutes an object's *state*.

Unlike relations in the relational model, an object also aims to model the dynamic *behavior* of the system. The behavior of an object is expressed as a set of operations that the object can perform under appropriate conditions. For example, the idea of a region is captured not just by specifying a set of points or curves giving the boundary of its extent (the data), but also the operations that we can expect a region to support. Such operations might include the calculation of the region's area and perimeter; plotting the region at different scales and levels of detail; the creation and deletion of regions from the system; and operations that

*object*

*state*

*behavior*

return the lineage of the region (when it was created, to what accuracy, and so forth). For the object-oriented approach, the key notion is that:

$$object = state + behavior.$$

Objects with state and behavior are the foundation of the object-oriented approach, common to all object-oriented systems. Objects interact by sending *messages* to each other, which can activate particular behaviors for another object. A particular behavior for an object is sometimes referred to as a *method*. Exactly how an object will respond to a message in any particular case is dependent on that object's state. An object is thus characterized by its behavior, which is the totality of its responses to messages. Objects with similar behaviors are organized into *classes*. The set of behaviors of a class of objects form an *interface*, defining how objects can interact with one another and with the external environment.

To illustrate, a **window** object might have different states, such as **opened** or **closed**, **broken** or **unbroken**. It may also have behaviors that affect the window's state, such as **open**, **shut**, or **hit**. The effect of a **golf ball** object hitting the **window** is modeled by the **golf ball** sending the **window** a message to activate the window's **hit** behavior. How the **window** responds to that message may depend on its state. If the **window** is already **opened**, the **window**'s state may be unaffected by the message (i.e., the golf ball flies right through the opened window). If the **window** is **closed**, the **window** will change its state from **unbroken** to **broken**, as in Figure 2.19.

*message*

*method*

*class*

*interface*

**Figure 2.19:** Objects send messages to activate other object's behaviors, sometimes changing their state

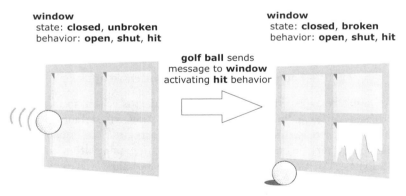

window
state: **closed, unbroken**
behavior: **open, shut, hit**

window
state: **closed, broken**
behavior: **open, shut, hit**

**golf ball** sends message to **window** activating **hit** behavior

From a modeling perspective, object-orientation can be summarized as offering four main features:

*Reduces complexity*: Decomposing complex phenomena into objects with state and behavior helps make conceptual models simpler and easier to manage. Organizing groups of objects with similar behaviors into classes further helps reduce model complexity.

*Combats impedance mismatch*: The object-oriented approach to systems may be applied at every level of the system life-cycle. For example, there are *object-oriented analysis* (OOA) techniques, *object-oriented design* (OOD) methodologies, and *object-oriented programming languages* (OOPLs) for implementation. Consistently applying the same object-oriented concepts throughout the system development process reduces impedance mismatch.

*Promotes reuse*: Systems can be developed more efficiently if they can be constructed from collections of well-understood and specified sub-components. Once a system has been successfully captured in an object-oriented model, it can reused as a sub-component within more complex models.

*Metaphorical power*: Treating complex phenomena as objects is a useful *metaphor* because we are very familiar with handling objects, like windows, in everyday life. However, the same object-oriented metaphors may equally be applied to virtual or digital objects, such as a window in a *graphical user interface* (e.g., Microsoft Windows) or abstract concepts, as in a "window of opportunity." (Object-oriented ideas are applied in many domains outside of computer-based information systems, including project management.)

### 2.4.2   Object-oriented constructs

In addition to the foundational ideas of object-orientation, described in the previous section, four other constructs are important in object-orientation: *identity*, *encapsulation*, *inheritance* and *polymorphism*, and *association*. These constructs are not unique to object-orientation, and not all object-oriented systems include all of these constructs. However, these features are closely associated with object-orientation and appear in some combination in almost all object-oriented systems.

*Object identity*

Objects have unique identities, independent of their attribute values. This corresponds to a natural idea, which can be simply expressed by means of an example.

object identity

> My name is Mike, but I am not my name. My name may change, indeed every cell in my body may change, but my identity remains the same.

When modeling entities using the E-R model, it was stipulated that each entity should be capable of unique identification. Attributes that identified entities were therefore always present, and included in the list of attributes for an entity. Each occurrence of that entity was pinpointed uniquely by giving the values for the identifying attributes. This is sometimes summed up by saying that the E-R model and the relational model are *value-based*. By contrast, the object model is not value-based. An object occurrence retains its identity even if all its attributes change their values. Each object has an internal identity, independent of any of its attribute values (and if the model is implemented, this identity will be stored in the computer). This identity is immutable; it is created when the object is created, never altered, and only destroyed when the object is destroyed. Thus, it is unnecessary to provide instance variables solely for the purpose of identifying an object.

*Encapsulation*

encapsulation

Encapsulation ensures that the internal mechanisms of an object's behavior are rigidly separated from the external access to, and effects of that behavior. To take an example from the real world, I usually do not care about the state of my car under the hood (internal state of instance of class **car**) provided that when I put my foot on the accelerator pedal (send message) the car's speed increases (activate the **accelerate** behavior leading to a change in the observable state of the **car** object).

Encapsulation helps further reduce complexity in the object-oriented model, by treating *what* behaviors an object exhibits separately from *how* an object achieves these behaviors (cf. the description of system analysis and design in section 1.3.4). Encapsulation also promotes reuse. The key to successful reuse is that sub-components should behave in a predictable way in any setting. This can only be achieved if an object's internals are encapsulated to insulate them from the external environment.

Encapsulation has important implications for object-oriented databases. Relational databases are based on the *call-by-value* principle, whereby tuples are accessed and connected to other tuples by means of their values. However, values contained within an object (part of the object's state) are encapsulated and only accessible indirectly via an object's behaviors. This level of indirection can make the performance of a DBMS based on an object-oriented data model unacceptable if not handled properly; we shall return to this issue later.

*Inheritance and polymorphism*

inheritance

Inheritance allows classes, and so objects, to share common properties. Just as for inheritance hierarchies in the EER model (section 2.3.1), inheritance acts in two ways: generalization and specialization, one the inverse of the other. A generalized class will exhibit properties common

subclass
superclass

to all its specialized *subclasses*. Conversely, classes specialize, adapt, and add to the properties of generalized *superclasses*.

Unlike inheritance hierarchies in the EER model, which only allow specialization or generalization of attributes, inheritance in an object-oriented setting also allows specialization or generalization of behaviors. For example, Figure 2.20 shows two classes, **triangle** and **rectangle**, that are subclasses (indicated by an arrow) of **polygon**. As subclasses they inherit all the attributes and behaviors of the superclass as well as adding their own. In the inheritance hierarchy in Figure 2.20, the generalized class **polygon** has behaviors (**draw**, **area**) common to the subclasses **triangle** and **rectangle**. In our example, **triangle** and **rectangle** also have specialized operations, **isEquilateral** and **isSquare**, returning Boolean objects that determine whether the shapes are equilateral or square, respectively. Diagrams like Figure 2.20 that show inheritance hierarchies and class properties are termed *class diagrams*. Inheritance relationships class diagram are often called "is a" relationships, so, for example, a **rectangle** "is a" "is a" relationship **polygon**.

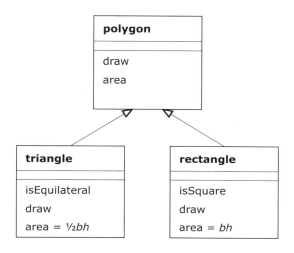

**Figure 2.20:** Simple class diagram

Inheritance enables objects to fulfill different roles in different contexts, a feature known as *polymorphism* (literally "many forms"). Specifi- polymorphism cally, an instance of a subclass can always be substituted for an instance of a superclass. For example, in an algorithm that calculates the total area of a set of **polygon** objects, **rectangle** or **triangle** objects can be substituted for **polygon** objects, because all subclasses of **polygon** possess an **area** behavior. This ability to substitute a subclass for a superclass is a particular type of polymorphism termed *inclusion polymorphism*. Another important type of polymorphism occurs when subclasses implement their own specialized algorithm for achieving generalized behavior, termed *overloading*. For example, **triangle** might implement the behavior **area** using the familiar formula $\frac{1}{2}bh$ (half the base $b$ multiplied by the height $h$ of the triangle), while **rectangle** might use the formula $bh$. The bibliographic notes contain further references to these and some other types of polymorphism.

One further feature of inheritance is worth mentioning. The inheritance hierarchy shown in Figure 2.20 is known as *single inheritance*, single inheritance meaning that a specialized object class inherits all the attributes and behavior of a single generic class (possibly adding some properties of its own). The notion of single inheritance may be extended to that of *multiple* multiple inheritance *inheritance*, meaning that a class is permitted to inherit properties from more than one superclass. Multiple inheritance introduces additional complexities into object-oriented models and systems, as a subclass may inherit different versions of same behavior from different superclasses. In such cases, there needs to exist some kind of protocol for resolving these conflicts, although here we do not go into the detail of precise rules for such conflict resolution.

*Association*

association    Related to the discussion of reuse above, an *association* groups objects together in order to model phenomena with complex internal structure. The relationship between an actor and a film, from the relational CINEMA database discussed earlier, is an example of an association. A special type

aggregation    of association is *aggregation*, which concerns part/whole associations between objects. For example, an object of class **car** is an aggregate of many component objects, including objects of class **seat**, **wheel**, **engine**, etc. These component objects may themselves be aggregated from other components (e.g., **engine** may be an aggregate of **piston**, **crankshaft**, and **cylinder** objects).

"part of"    Aggregation relationships are sometimes called "part of" relation-
relationship    ships, because, for example, an **engine** is "part of" a **car**. Aggregation is *antisymmetric* (if $A$ is part of $B$ it means $B$ cannot be part of $A$). A

partonomy    hierarchy of aggregation relationships is termed a *partonomy* or sometimes a *mereological hierarchy* (in contrast to an inheritance hierarchy,

taxonomy    also termed a *taxonomy*). There are at least two more specialized types of association that are common:

homogeneous    • An association is said to be *homogeneous* if it is formed from a
association       set of objects, *all of the same class*. For example, a **soccer team** is a homogeneous association (aggregation) of **soccer players** (i.e., in this definition, the manager, supporters, etc. are not part of the team).

ordered    • An *ordered association* is an association where the ordering of
association       component objects is important. For example, **point_sequence** might be an object structured as a linear ordering of **point** objects.

### 2.4.3   Object-oriented modeling

The modeling tools associated with relational technology are sufficient for analysis of the kind of information structures that exist in many traditional applications. However, as the technology is applied to wider classes of applications, shortcomings in the E-R model approach become apparent. This is particularly true for applications, like GIS, where entities may have a rich substructure or organization (like partonomies or taxonomies).

An important initial task in object-oriented modeling is defining the classes for the application, and the attributes, behaviors, associations, and inheritance of those classes. Attributes for each class are defined in a similar way to attributes in E-R modeling. For example, the class **car** might have attributes such as **make**, **color**, and **year**. An individual

**Constructors** *In addition to basic constructors, which create a new object based on specification of its class and attributes, constructors may also utilize more complex references to other objects. For example:*

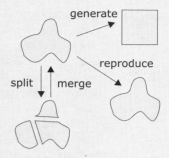

- generate *creates an object that will depend on the referenced object with regard to type, state, or behavior.*
- reproduce *creates an identical copy of the referenced object with regard to type, state, and behavior (i.e., everything except identity).*
- split *creates a set of objects whose composition (aggregation) is the referenced object.*
- merge *creates a single object that is the composition (aggregate) of the referenced objects.*

*From the definitions above, illustrated above right, it should be clear that reproduce is a special case of generate, while split and merge are inverse to each other.*

object of class **car** will assign particular values to these attributes, which constitutes the object's state (e.g., for the object 'my_car' the **make** is 'Chevrolet', **color** is 'tan', **year** is '1993'). Behaviors for a class can be separated into three different categories.

- *Constructors* are behaviors that are activated when objects are created, while *destructors* are activated when an object is deleted. Constructors and destructors are particularly useful for ensuring data integrity. For example, a **road** object should always have a **centerline**, which describes the geometry of the road. A constructor could ensure that a new **road** can only be created by specifying what that geometry is. Some other types of constructor are illustrated in the "Constructors" inset, on this page. *(constructor, destructor)*

- *Accessors* are behaviors that may be used to examine the state of an object. For example, the total length of a **road** object might be accessed by the behavior **length**. *(accessor)*

- *Transformers* are behaviors that change the state of objects. For example, traffic-calming measures to reduce the maximum speed of vehicles might result in the **speed** attribute of a **road** object being reduced. This change may be communicated to the appropriate object by means of the behavior **set_speed**. *(transformer)*

Defining association and inheritance relationships between classes is typically an iterative and application-dependent process. A good rule of thumb for detecting inheritance and aggregation relationships is to try to use "is a" and "part of" in a sentence connecting different classes. For example, the sentence "An expressway 'is a' road" makes sense, suggesting that **expressway** may be a subclass of **road**. In contrast, the sentences "A road is 'part of' an expressway" or even "A road 'is an' expressway" make less sense, and suggest that the corresponding aggregation and inheritance relationships are not correct.

Class diagrams, such as that in Figure 2.20, are important tools in developing and communicating object-oriented models. Figure 2.21 shows an example of a class diagram that includes a variety of attributes, behaviors, and association relationships, as well as inheritance. There exist several different graphical notations for class diagrams; the one used in this book is based on the most common notation, Unified Modeling Language (UML). As in Figure 2.20, Figure 2.21 shows the classes (**point**, **centerline**, **road**, and **expressway**) as labeled boxes. The inheritance relationship between **road** and **expressway** (a road "is an" expressway) is indicated by an arrow emanating from the subclass and pointing to the superclass.

Association is indicated using a line between two classes. As in E-R diagrams, the cardinality of the association can be added to the line. Each **road** is associated with one **centerline**, and a **centerline** may be associated with zero or one **road** objects. Aggregation is indicated by a diamond headed arrow pointing from the part to the whole. In the example in Figure 2.21, a **point** is part of a **centerline**. Again, we may indicate cardinality of aggregation, so that a centerline is composed of two or more points, and a point is part of zero or more centerlines. We may add further detail by annotating the aggregation line with the word "ordered" to indicate that the ordering of points in a centerline is significant.

A class diagram also lists the attributes for each class, if any, and then each class's behaviors, if any. In Figure 2.21, **road** has attributes **name** and **speed**. An important aspect of object-oriented modeling is that attribute values are themselves objects. For example, the **name** of a **road** (such as 'A500', see Figure 1.2) is itself an object belonging to the class **string**. The class of an attribute is indicated next to the name of the attribute in a class diagram.

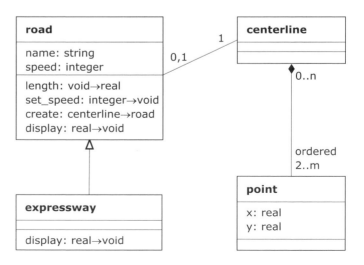

**Figure 2.21:**
Road class
diagram

Finally, a class diagram lists the behaviors associated with each class. For example, the class **road** has the behavior **length**, which returns the length of the road (an object of class **real**). The word "void" is used in the definition of **length** to indicate that no information is required as input to this behavior, because the road length can be calculated via the association between the **road** and its **centerline** (using some procedure not yet defined). Behaviors **length** and **display**, which display the road at a certain scale, are examples of accessor operations. The behavior **set_speed** is a transformer operation, changing the **speed** attribute. The **set_speed** behavior requires an **integer** object as input (the new speed), but returns no value.

The **create** behavior is an example of a constructor. The **create** behavior requires an object of class **centerline** as input. This ensures that a **road** object cannot be created without first creating a **centerline** object. As indicated above, such constructors are useful for ensuring data integrity. Constructors and destructors that do not perform some integrity function are usually omitted from class diagrams. Most of the behaviors and attributes inherited by the subclass **expressway** from **road** have also been omitted; they are superfluous because a subclass must possess all the behaviors of a superclass. An exception to this rule is the **expressway** behavior **display**, which is included because it specializes the **road** version of this behavior, resulting in a expressway being displayed in a different style from other roads (see, for example, Figure 1.2, where a "motorway" has similarities to an expressway). The multiple usage of **display** is an illustration of polymorphism, discussed earlier.

Class diagrams, such as those above, can help communicate concepts from one stage of the life-cycle to the next, because same object-oriented concepts pervade the entire object-oriented system life-cycle. Just as E-R diagrams can be easily implemented in a RDBMS, so class diagrams can be easily implemented using an OOPL or object-oriented database management system. In addition to the class diagrams illustrated above, other diagrams can also be important in object-oriented system development. *Instance* or *object diagrams* show the relationship between objects (rather than classes) within a particular context. *Interaction diagrams* are used to illustrate the execution of objects' behaviors in response to particular messages. The references in the bibliography include literature with more detailed description of object-oriented analysis and design.

object diagram

interaction diagram

### 2.4.4 Object-oriented DBMS

A DBMS that utilizes an object-oriented data model is termed an *object-oriented database management system* (OODBMS). In addition to supporting the object constructions described above, several other features are needed in an OODBMS. These include:

OODBMS

- *scheme management*, including the ability to create and change class schemes;

- automatic *query optimization* of queries formulated within a declarative *query language*, as part of a usable *query environment*;

- *storage* and *access management*; and

- *transaction management*, including control of concurrent access, data integrity, and system security.

Unfortunately there are technological problems in achieving some of these facilities with an OODBMS. For example, the provision of query optimization is made difficult by the complexity of the object classes in the system. No longer do we have a small number of global and well-defined operations, such as the relational operations of project, select, and join, for which it is possible to estimate the cost of execution and so choose between different strategies for executing a query. In an object-oriented system there may be a multitude of behaviors, each without any measure of implementation cost. Another problem is the seeming incompatibility of indexing with the notion of encapsulation. Here, the difficulty is that indexes rely on direct access to attribute values, but an object is only accessible via messages activating accessor behaviors. Finally, transactions in an object-oriented database may also be of a higher level of complexity than simple transactions with a relational database. Due to the hierarchical nature of much object data, transactions may cascade downward and affect many other objects, decreasing performance.

Performance has always been lower for OODBMSs than RDBMSs. However, even with lower performance, OODBMSs have the great advantage that impedance mismatch is lower than for RDBMSs, because the user's conception of the application domain can be closer to the conceptual data model, logical data model, and implementation. Traditional systems often force users to adopt highly system-oriented rather than application-oriented concepts. The problem is rather like playing a Beethoven piano sonata on the trumpet: the instrument does not match the conception, and much of the intended meaning will be lost in the execution.

ORDBMS Some DBMS software companies have attempted to achieve the best of both worlds by utilizing a hybrid model, with an RDBMS as the core DBMS and an object-oriented "shell" mediating user access to the core DBMS. These hybrid databases are called *object-relational database management systems* (ORDBMSs). ORDBMSs have the advantage of good performance with support for user-defined data types and expressive object-oriented user models. While this approach does insulate the user from exposure to too many system-oriented concepts, it does not solve the problems of impedance mismatch, nor can it solve some of the other difficulties posed by geospatial information to RDBMSs, as we shall see in later chapters.

## BIBLIOGRAPHIC NOTES

2.1 Overview books on general database technology are plentiful. Date (2003) is a classic text now in its eighth edition. Elmasri and Navathe (2003) is wide-ranging and full of information on relational and object-oriented databases. Connolly and Begg (1999) and Ramakrishnan and Gehrke (2000) are also highly recommended general database texts. Aside from Date (2003), which is a more advanced text focusing primarily on the fundamental theory of relational databases, all of these books provide more detail on every aspect of databases introduced in this chapter.

2.2 The relational model for databases was constructed by Codd (1970). Codd worked at IBM San Jose, and his colleagues Chamberlin and Boyce (1974) introduced SEQUEL, the forerunner to SQL. Codd's original 1970 paper is quite readable and gives a good account of the first stages in the construction of the relational model. At the time of writing, this paper is freely available from the ACM classic papers website (ACM, 2003). Codd's later work (1979) extended the relational model to incorporate null values and extra semantics. Abiteboul et al. (1995) provides an advanced and theoretical database text, mainly focused on the relational model.

2.2.2 Chapter 18 of Connolly and Begg (1999) provides a clear and thorough introduction to query optimization and relational algebra transformation rules.

2.3 The entity-relationship model was introduced by Chen (1976). Constructs that enhanced Chen's original E-R model were developed by several authors: for example, generalization and aggregation in Smith and Smith (1977) and the lattice of subtypes in Hammer and McLeod (1981). An interesting and extensive discussion of the enhanced E-R model is given by Teorey et al. (1986). The best general modern treatment of these ideas is the already recommended general database text of Elmasri and Navathe (2003).

2.3.3 Modeling approaches other than E-R and EER are described in Brodie et al. (1984). A survey of modeling techniques emphasizing the importance of modeling application semantics is Hull and King (1987). Fernández and Rusinkiewicz (1993) apply the EER modeling approach to the development of a soil database for a GIS.

2.4 Booch (1993), Rumbaugh et al. (1990), Coad and Yourdon (1991a), and Coad and Yourdon (1991b) are standard texts on the object-oriented approach to systems analysis and design.

Abadi and Cardelli (1996) is an advanced text that attempts to provide a solid theoretical underpinning to object-oriented systems. Wood (2002) shows how object-oriented ideas can be applied to spatial information, using the object-oriented programming language, Java.

2.4.2 The different types of polymorphism are classified and discussed by Cardelli and Wegner (1985).

2.4.3 UML and related object-oriented modeling techniques are described by Booch et al. (1998).

# Fundamental spatial concepts

3

**Summary**

Many different representations of space are commonly used within a GIS, and this chapter discusses the concepts underlying them. A familiar representation is **Euclidean geometry**, in which distances, angles, and coordinates may be defined. However, other representations are also important. **Sets** of **elements** provide a much simpler representation of space. **Topology** can be constructed based on the concept of a **neighborhood**. A **graph** comprises **nodes** connected by **edges**, and may be used to represent network spaces, such as a road network. A **metric space** formalizes the concept of distance between points in space.

The term "space" is difficult to define (see "What is space?" inset, page 84). We all have an intuitive idea about the concrete space in which our bodies move. In the context of GIS, we normally use the term "space" to refer to "geographic space": the structure and properties of the relationships between locations at the Earth's surface. In this chapter we examine space more carefully by considering the different ways of representing and reasoning about geographic space.

A fundamental concept underlying these different representations is that of *geometry*. A geometry provides a formal representation of the abstract properties and structures within a space. Modern treatments of geometry are founded on the notion of *invariance*: geometries can be classified according to the group of transformations of space under which their propositions remain true. This idea was first proposed in 1872, by the German mathematician Felix Klein in his inaugural address to the University of Erlangen (known as the Erlangen Program).

geometry

invariance

To illustrate, consider a space of three dimensions and our usual notion of distance between two points. Then a geometry is formed by the set of all transformations that preserve distances (that is, for which distance is an *invariant*). Into this set would fall translations and rotations,

**What is space?**  *The notion of space is not easy to define. Gatrell (1991) defines space as "a relation defined on a set of objects," which includes just about any structured collection. This definition is too general to describe the spaces of interest in a GIS. A distinction is often made between the space we can apprehend using our visual perception (termed perceptual or small-scale space) and space that is too big for humans to observe all at once (termed transperceptual, large-scale space, or geographic space, see Kuipers, 1978; Mark and Frank, 1996; Montello, 1993). Zubin (1989) provides more detail and distinguishes four types of space: A-space contains manipulatable everyday objects, like phones and books; B-space contains objects larger than humans, but still observable from a single perspective, like buildings and buses; C-space contains geographic scenes that are too large to apprehend at one time, like landscapes; D-space contains objects that are too large for any human to truly experience, like the solar system or the galaxy. Freundschuh and Egenhofer (1997) give a full overview and synthesis of the different classifications of space that have been proposed. In GIS, we are primarily interested in geographic spaces, although there is also much interest in the relationship between the geographic spaces of GIS and other more general types of information space, such as the Internet and cyberspace (sometimes termed "cybergeography," see Dodge and Kitchin, 2002; Fabrikant and Buttenfield, 2001).*

because the distance between two points is the same before and after a translation or rotation is effected. Scalings (enlargements) would not be members because they usually change distance. Looking at this the other way round provides us with a definition of a geometry as the study of the invariants of a set of transformations. Thus the invariants of the set of translations, rotations, and scalings include angle and parallelism, but not distance.

Coordinatized Euclidean geometry provides a view of space that is intuitive, at least in Western culture, so this is our starting point (section 3.1). In later sections we discuss the most primitive space of all, just collections of objects with no other structure (section 3.2), and proceed to build up to richer geometries (sections 3.3–3.5). Finally, in section 3.6 we introduce fractal geometry, which is concerned with invariance in the scaling properties of objects. Because of the nature of the topic, the treatment will of necessity be sometimes abstract and formal, but examples are provided along the way. Some of the texts in the bibliography provide further background.

## 3.1   EUCLIDEAN SPACE

Euclidean space

Geospatial phenomena are commonly modeled as embedded in a coordinatized space, which enables measurements of distances and bearings between points according to the usual formulas (given below). This section describes this coordinatized model of space, called *Euclidean space*, which transforms spatial properties into properties of tuples of real numbers. We assume for simplicity a two-dimensional model, although all the concepts in this section can be generalized to higher dimensional spaces. For the Euclidean plane, we can set up a coordinate frame consisting of

a fixed, distinguished point (*origin*) and a pair of orthogonal lines (*axes*),      origin
intersecting in the origin.                                                           axis

### 3.1.1  Point objects

A *point* in the plane of the axes has associated with it a unique pair of real       point
numbers $(x, y)$ measuring its distance from the origin in the direction of
each axis, respectively. The collection of all such points is the *Cartesian*         Cartesian plane
*plane*, often written as $\mathbb{R}^2$. It is often useful to view Cartesian points $(x, y)$
as *vectors*, measured from the origin to the point $(x, y)$, having direction        vector
and magnitude and denoted by a directed line segment (see Figure 3.1).
Thus they may be added, subtracted, and multiplied by scalars according
to the rules:

$$(x_1, y_1) + (x_2, y_2) = (x_1 + x_2, y_1 + y_2)$$
$$(x_1, y_1) - (x_2, y_2) = (x_1 - x_2, y_1 - y_2)$$
$$k(x, y) = (kx, ky)$$

Given a point vector, $a = (x, y)$, we may form its *norm*, defined as               norm
follows:
$$\|a\| = \sqrt{(x^2 + y^2)}$$

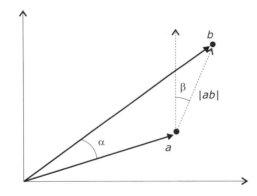

**Figure 3.1:**
Distance,
angle, and
bearing
between points
(vectors) in the
Euclidean
plane

In a coordinatized system, measures of distance may be defined in a
variety of ways (see section 3.5 on metric spaces). A *Euclidean plane* is a          Euclidean plane
Cartesian plane with the particular measures of distance and angle given
below. These measures form the foundation of most school geometry
courses and refer to the "as the crow flies" concept of distance. Given
points (vectors) $a, b$ in $\mathbb{R}^2$, the distance from $a$ to $b$, $|ab|$ (see Figure 3.1)
is given by:
$$|ab| = \|a - b\|$$

Suppose that the points $a, b$ in $\mathbb{R}^2$ have coordinates $(x_a, y_a)$ and
$(x_b, y_b)$, respectively. Then the distance $|ab|$ is precisely the Pythagorean
distance familiar from school days, given by:
$$|ab| = \sqrt{(x_b - x_a)^2 + (y_b - y_a)^2}$$

angle    The *angle* $\alpha$ (see Figure 3.1) between vectors $a$ and $b$ is given as the solution of the trigonometrical equation:

$$\cos\alpha = \frac{x_a x_b + y_a y_b}{\|a\| \cdot \|b\|}$$

bearing    The *bearing* $\beta$ (see Figure 3.1) of $b$ from $a$ is given by the unique solution in the interval $[0, 360[$ of the simultaneous trigonometrical equations:

$$\sin\theta = \frac{x_b - x_a}{|ab|}$$

$$\cos\theta = \frac{y_b - y_a}{|ab|}$$

### 3.1.2   Line objects

Line objects are very common spatial components of a GIS, representing the spatial attributes of objects and their boundaries. The definitions of some commonly used terms related to straight lines follow.

line
- Given two distinct points (vectors) $a$ and $b$ in $\mathbb{R}^2$, the *line* incident with $a$ and $b$ is defined as the point set $\{\lambda a + (1 - \lambda)b | \lambda \in \mathbb{R}\}$.

line segment
- Given two distinct points (vectors) $a$ and $b$ in $\mathbb{R}^2$, the *line segment* between $a$ and $b$ is defined as the point set $\{\lambda a + (1 - \lambda)b | \lambda \in [0, 1]\}$.

half line
- Given two distinct points (vectors) $a$ and $b$ in $\mathbb{R}^2$, the *half line* radiating from $b$ and passing through $a$ is defined as the point set $\{\lambda a + (1 - \lambda)b | \lambda \geq 0\}$.

These definitions represent different types of lines in a parameterized form. The parameter $\lambda$ is constrained to vary over a given range (the range depending upon the object type). As $\lambda$ varies, so the set of points constituting the linear object is defined. Figure 3.2 shows some examples.

**Figure 3.2:**
Parameterized
representation
of linear
objects

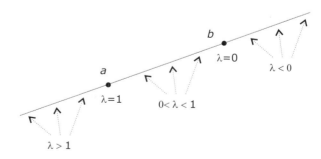

Not only straight lines are of interest for GIS. Straight lines may be specified by a single bivariate polynomial equation of degree one ($ax + by = k$). Higher-degree bivariate polynomials specify further classes of one-dimensional objects. Thus, polynomials of degree two (quadratics of the form $ax^2 + bxy + cy^2 = k$) specify conic sections, which could be circles, ellipses, hyperbolas, or parabolas. Cubic polynomials are used extensively in graphics to specify smooth curves (see Chapter 4).

### 3.1.3 Polygonal objects

A *polyline* in $\mathbb{R}^2$ is defined to be a finite set of line segments (called *edges*) such that each edge end-point is shared by exactly two edges, except possibly for two points, called the *extremes* of the polyline. If, further, no two edges intersect at any place other than possibly at their end-points, the polyline is called a *simple* polyline. A polyline is said to be *closed* if it has no extreme points. A (*simple*) *polygon* in $\mathbb{R}^2$ is defined to be the area enclosed by a simple closed polyline. The polyline forms the *boundary* of the polygon. Each end-point of an edge of the polyline is called a *vertex* of the polygon. Some possibilities are shown in Figure 3.3. An extension to the definition would allow a general polygon to contain holes, islands within holes, etc.

polyline

simple<br>polygon

boundary

vertex

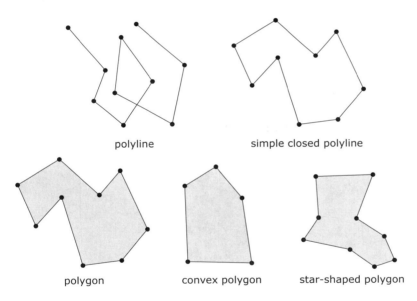

polyline     simple closed polyline

polygon     convex polygon     star-shaped polygon

**Figure 3.3:** Polylines and types of polygons

Many different types of polygon have been defined in the computational geometry literature. A useful category is the set of *convex polygons*, each of which has an interior that is a convex set and thus has all its internal angles not greater than $180°$. For a convex polygon, every interior point is visible from every other interior point in the sense that the line of sight lies entirely within the polygon. A *star-shaped polygon* has the weaker property that there need exist only at least one point which is visible from every point of the polygon (see Figure 3.3). Convexity is discussed further in section 3.2.4.

convex polygon

star-shaped<br>polygon

The definition of *monotone polygons* depends upon the concept of *monotone chain*. Let chain $C = [p_1, p_2, ..., p_n]$ be an ordered list of $n$

monotone

points in the Euclidean plane. Then $C$ is *monotone* if, and only if, there is some line in the Euclidean plane such that the projection of the vertices onto the line preserves the ordering of the list. Figure 3.4 shows monotone and non-monotone chains.

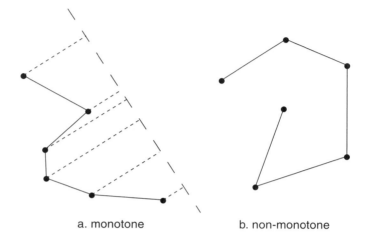

**Figure 3.4:**
Monotone and
non-monotone
chains

a. monotone                    b. non-monotone

monotone
polygon

A polygon is a *monotone polygon* if its boundary can be split into two polylines, such that the chain of vertices of each polyline is a monotone chain. Clearly, every convex polygon is monotone. However, the converse of this statement is not true. It is not even true that every monotone polygon is star shaped, as may be seen from the example in Figure 3.5.

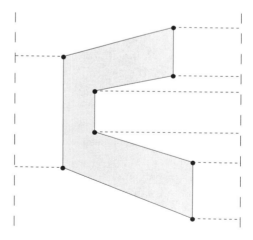

**Figure 3.5:**
A monotone
but not
star-shaped
polygon

triangulation

A *triangulation* of a polygon is a partition of the polygon into triangles that intersect only at their mutual boundaries. It is not too hard to show that a triangulation of a simple polygon with $n$ vertices that introduces an extra $m$ internal vertices (due to the triangulation) will result in exactly

**Diagonal triangulation**  *We can show that a diagonal triangulation of a polygon with $n$ vertices results in $n - 2$ triangles by using an inductive argument on $n$. For $n = 3$, there is but a single triangle and the result is clearly true. For the inductive step, take a general polygon $P$ with an arbitrary number $k > 3$ vertices. Suppose that the result holds for all polygons with less than $k$ vertices. Split $P$ down one of its triangulation edges into two smaller polygons $P'$ and $P''$ (this can be done because $k > 3$). Suppose that $P'$ and $P''$ have $k'$ and $k''$ vertices, respectively. Then:*

1. *$k' + k'' - 2 = k$ (because two vertices will be counted twice in adding up the vertices of $P'$ and $P''$).*

2. *The number of triangles in $P'$ is $k' - 2$ (because we are assuming the result holds for $P'$).*

3. *The number of triangles in $P''$ is $k'' - 2$ (because we are assuming the result holds for $P''$).*

*Adding the numbers of triangles given by 2 and 3, the number of triangles in $P$ is $k' - 2 + k'' - 2$, which simplifies to $k' + k'' - 2 - 2$, and this is $k - 2$ (from 1). This is what we wanted to show.*

$n + 2m - 2$ triangles. If no internal vertices (sometimes called *Steiner points*) are introduced, the triangulation is called a *diagonal triangulation* and results in $n - 2$ triangles (see "Diagonal triangulation" inset, on this page). Figure 3.6 shows a diagonal and non-diagonal triangulation of a polygon with 11 vertices. As predicted by the formula, the diagonal triangulation results in nine triangles and the non-diagonal triangulation that introduces three internal points results in 15 triangles.

Steiner point

diagonal
triangulation

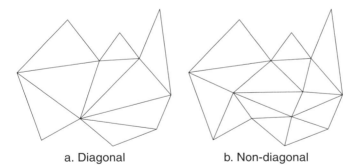

a. Diagonal          b. Non-diagonal

**Figure 3.6:**
A diagonal and
non-diagonal
triangulation of
a polygon

### 3.1.4    Transformations of the Euclidean plane

*Translations, reflections, and rotations*

This section describes some common transformations of the Euclidean plane. A *transformation* of $\mathbb{R}^2$ is a function from $\mathbb{R}^2$ to itself. Thus every point of the plane is transformed into another (maybe the same) point. Some transformations preserve particular properties of embedded objects:

transformation

*Euclidean transformations*: (also termed congruences) preserve the shape and size of embedded objects. An example of a Euclidean transformation is a translation.

*Similarity transformations*: preserve the shape but not necessarily the size of embedded objects. An example of a similarity transformation is a scaling; all Euclidean transformations are also similarities.

*Affine transformations*: preserve the affine properties of embedded objects, such as parallelism. Examples of affine transformations are rotations, reflections, and shears; all similarity transformations are also affine.

*Projective transformations*: preserve the projective properties of embedded objects. An intuitive idea of a central projection is the action of a point light source, sending a figure to its projection upon a screen. For example, the projection of a circle may result in an ellipse. All affine transformations are also projective.

*Topological transformations*: (homeomorphisms or bicontinuous maps) preserve topological properties of embedded objects. We shall study this class in detail later.

For some classes of transformations, formulas may be provided.

*Translation*: through real constants $a$ and $b$
$$(x, y) \rightarrow (x + a, y + b)$$

*Scaling*: by real constants $a$ and $b$
$$(x, y) \rightarrow (ax, by)$$

*Rotation*: through angle $\theta$ about origin
$$(x, y) \rightarrow (x \cos \theta - y \sin \theta, x \sin \theta + y \cos \theta)$$

*Reflection*: in line through origin at angle $\theta$ to $x$-axis
$$(x, y) \rightarrow (x \cos 2\theta + y \sin 2\theta, x \sin 2\theta - y \cos 2\theta)$$

*Shear*: parallel to the $x$-axis with real constant $a$
$$(x, y) \rightarrow (x + ay, y)$$

## 3.2   SET-BASED GEOMETRY OF SPACE

The Euclidean plane is a highly organized kind of space, with many well-defined operations and relationships that can act upon objects within it. This section retreats to the much more rarefied realm of set-based space, and then gradually builds up more structure with topological and metric spaces.

### 3.2.1 Sets

The set-based model of space does not have the rich set of constructs of the Euclidean plane. The set-based model simply involves:

- The constituent objects to be modeled, called *elements* or *members*.  element

- Collections of elements, called *sets*. For computer-based models, such collections are usually finite, or at least countable.  set

- The relationship between the elements and the sets to which they belong, termed *membership*. We write $s \in S$ to indicate that an element $s$ is a member of the set $S$.  membership

The set-based model is abstract and provides little in the way of constructions for modeling spatial properties and relationships. Nevertheless, sets are a rich source of modeling constructs (see "Russell's paradox" inset, on the following page) and are fundamental to the modeling of any geospatial information. For example, relationships between different base units of spatial reference may be modeled using set theory. States or counties may be contained within (members of) countries, which may themselves be contained within continents. Cities may be elements of countries. Such hierarchical relationships are adequately modeled using set theory. Sometimes, areal units are not so easily handled. For example, in the UK there is no simple set-based relationship between the postcoded areas, used for the distribution of mail, and the administrative units such as ward, district, and county. This mismatch causes considerable problems when data referenced to one set of units is compared or combined with data referenced to a different set.

In classical set theory an object is either an element of a particular set or it is not. There is no halfway house or degree of membership. If the binary on-off nature of the membership condition is relaxed, then it is possible to arrive at some more expressive models. Some of these models are important for modeling uncertainty in geospatial information, a topic tackled in Chapter 9.

From the basic constructs of element, set, and membership, a large number of modeling tools may be constructed. We shall consider just a few here.

- *Equality* is defined as a relationship between two sets that holds when the sets contain precisely the same members.  equality

- The relationship between two sets where every member of one set is a member of the second is termed *subset*. The relationship that set $S$ is a subset of set $T$ is denoted $S \subseteq T$.  subset

- A *power set* is the set of all subsets of a set. The power set of set $S$ is denoted $\mathcal{P}(S)$.  power set

- The *empty set* is the set containing no members, denoted $\varnothing$.  empty set

- The number of members in a set is termed *cardinality*. The cardinality of set $S$ is denoted $\#S$.  cardinality

**Russell's paradox** *Even though the set concept expresses the most rudimentary relation-ships and structuring of space, it is surprisingly difficult to capture the essence of a set in a few words. A first attempt might be "a set is any collection of objects," but this lays itself open to Russell's paradox (after the British mathematician Bertrand Russell). Russell's paradox is sometimes explained in terms of a library index. Imagine that an assiduous librarian wishes to index every book in the library that does not contain a reference to itself (unlike this book, Worboys and Duckham, 2004). The librarian can write down the names of books that do not refer to themselves in a new index book. To be complete, this index should list itself, as it does not contain a self reference. But as soon as the librarian adds a self reference to the index, that reference should be removed, because now the index book does cite itself! We can express this tortuous problem mathematically by considering the set S of all sets that are not members of themselves. If S is a member of itself, then by definition it is not a member of itself; on the other hand, if S is not a member of itself, then it must be a member of itself. Either way, we arrive at a contradiction. This paradox spurred considerable efforts in the 20th century to find a more adequate definition of a set, but no definition has proved completely satisfactory.*

intersection
- The *intersection* operation is a binary operation that takes two sets and returns the set of elements that are members of both the original sets. The intersection of sets $S$ and $T$ is denoted $S \cap T$.

union
- The *union* operation is a binary operation that takes two sets and returns the set of elements that are members of at least one of the original sets. The union of sets $S$ and $T$ is denoted $S \cup T$.

difference
- The *difference* operation is a binary operation that takes two sets and returns the set of elements that are members of the first set but not the second set. The difference of sets $S$ and $T$ is denoted $S \backslash T$.

complement
- The *complement* operation is a unary operation that when applied to a set returns the set of elements that are not in the set. The complement is always taken with reference to an (implicit) universal set. The complement of set $S$ is denoted $S'$.

Figure 3.7 shows by shading on set diagrams some of the Boolean set operations described above.

**Figure 3.7:**
Set
intersection,
union, and
difference

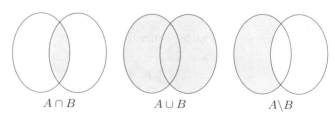

$$A \cap B \qquad\qquad A \cup B \qquad\qquad A \backslash B$$

Some sets, particularly sets of numbers, are used so often that they have a special name and symbol. Some of these are listed in Table 3.1. The Boolean set $\mathbb{B}$ is used whenever there is a two-way choice to be made. This choice may be viewed as between on and off, true and false, or one and zero, depending upon the context. The set of integers $\mathbb{Z}$ is used in discrete models. Sometimes, only the positive integers are needed, written $\mathbb{Z}^+$.

For continuous models, the real numbers $\mathbb{R}$ are required. In fact, it is provably impossible to capture the reals completely using a computer; therefore rational numbers are used in practice. Rational numbers are real numbers that can be expressed as a ratio of integers, e.g., 123/46. Like real numbers, rational numbers have the property that they are *dense*: that is, between any two rationals $a$ and $b$, where $a$ is greater than $b$, no matter how close $a$ and $b$ are, it is always possible to find a third rational $c$ such that $a < c < b$ (see Figure 3.8). In this way, the rationals are a useful approximation for modeling continuous processes. Any particular computer implementation will place a restriction upon the precision of the rationals.

*dense set*

| Name | Symbol | Description |
|------|--------|-------------|
| Booleans | $\mathbb{B}$ | Two-valued set of true/false, 1/0, or on/off |
| Integers | $\mathbb{Z}$ | Positive and negative numbers, including zero |
| Reals | $\mathbb{R}$ | Measurements on the number line |
| Real plane | $\mathbb{R}^2$ | Ordered pairs of reals (often the Euclidean plane when viewed as a point in space) |
| Closed interval | $[a, b]$ | All reals between $a$ and $b$ (including $a$ and $b$) |
| Open interval | $]a, b[$ | All reals between $a$ and $b$ (excluding $a$ and $b$) |
| Semi-open interval | $[a, b[$ | All reals between $a$ and $b$ (including $a$ and excluding $b$) |

**Table 3.1:** Some distinguished sets

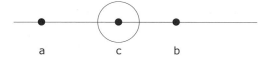

a    c    b

**Figure 3.8:** The rationals are dense

Some subsets of the reals are particularly useful. The *intervals* are connected sets of real numbers. They may or may not contain their endpoints and are then called *closed* or *open*, respectively. It is also possible for intervals to be closed at one end and open at the other. Such intervals are called *semi-open* (or *semi-closed*). Thus the closed interval $[2, 5]$ denotes the set of all real numbers not less than two and not greater than five. The semi-open interval $]2, 5]$ denotes the set of all real numbers greater than two but not greater than five.

*interval*

### 3.2.2   Relations

Sets on their own are limited in their application to modeling. Life becomes more interesting when relationships between two or more sets are modeled. In order to provide these tools, a further set-based operation is defined.

product
- The binary operation *product* returns the set of ordered pairs, whose first element is a member of the first set and second element is a member of the second set. The product of sets $S$ and $T$ is denoted $S \times T$.

An example of a product set is the set of points in the Cartesian plane, introduced in section 3.1.1. Each point in the Cartesian plane is represented as an ordered pair of real numbers, measuring the point's distance from a given origin in the direction of the two axes. In set-theoretic terms, the collection of all such points is a product set, being the product of the set of real numbers with itself. This set is denoted by $\mathbb{R} \times \mathbb{R}$ or $\mathbb{R}^2$. This notion may be generalized to Cartesian 3-space $\mathbb{R}^3$ or indeed Cartesian $n$-space $\mathbb{R}^n$. A second example would be the set of points in the *unit square*, the square with vertices $(0,0)$, $(0,1)$, $(1,0)$, and $(1,1)$ in the Cartesian plane. This set is the product of two intervals, $[0,1] \times [0,1]$. It is the case that $[0,1] \times [0,1] \subseteq \mathbb{R}^2$.

Product spaces provide a means of defining relationships between objects. Hence the following construction:

binary relation
- A *binary relation* is a subset of the product of two sets, whose ordered pairs show the relationships between members of the first set and members of the second set.

Suppose that $S = \{$Fred, Mary$\}$ and $T = \{$apples, oranges, bananas$\}$. Then the formal way of expressing the relationship "likes" between people and fruit is by means of the relation, constructed as a set of ordered pairs $\{($Fred, apples$)$, $($Fred, bananas$)$, $($Mary, apples$)\}$. This set of ordered pairs is a subset of the entire product space $S \times T$, containing six pairs.

Relations may in general apply to the product of more than two sets (as in the relations in the relational database model). However, even binary relations have many modeling applications. For example, suppose that we are given two sets: the places of interest in the Potteries $P$ and town centers in the Potteries $T$. A relation $R_1 \subseteq P \times T$ might provide for each a place of interest $p \in P$ the nearest town center $t \in T$ to $p$. Figure 3.9 shows the relation $R_1$ diagrammatically. The fact that Hanley is the nearest town center to the City Museum is indicated in Figure 3.9 with a line joining these two locations, and would be written $($City Museum, Hanley$) \in R_1$.

As we have seen with the relation $\mathbb{R}^2$, a relation may act upon a single set. Suppose for the set of places $P$, a place of interest $p \in P$ is related to $p' \in P$ if $p'$ is the nearest place of interest to $p$. This relation $R_2 \subseteq P \times P$

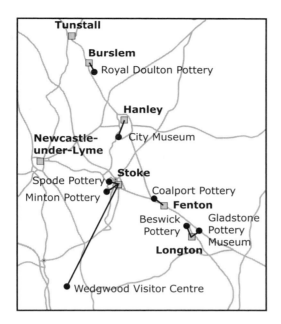

**Figure 3.9:**
Binary relation
on the sets of
Potteries places
of interest and
town centers

is shown in Figure 3.10. The fact that Spode Pottery is the nearest place
of interest to the City Museum is indicated with an arrow between these
locations, and would be written (City Museum, Spode Pottery) $\in R_2$.

Some binary relations between objects of the same set have special
properties.

- A relation where every element of the set is related to itself is
  termed a *reflexive* relation.                                          reflexive

- A relation where if $x$ is related to $y$ then $y$ is related to $x$ is termed
  a *symmetric* relation.                                                 symmetric

- A relation where if $x$ is related to $y$ and $y$ is related to $z$ then $x$ is
  related to $z$ is termed a *transitive* relation.                       transitive

In the example of Figure 3.10 there is an implicit assumption that a
place cannot be related to itself; therefore the relation is not reflexive. It
is less obvious whether or not the relation "has nearest place of interest"
is symmetric. For example, Spode Pottery is nearest to Minton Pottery
and vice versa; hence there is no arrow on the line between these two
locations. However, the nearest place to the City Museum is also Spode
Pottery while the converse is not true, so the relation "has nearest place of
interest" cannot be symmetric. Also, the relation is not transitive because
if $p$ has nearest place of interest $q$ and $q$ has nearest place of interest $r$
then it is unlikely that $p$ has nearest place of interest $r$.

A binary relation that is reflexive, symmetric, and transitive is termed
an *equivalence relation*. Another useful class of relations is *order* rela-    equivalence
tions, which satisfy the transitive property and are also irreflexive and        relation

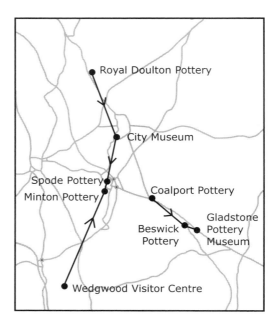

**Figure 3.10:**
Binary relation
on the sets of
Potteries places
of interest

antisymmetric. These further properties and partial orders are discussed
in section 3.4.1.

### 3.2.3 Functions

function

A *function* is a special type of relation which has the property that each
member of the first set relates to exactly one member of the second
set. Thus a function provides a rule that transforms each member of the

domain

first set, called the *domain*, into a member of the second set, called the

codomain

*codomain*. We use the notation:

$$f : S \rightarrow T$$

to mean that $f$ is a function, $S$ is the domain, and $T$ is the codomain. If the
result of applying function $f$ to element $x$ of $S$ is $y$, we write $y = f(x)$
or $f : x \mapsto y$. Figure 3.11 shows schematically the relationship between
function, domain, codomain, and image (defined shortly).

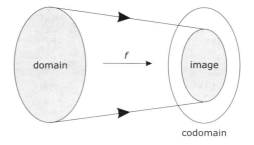

**Figure 3.11:**
An abstract
function $f$

For example, suppose that $S$ is the set of points on a spheroid and $T$ the set of points in the plane. A map projection is essentially a function with domain $S$ and codomain $T$. Usually, map projection functions are constructed to leave some properties of the sets invariant, for example, lengths, angles, or areas. The familiar Universal Transverse Mercator (UTM) is a function that preserves angles. The UTM projection has a further property that any two different points in the domain are transformed to two distinct points in the codomain. Such a function is said to be an *injection*.

Not all functions are injections. Consider, for example, the function that computes the square of any integer. Then, because both 2 and $-2$ square to the same number, 4, the square function is not an injection. The two relations described in the previous section (Figures 3.9 and 3.10) are also functions, but not injections.

The set of outputs from the application of a function to points in its domain will form a subset of the codomain. Thus the UTM will project a spheroid onto a finite subset of the plane. The set of all possible outputs is called the *image* or *range* of the function. If the image actually equals the codomain, then the function is termed a *surjection*. The UTM projection is not a surjection, because the spheroid projects onto only a finite portion of the plane and not the whole plane. A function that is both a surjection and an injection is termed a *bijection*.

Injective functions have the special property that they have inverse functions. Consider again the UTM projection from the spheroid to the plane. Given a point in the plane that is part of the image of the transformation, it is possible to reconstruct the point on the spheroid from which it came. This reversal of the process allows us to form a new function whose domain is the image of the UTM, and which maps the image back to the spheroid. This is called the *inverse* function.

*injection*

*image*
*surjection*

*bijection*

*inverse function*

### 3.2.4 Convexity

Convexity has already been discussed for polygons and is now generalized to the same property for arbitrary point-sets in the Euclidean plane. The same notion is also meaningful in Euclidean 3-space, and the definition is easily extensible to this case. The essential idea is that a set is convex if every point is visible from every other point within the set. To make this idea precise, we define visibility and then convexity.

- Let $S$ be a set of points in the Euclidean plane. Then point $x$ in $S$ is *visible* from point $y$ in $S$ if either $x = y$ or it is possible to draw a straight-line segment between $x$ and $y$ that consists entirely of points of $S$.

- Let $S$ be a set of points in the Euclidean plane. The point $x$ in $S$ is an *observation point* for $S$ if every point of $S$ is visible from $x$.

- Let $S$ be a set of points in the Euclidean plane. The set $S$ is *semi-convex* (*star-shaped* if $S$ is a polygonal region) if there is some observation point for $S$.

convex

- Let $S$ be a set of points in the Euclidean plane. The set $S$ is *convex* if every point of $S$ is an observation point for $S$.

Figure 3.12 shows the visibility relation within a set between three points $x$, $y$, and $z$. Points $x$ and $y$ are visible from each other, as are points $y$ and $z$. But, points $x$ and $z$ are not visible from each other. The visibility relation is reflexive and symmetric, but not transitive. Also, observe that any convex set must be semi-convex (but not conversely). Figure 3.13 gives some examples of sets that are not semi-convex, semi-convex but not convex, and convex, respectively.

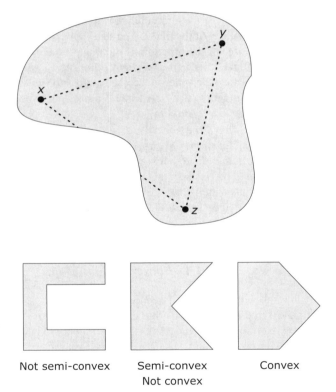

**Figure 3.12:**
Visibility
between points
$x$, $y$, and $z$

**Figure 3.13:**
Degrees of
convexity in
point sets

Not semi-convex        Semi-convex        Convex
                       Not convex

convex hull

The intersection of a collection of convex sets is also convex, and therefore any collection of convex sets closed under intersection has a minimum member. This leads to the definition of a *convex hull* of a set of points $S$ in $\mathbb{R}^2$ as the intersection of all convex sets containing $S$. From above, the convex hull must be the unique smallest convex set that contains $S$ (Figure 3.14). A convex hull of a finite set of points is always a polygonal region.

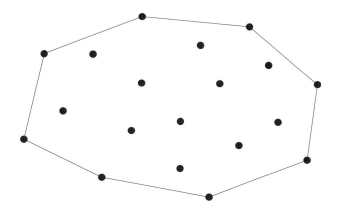

**Figure 3.14:**
Convex hull of
a point set

## 3.3   TOPOLOGY OF SPACE

The word "topology" derives from Greek and literally translates to "study of form." This may be contrasted with the word "geometry," which translates to "measurement of the Earth." *Topology* is a branch of geometry, concerned with a particular set of geometrical properties—those that remain invariant under topological transformations. So, what is a topological transformation? In fact, topology is a general notion that can be applied to many different kinds of space.

topology

### 3.3.1   **Topological spaces**

To gain some topological intuition, imagine the Euclidean plane to be an unbounded sheet of fine-quality rubber that has the ability to stretch and contract to any desired degree. Imagine a figure drawn upon this rubber sheet. Allow this sheet to be stretched but not torn or folded. Certain properties of the original figure will remain while others will be lost. For example, if a polygon were drawn upon the sheet and a point was drawn inside the polygon, then after any amount of stretching the point would still be inside the polygon; on the other hand, the area of the polygon may well have changed. The property of "insideness" is a topological property (because it is invariant under rubber sheet transformation) while "area" is not a topological property. The transformation induced by stretching a rubber sheet is called a *topological transformation* or *homeomorphism*. Thus we have the following definitions:

*Topological properties*: properties that are preserved by topological transformations of the space.

*Topology*: the study of topological transformations and the properties that are left invariant by them.

Table 3.2 lists some topological and non-topological properties of objects embedded in the Euclidean plane. For object types we take points, arcs, and area. Later, these types will be defined more carefully; for now, assume the obvious meanings to arc (possibly curved linear object) and area (two-dimensional piece of plane, possibly with holes and islands).

**Table 3.2:**
Topological and non-topological properties of objects in the Euclidean plane with the usual topology

| Topological | A point is at an end-point of an arc |
|---|---|
| | A point is on the boundary of an area |
| | A point is in the interior/exterior of an area |
| | An arc is simple |
| | An area is open/closed/simple |
| | An area is connected |
| Non-topological | Distance between two points |
| | Bearing of one point from another point |
| | Length of an arc |
| | Perimeter of an area |

point-set topology

combinatorial topology

The discussion will cover two branches of topology: point-set (or analytic) topology and combinatorial (or algebraic) topology. In *point-set topology*, the focus, as one would expect, is on sets of points and in particular on the concepts of neighborhood, nearness, and open set. We shall see that several important spatial relationships, such as connectedness and boundary may be expressed in point-set topological terms. The other important branch of topology, which has been applied to spatial data modeling, is *combinatorial topology*, in particular the theory of simplicial complexes. Even though these ideas may at times seem rarefied and far removed from spatial databases, in fact they do form the basis of several prominent conceptual models for spatial systems and GIS. It is certainly true that the construction of sound and lasting generic spatial models relies on knowledge of the material that is introduced here. The reader is encouraged to gain further understanding of this area by sampling material from the bibliographic notes provided at the end of the chapter.

### 3.3.2   General point-set topology

It is possible to define a topological space in several different ways. The definition below is based upon a single primary notion, that of *neighborhood*. A set upon which a well-defined notion of neighborhood is provided is then a topological space. It turns out that all the familiar topological properties are definable in terms of the single concept of neighborhood. Given any set, the approach is to define a collection of its subsets, constituting the neighborhoods, and thus provide a neighborhood topology on the set. The formal definition is now given.

topological space
neighborhood

- Let $S$ be a given set of points. A *topological space* is a collection of subsets of $S$, called *neighborhoods*, that satisfy the following two conditions.

$T1$ Every point in $S$ is in some neighborhood.

$T2$  The intersection of any two neighborhoods of any point $x$ in
       $S$ contains a neighborhood of $x$.

Figure 3.15 shows the two conditions of a topological space in action.
Neighborhoods are shown surrounding each point in the set and two
neighborhoods are shown overlapping and containing in their intersection
another neighborhood.

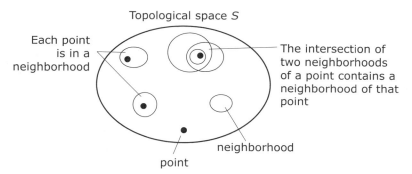

**Figure 3.15:**
Points and
neighborhoods
in a topological
space

By far the most important example of a topological space, for our
purposes, is the *usual topology* for the Euclidean plane. The usual topology is so called because it is the topology that naturally comes to mind
with the Euclidean plane, and corresponds to the rubber-sheet topology
introduced earlier. It is possible to define other (unusual?) topologies on
the Euclidean plane (see "Topological spaces" inset, on the next page),
and one important example of these, travel time topology, follows the
usual topology.

*The usual topology of the Euclidean plane*

Define an *open disk* to be a set of points bounded by a circle in the
Euclidean plane, but not including the boundary. An example is given
in Figure 3.16. The convention is that a hatched line at the boundary
indicates that the boundary points are excluded, whereas a continuous
line indicates that boundary points are included.

open disk

**Figure 3.16:**
An open disk in
the Euclidean
plane with the
usual topology

Define a neighborhood of a point $x$ in $\mathbb{R}^2$ to be any open disk that has
$x$ within it (see Figure 3.17). We now show that, under this definition of
neighborhood, $\mathbb{R}^2$ is a topological space. To check that condition $T1$ for
a topological space holds, it is sufficient to observe that every point in $\mathbb{R}^2$
can certainly be surrounded by an open disk. For $T2$, take any point $x$ in

**Topological spaces** *There are many common topological spaces other than the usual topology. For example, let S be any set, and define the neighborhoods to be all the subsets of S. It is easy to confirm that this neighborhood structure defines a topological space (by checking conditions T1 and T2). The space is called the discrete topology, because the smallest neighborhood of each point x in S is {x}, so each point in S is separated by a neighborhood from every other point. Another example of an extreme topology occurs if we let S be any set, and define the only neighborhood to be the set S itself. Again, this may easily be verified to be a topological space, called the indiscrete topology. The usual topology of the Euclidean plane may be scaled up or down to Euclidean space of any dimension. This is illustrated with the usual topology on the Euclidean line. For any real number $x \in \mathbb{R}$, define a neighborhood of x to be any open interval containing x. This is the one-dimensional equivalent of the usual topology on the Euclidean plane, and can be shown to satisfy properties T1 and T2 in a similar way.*

$\mathbb{R}^2$ and surround it by two of its neighborhoods (open disks with $x$ inside), $N_1$ and $N_2$. Now, $x$ will lie in the intersection of these two neighborhoods and it is always possible to surround $x$ with an open disk entirely within this intersection. To see this, let $d_1$ be the minimum distance of $x$ from the boundary of $N_1$ and $d_2$ be the minimum distance of $x$ from the boundary of $N_2$. Then the open disk with center $x$ and radius the minimum of $d_1$ and $d_2$ will contain $x$ and lie entirely in the intersection of $N_1$ and $N_2$ (see Figure 3.18). Thus $T1$ and $T2$ are satisfied, and under this definition

usual topology

of neighborhood, $\mathbb{R}^2$ is a topological space, called the *usual topology* for $\mathbb{R}^2$.

**Figure 3.17:**
A neighborhood of $x$ in $\mathbb{R}^2$ with the usual topology

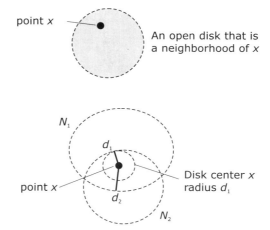

**Figure 3.18:**
Condition $T2$ is satisfied for $\mathbb{R}^2$ with the usual topology

*Travel time topology*

travel time topology

Another important topology for geospatial information is *travel time topology*. Let $S$ be the set of points in a region of the plane. Suppose that the region contains a transportation network and that we know the average travel time between any two points in the region using the network, following the optimal route. For the purposes of this example, we need to assume that the travel time relation is symmetric: that is, it must always be the case that the travel time from $x$ to $y$ is equal to the travel time from $y$ to $x$. For each time $t$ greater than zero, define a $t$-zone around point $x$ to be the set of all points reachable from $x$ in less than time $t$. As an illustration, Figure 3.19 shows a 5-zone, 10-zone, and 15-zone around the Spode Pottery. Let the neighborhoods be all $t$-zones (for all times $t$) around all points. Then, clearly $T1$ is satisfied, since each point will have some $t$-zones surrounding it. The argument that $T2$ is satisfied is similar to that used for the usual topology of the Euclidean plane and is omitted. The symmetry of the travel time relation is required at a critical stage in the argument. The travel time measure between two points is an example of a *metric* (discussed later in section 3.5) and the travel time topology is a special case of the topology that can be induced by any metric on a space.

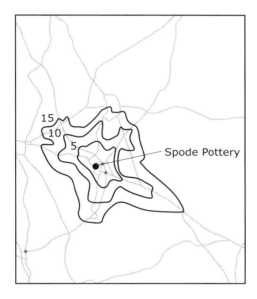

**Figure 3.19:** Travel-time topology example, showing some neighborhoods ($t$-zones)

### 3.3.3   Properties of a topological space

It is surprising that out of the single primitive notion of neighborhood it is possible to construct all the features and properties of a topological space. This section describes some of these constructions, in a similar way to Henle (1979), beginning with the definition of "nearness." Many topologists use the phrase "limit point" to replace our use of "near point."

- Let $S$ be a topological space. Then $S$ has a set of neighborhoods associated with it. Let $X$ be a subset of points in $S$ and $x$ an individual point in $S$. Define $x$ to be *near* $X$ if every neighborhood of $x$ contains some point of $X$.

For example, in the Euclidean plane with the usual topology, let $C$ be the open *unit disk*, centered on the origin, $C = \{(x,y)|x^2 + y^2 < 1\}$. Then the point $(1,0)$, although not a member of set $C$, is near to $C$, because any open disk (no matter how small) that surrounds $(1,0)$ must impinge into $C$. In fact, any point on the circumference of $C$ is near to $C$, as indeed is any point inside $C$. However, any point exterior to $C$ and not on the circumference is not near $C$, since it will always be possible to surround it with a neighborhood that separates it from $C$ (see Figure 3.18).

**Figure 3.20:** Points near and not near to the open unit disk $C$ in the Euclidean plane with the usual topology

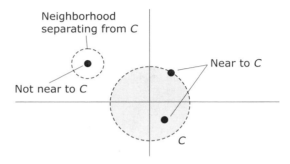

A fundamental topological invariant of any set is its boundary. This notion can be constructed out of our primitives as follows. First, open and closed sets are introduced using the neighborhood idea. It will emerge that an *open set* is a set that does not contain its boundary, whereas a *closed set* is a set that contains all its boundary.

open set

closed set

- Let $S$ be a topological space and $X$ be a subset of points of $S$. Then $X$ is *open* if every point of $X$ can be surrounded by a neighborhood that is entirely within $X$.

- Let $S$ be a topological space and $X$ be a subset of points of $S$. Then $X$ is *closed* if it contains all its near points.

The open unit disk $C$ above is (obviously) open, because any point, no matter how close to the circumference, may be surrounded by a neighborhood made sufficiently small that it is entirely within $C$. $C$ is not closed, because points on the circumference are near points to $C$ but not contained in $C$. For $C$ to be closed, it would have to include its circumference. This leads us to the related definition of *closure*.

Note that it is possible for a topological space to contain sets that are both open and closed or neither open nor closed (see "Topological spaces, open and closed sets" inset, on the facing page).

closure

- Let $S$ be a topological space and $X$ be a subset of points of $S$. Then the closure of $X$ is the union of $S$ with the set of all its near points. The closure of $X$ is denoted $X^-$.

Clearly the closure of a set is itself closed; in fact the closure of set $X$ is the smallest closed set containing $X$. In our example, the closure

**Topological spaces (open and closed sets)**  *Recall from the previous "Topological spaces" inset, on page 102, that the discrete topology on a space S defined the neighborhoods to be all the subsets of S. Let X be any subset of S, then the only near points of X are the points of X itself (since any point not in X may be surrounded by the neighborhood containing just that point). Thus X is closed. X is also open, because we can surround any point of X with a neighborhood entirely in X containing just that point. The discrete topology is therefore odd in that every set within it is both open and closed. This is not the case for the indiscrete topology. Let S be a set upon which the indiscrete topology is defined and let X be any subset of S. Then, the only neighborhood in S is S itself. Therefore every point in S is a near point of X. Thus, unless X is either empty or equal to S, it is neither open nor closed. For the travel time topology, an example open set is the set of all points less than 1 hour's traveling time from a specified point, say, Spode Pottery. This set has as its boundary the set of all points that are exactly 1 hour from Spode Pottery, and as its closure the set of all points having travel time from Spode Pottery not greater than 1 hour.*

of the open unit disk $C$ is formed by annexing its circumference. Thus $C = \{(x, y)|x^2 + y^2 \leq 1\}$. This set is called the *closed unit disk*. We may also force a set to be open by stripping away unwanted near points, constructing the *interior* of a set as follows:

- Let $S$ be a topological space and $X$ be a subset of points of $S$. Then the *interior* of $X$ consists of all points which belong to $X$ and are not near points of $X'$, the complement of $X$. The interior of set $X$ is denoted $X^\circ$.

interior

Notice that for a point $x$ to be near to the complement of set $X$, it must be the case that each neighborhood of $x$ impinges upon $X'$. Therefore, a point in $X$ which is not a near point of $X'$ has at least one neighborhood of it that is entirely within $X$. Thus, the interior of a set is open. In fact, the interior of a set $X$ is the largest open set contained in $X$. As an example, if $D$ is the closed unit disk, then $D^\circ$ is the open unit disk. Figure 3.21 shows a further example of an open and closed set in the Euclidean plane with the usual topology.

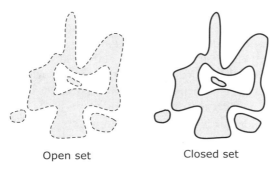

Open set                     Closed set

**Figure 3.21:** Open and closed sets in the Euclidean plane with the usual topology

We are now sufficiently prepared to define the *boundary* of a set in purely topological terms.

boundary

- Let $S$ be a topological space and $X$ be a subset of points of $S$. Then the boundary of $X$ consists of all points which are near to both $X$ and $X'$. The boundary of set $X$ is denoted $\partial X$.

Let point $x$ be a member of $\partial X$. Since $x$ is near to $X$, then $x$ must be in $X^-$. Since $x$ is near to $X'$, then $x$ cannot be in $X^\circ$. Thus $\partial X$ is the set difference of $X^-$ and $X^\circ$. In the case of the unit disk $C$ in the Euclidean plane with the usual topology, $\partial X$ is the circumference of $C$, as we would expect. As a further example using the Euclidean plane with the usual topology, suppose that $S$ is the connected region of the plane containing a single hole shown in Figure 3.22. We see that the outer boundary of $S$ is excluded from $S$, but the inner boundary of $S$ (i.e., the boundary of the hole) is contained in $S$. The interior, closure, and boundary of $S$ are as shown in Figure 3.22. $S^\circ$ contains all the points of $S$ excluding its inner boundary. $S^-$ includes both inner and outer boundaries. $\partial S$ is the union of the inner and outer boundaries of $S$.

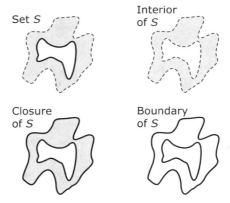

**Figure 3.22:** Interior, closure, and boundary of a region in the Euclidean plane with the usual topology

It is important to realize that it is not possible to consider the topological properties of sets in exclusion from the larger spaces in which they are embedded. To illustrate this point, consider a finite length of straight line. If we take the line to be embedded in a two-dimensional (or higher-dimensional) Euclidean space, as shown in Figure 3.23a, then its interior is the empty set, and its boundary and closure are both the line itself. On the other hand, if the same line is embedded in a one-dimensional Euclidean space, that is, in the real number line, as shown in Figure 3.23b, then its interior is the line excluding its end-points, its closure is the line itself, and its boundary consists of the end-points of the line.

We conclude this brief excursion into point-set topology by considering the notion of *connectedness*. In fact, point-set topology recognizes several different kinds of connectedness. This section defines a simple form based directly upon the neighborhood properties of a topological space. The next section includes a description of further forms, including weak, strong, and path-connectedness.

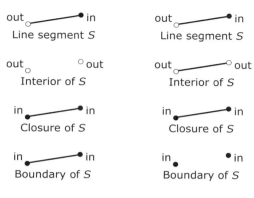

**Figure 3.23:** Interior, closure, and boundary of a line segment in Euclidean 2-space and 1-space with the usual topology

a. Topology of a line in Euclidean 2-space

b. Topology of a line in Euclidean 1-space

- Let $S$ be a topological space and $X$ be a subset of points of $S$. Then $X$ is *connected* if whenever it is partitioned into two non-empty disjoint subsets, $A$ and $B$, then either $A$ contains a point near $B$, or $B$ contains a point near $A$, or both.

connected

Consider the three sets shown in Figure 3.24. In the case of Figures 3.24a and b, no matter how we choose to divide them into two, the partition will always satisfy the condition of the definition of connectedness above. Even if the set in Figure 3.24a is partitioned into its upper and lower disks, with the point of intersection included in the upper disk, then this point is certainly a near point of the lower disk. Therefore the sets in Figures 3.24a and b are connected. Figure 3.24c shows a set that is not connected. To see this, partition the set in Figure 3.24c into its upper and lower component disks. Then no point of the upper disk is a near point of the lower disk and no point of the lower disk is a near point of the upper disk. This example shows that the topological definition of connectedness accords with our intuition.

### 3.3.4 Point-set topology of the Euclidean plane

The Euclidean plane with the usual topology provides by far the most important example of a topological space, for the purposes of GIS. A *homeomorphism* (or *topological transformation*) of $\mathbb{R}^2$ is a bijection of the plane that transforms each neighborhood in the domain to a neighborhood in the image. Furthermore, any neighborhood in the image must be the result of the application of the transformation to a neighborhood in the domain. Put more simply and intuitively, a homeomorphism corresponds to the notion of a rubber sheet transformation, which stretches and distorts the plane without folding or tearing.

homeomorphism

If the result of applying a homeomorphism to a point-set $X$ is point-set $Y$, we say that $X$ and $Y$ are *topologically equivalent*. Thus, in Figure

topological equivalence

**Figure 3.24:**
Connected (a,
b) and
disconnected
(c) sets in the
Euclidean
plane with the
usual topology

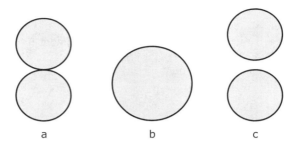

a                    b                    c

3.25, the disk $S$ and the area $T$ are topologically equivalent, but neither is topologically equivalent to the area $U$. From an intuitive point of view, it is clearly possible (at least, in the mind's eye, with a lot of stretching) to transform $S$ to $T$ (and back again) by stretching and contracting a rubber sheet. However, the only option to arrive at the area $U$ from $S$ is to tear the sheet, so as to form the hole in $U$. Tearing is not allowed and therefore $S$ and $U$ are not topologically equivalent. Set $V$, which is formed by gluing the disk to itself at a single point, is homeomorphic to none of the sets $S$, $T$, or $U$.

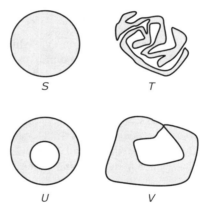

**Figure 3.25:**
Topologically
equivalent and
inequivalent
planar objects

Many mapping ideas are based upon the idea of homeomorphism. For example, Figures 3.26 and 3.27 show two maps of the Potteries bus routes, first (Figure 3.26) as a similarity transformation of the actual routings (neglecting the Earth's curvature), and second (Figure 3.27) as a topological transformation of the actual routings. The two bus route maps are topologically equivalent.

Properties that are preserved by homeomorphisms are called *topological invariants*. From the definition of a homeomorphism, it is clear that the configuration of neighborhoods of a space is a topological invariant. Other constructs, such as open set, closed set, and boundary, are also topological invariants because they are defined purely in terms of neighborhoods.

topological
invariant

The paradigm for all open sets in the Euclidean plane is the unit open disk, with its center at the origin. Therefore, all topologically equivalent

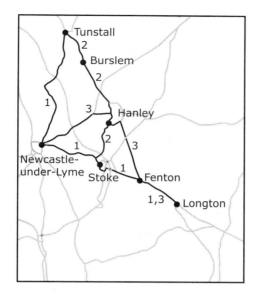

**Figure 3.26:** Map (similarity transformation) of Potteries bus routes

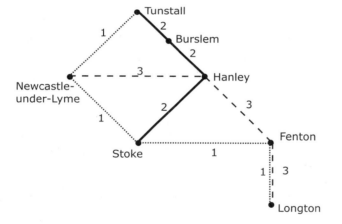

**Figure 3.27:** Map of Potteries bus routes, topologically equivalent to Figure 3.26

areas to this are open as well. Any set homeomorphic to the unit open disk we term an *open cell*. Any set homeomorphic to the closed unit disk is a *closed cell* (see "Brouwer's fixed point theorem" inset, on the following page). An open cell is clearly an open set, and similarly a closed cell is clearly a closed set. In general, if a set of points owns its entire boundary then it is closed, and if it owns none of its boundary then it is open.

Connectedness has already been defined in the context of a general topological space. Connectedness is defined purely in terms of the topology, so it is a topological invariant. Essentially, a connected set is "all of a piece," such as a line, a cell, or an annulus (a ring). Connected sets are not necessarily homeomorphic to each other; a cell is not homeomorphic to an annulus, because the rubber sheet would need to be torn to make the hole. Connected sets that do not have holes are called *simply connected*. Being

open cell

closed cell

simply connected

**Brouwer's fixed point theorem**  *Cells, though basic to topological work, are the subject of one of the most intriguing of results in introductory topology. Imagine a cell laid out as a rubber shape on the plane. Now deform the cell by twisting, stretching, even folding, but not tearing, and replace the cell on the plane in such a way that it is entirely within its original outline. Then, there will always be at least one point that has not moved from its original position! Such a point is called a fixed point and the result is known as Brouwer's fixed point theorem. Brouwer's fixed point theorem holds for Euclidean n-space and has many intriguing and counter-intuitive implications. As a three-dimensional example, assume the water within a lake is homeomorphic to the closed unit sphere. Assume further that the currents within the lake operate as a continuous rubber sheet transformation of the water. Then according to Brouwer's fixed point theorem for any two points in time there must always be some water that is in exactly the same location within the lake at both points in time (although it may have moved in between). Fixed point theorems have played an important role in computer science, in particular underpinning the theory of recursion.*

simply connected is a topological invariant. Cells are simply connected, annuli are not simply connected.

Two basic one-dimensional object types are the straight-line segment and the circle (a circle is the boundary of a disk). Using the notion of homeomorphism, we may generalize these two basic object types as follows. A *simple arc* is topologically equivalent to a straight-line segment: it is clearly connected. The homeomorphism from a straight-line segment to a simple arc is a bijection, so it cannot be possible for a simple arc to cross over itself or for its end-points to be coincident. If the condition on no self-crossings is relaxed, then the resulting one-dimensional object is termed an *arc*. If the end-points are coincident and self-crossings are not allowed, then the object is termed a *simple loop*. If the end-points are coincident and self-crossings are allowed, then the object is termed a *loop*. A simple loop is topologically equivalent to a circle. Examples of some one-dimensional planar objects are shown in Figure 3.28.

*simple arc*

*arc*
*simple loop*

*loop*

**Figure 3.28:**
Some one-dimensional planar objects

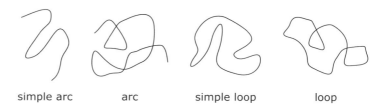

simple arc            arc            simple loop            loop

Moving up a dimension, we have already defined the cell as the primary two-dimensional topological object, topologically equivalent to the disk and simply connected. Another notable class of areal objects is the annuli (cells with a single hole). We can also allow the holes to be occupied by further objects (islands) and so on (see Figure 3.29).

The next paragraphs explore further the concept of connectedness, particularly as it relates to areal objects. The topological property of con-

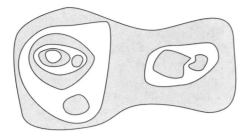

**Figure 3.29:**
Holes within
holes within
holes ...

nectedness was defined earlier, in section 3.3.2. We begin by providing a different definition.

- A set in a topological space is *path-connected* if any two points in the set can be joined by a path that lies wholly in the set.

path-connected

This definition relies on the definition in topological terms of the notion of "path," and while this is not hard to do, it takes us away from our main themes. Instead we suggest some intuitive ideas about a path as an unbroken and "well-behaved" curve (in the sense of not doing odd things like getting infinitely tangled anywhere). In the Euclidean plane with the usual topology, for "path" we may read "simple arc."

A natural first question is: Given that there are two notions of connectedness, namely, connectedness (as defined earlier) and path-connectedness, do these notions define the same property? The answer to this is, in general, no. Although it is possible to show that every path-connected set is connected, there are sets that are connected but not path-connected. However, in the special case of the Euclidean plane with the usual topology, each example of a connected but not path-connected set is pathological, involving an infinite number of twists and turns. Therefore, for practical purposes we may identify notions of connectedness and path-connectedness, certainly for the areal objects that we will define shortly. Path-connectedness is a more intuitive notion than pure connectedness, and can therefore be used as a test for connectedness in practical cases. To summarize, test for connectedness by asking the question: Given any two points in the set, is it possible to move from one point to the other along a path entirely within the set?

Many applications of spatial analysis require classes of planar objects that are purely areal, that is, not mixtures of points, lines, and areas. Also they do not have isolated missing points (*punctures*) or arcs (*cuts*). Interestingly, it is possible to define the notion of a purely areal object using only topological notions.

- Let $X$ be a set of points in the Euclidean plane under the usual topology. Then define the *regularization* of $X$ to be the closure of the interior of $X$, that is, $reg(X) = X^{\circ-}$.

regularization

The regularization process has the effect of eliminating from a set any pathological and non-areal features. Consider the example shown in Figure 3.30a, which is an amalgamation of a punctured and cut cell with some arcs and isolated points. The regularization of the set in Figure 3.30b removes all cuts, punctures, extraneous arcs, and isolated points. Regularization first finds the interior of the object, which will remove exterior arcs and points. Taking the closure will then remove cuts and punctures. What remains is always a closed, purely areal object. In our example, shown in Figure 3.30b, the result is a cell.

**Figure 3.30:**
Spatial object
$X$ comprising
an area with
cut, puncture,
arcs, and
points, and
$reg(X)$, its
regularization

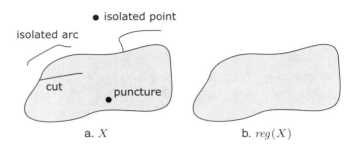

a. $X$                                      b. $reg(X)$

The regularization concept can now be used to characterize pure area. If an object is already purely areal, then regularizing it will have no effect. An object for which regularization has no effect is termed *regular closed*. The regular closed sets are exactly the purely areal objects that we require. The formal definition follows.

regular closed

- Let $X$ be a set of points in the Euclidean plane under the usual topology. Then $X$ is *regular closed* if and only if $X^{\circ-} = X$

Having finally arrived at a topological characterization of objects that are purely areal, we now reconsider in more detail the notion of connectedness. Figure 3.31 shows three connected sets.

**Figure 3.31:**
Three
connected sets

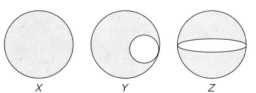

X               Y               Z

There are some clear differences in the kinds of connectedness here. In the first two cases $X$ and $Y$, given any two points in each set, there are few constraints upon the path of connection from the first point to the second. All that is required is that the path starts at the first point, stays within the set, and ends at the second point. However, in the case of set $Z$, if the two points are in the upper portion and lower portion respectively, then the path is constrained to pass through one of the two points on the horizontal diameter. This difference is expressed by saying that $X$ and $Y$ are *strongly connected*, but that $Z$ is *weakly connected*. To arrive at a formal definition, note that $Z$ may be made disconnected by removing a finite number of points (in fact, the two points on its horizontal diameter).

However, no matter how large a finite number of points we remove from $X$ and $Y$, they will remain connected.

- A set $X$ in the Euclidean plane with the usual topology is *weakly connected* if it is possible to transform $X$ into an unconnected set by the removal of a finite number of points.

  weakly connected

- A set $X$ in the Euclidean plane with the usual topology is *strongly connected* if it is not weakly connected.

  strongly connected

Figure 3.32 shows some more strongly and weakly connected sets. The notions of strong and weak connectedness play an important role later in the categorization of planar objects in the object-based approach to spatial modeling.

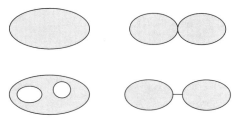

**Strongly connected      Weakly connected**

**Figure 3.32:** More strongly and weakly connected sets

### 3.3.5   Combinatorial topology of the Euclidean plane

The other topological area considered in this chapter is combinatorial topology. In some ways, combinatorial topology is more pertinent to computer-based models than point-set topology, because the often finite and discrete structures that arise in combinatorial topology are highly suitable for representation in computer-based data structures. A typical result of combinatorial topology is Euler's famous formula:

Euler's formula

- Given a polyhedron with $f$ faces, $e$ edges, and $v$ vertices, then $f - e + v = 2$.

For example, Figure 3.33a shows a cube (six faces, 12 edges, and eight vertices). A very similar formula applies to an arrangement of cells in the plane. Remove a single face from a polyhedron (for example, in Figure 3.33a remove face $F$) and apply a 3-space homeomorphism to flatten the shape onto the plane. What results is a configuration of cells, with arcs forming common boundaries and nodes forming the intersection points of the arcs. Flattening the cube in Figure 3.33a results in the planar configuration in Figure 3.33b. Since we have removed a face from the polyhedron (it has actually become the exterior to the cellular configuration, as in Figure 3.33b), we may simply modify Euler's formula for the sphere to derive Euler's formula for the plane.

**Figure 3.33:**
Example
polyhedron and
3-space
homeo-
morphism

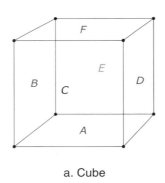 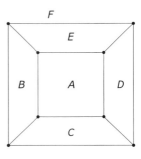

a. Cube                    b. Planar configuration
                              homeomorphic to cube

- Given a cellular arrangement in the plane, with $f$ cells, $e$ edges, and $v$ vertices, $f - e + v = 1$.

Figure 3.34 shows an example of a planar configuration of cells, with seven faces, 17 edges, and 11 nodes.

**Figure 3.34:**
Planar cellular
arrangement
where $f = 7$,
$e = 17$, and
$n = 11$

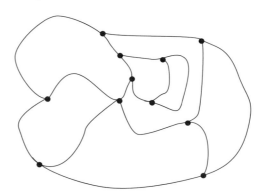

The topological content of these results becomes clear when we observe that no matter how the surface of the sphere is divided into polyhedral arrangements, the result of $f - e + v$ is always two, and for planes the result is always one. Thus the number two characterizes a sphere and distinguishes it from a plane. If we were to perform the same exercise on the surface of a torus (doughnut shape), the result is always Euler zero. The result of $f - e + v$ is called the *Euler characteristic* of a surface.

Euler
characteristic

*Simplexes and complexes*

As the next chapter will discuss, much of the current work on generic models of space is done using variations of the planar cellular arrangements described above. The most fundamental formal model uses the notion of a *simplicial complex*. In the two-dimensional case, simplicial complexes are simple triangular network structures in the Euclidean plane. The constructions performed in this section are all planar, but the ideas can be generalized to higher dimensional structures.

- A *0-simplex* is a set consisting of a single point in the Euclidean plane.

- A *1-simplex* is a closed finite straight-line segment.

- A *2-simplex* is a set consisting of all the points on the boundary and in the interior of a triangle whose vertices are not collinear.

An $n$-simplex is said to have *dimension* $n$. The *vertices* of a simplex are defined as: for a 0-simplex the point itself, for a 1-simplex its endpoints, and for a two-simplex the vertices of the triangle. A *face* of a simplex $S$ is a simplex whose vertices form a proper subset of the vertices of $S$. Figure 3.35 shows examples of 0-, 1-, and 2-simplexes. Of course, the dimensionality can be extended beyond two. Thus a 3-simplex would be a tetrahedron. The *boundary* of a simplex $S$, written $\partial S$, is the union of all its faces. For example, suppose that a 2-simplex $S$ has vertices $x$, $y$, and $z$. Then the faces of $S$ are the three 1-simplexes $xy$, $xz$, $yz$ and the three 0-simplexes $x$, $y$, $z$. The boundary of $S$ is the union of these faces, thus corresponding to the usual point-set topological definition of boundary.

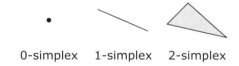

0-simplex    1-simplex    2-simplex

**Figure 3.35:** Examples of 0-, 1- and 2-simplexes

Simplexes are the building blocks of larger structures, called simplicial complexes. Complexes are built out of simplexes in a way that is now made precise. A simplicial complex $C$ is a finite set of simplexes satisfying the properties:

1. A face of a simplex in $C$ is also in $C$.

2. The intersection of two simplexes in $C$ is either empty or is also in $C$.

Figure 3.36 shows examples of configurations, two of which are simplicial complexes and two are not. In fact, the complexes on the right are formed by adding sufficient nodes and edges to the configurations on the left to make them satisfy the simplicial complex formation rules, a form of "completing the topology." In the case of the 2-simplexes $abc$ and $def$, their intersection is not a face of either simplex. This is rectified by adding nodes $k$ and $l$ and decomposing the original simplexes $abd$ and $def$ into simplexes $akf$, $afb$, $bfc$, $cfl$, $fkl$, $dkl$, and $del$. These simplexes, along with their faces $ab$, $af$, $ak$, $bc$, $bf$, $cf$, $cl$, $dk$, $dl$, $de$, $el$, $fl$, $fk$, $kl$ and vertices $a$, $b$, $c$, $d$, $e$, $f$, $k$, $l$, form a simplicial complex. The 1-simplexes

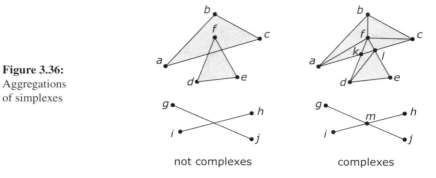

**Figure 3.36:**
Aggregations
of simplexes

                                not complexes            complexes

in the lower part of the figure are enhanced in a similar way by adding
vertex $m$.

The set of points contained in the constituent simplexes of a simplicial
planar embedding    complex is a set in the Euclidean plane, called the *planar embedding*
of the complex. Figure 3.37 shows the planar embedding of the upper
complex in Figure 3.36.

**Figure 3.37:**
Planar
embedding of a
simplicial
complex from
Figure 3.36

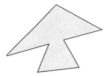

Even though we have given an abstract presentation of simplicial
complexes, such structures are common in the spatial sciences. The
*dimension* of a simplicial complex is the maximum dimension of its
constituent simplexes. For an $n$-complex, $C$, the boundary of $C$, $\partial C$, is
a simplicial complex of dimension $n - 1$. One-dimensional complexes
are *graphs*, covered in section 3.4.1. Two-dimensional complexes may be
used to model the triangulated irregular networks (TIN) used in terrain
modeling, or indeed any areal objects.

A different approach to combinatorial topology is known as the
combinatorial    *combinatorial map*. Most GIS practitioners are familiar with the standard
map    NAA (node-arc-area) representation of planar configurations, discussed
in more detail in Chapter 5. This representation begins to give a topological description of a planar object. In the NAA representation, arcs
have associated with them left and right polygons. However, the NAA
representation falls short of a full topological representation in at least
two ways.

1. The more detailed connectivity of the object is not explicitly given.
   Thus there is no explicit representation of weak, strong, or simple
   connectedness.

2. The representation is not *faithful*, in the sense that two different topological configurations may have the same representation.

A combinatorial map goes some way to meeting these two shortcomings. The problem is that the same shape may be viewed in different ways. Consider the weakly connected cellular arrangement in Figure 3.38. Is it a disk with an ellipse removed or the union of two lunes?

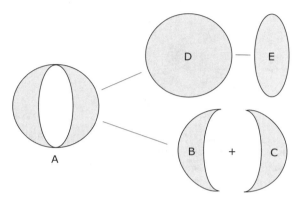

**Figure 3.38:** Ambiguity in representing a planar object

The notion of a combinatorial map helps in this regard. Assume that the boundary of a cellular arrangement is decomposed into simple arcs and nodes that form a network (networks are discussed further in the next section). Next, give a direction to each arc so that traveling along the arc, the object bounded by the arc is to the right of the directed arc. Thus the object above has a boundary network as shown in Figure 3.39. The final step is to provide a rule for the order of following the arcs: after following an arc into a node, move counterclockwise around the node and leave by the first unvisited outward arc encountered. Thus in our example, suppose the traversal begins at $x$ and moves along arc 1. When arriving at $y$ we rotate counterclockwise around $y$ in our search for the first outbound arc. This is arc 2. Now arriving back at $x$, we choose arc 1. Arc 1 is already in our list (cycle), so the process moves on. We start with an arc not already traversed and repeat the cycle-generating process. When there are no more arcs to be traversed, the process halts. Thus we have found two cycles $[1, 2]$ and $[3, 4]$.

The weakly connected object in Figure 3.39 has been represented as a union of lunes, each homeomorphic to a cell. It is possible to show that for any cellular configuration, the combinatorial map provides a set of cycles, each corresponding to a component cell of the configuration. Thus the representation is unique and unambiguous.

## 3.4   NETWORK SPACES

A great many geographic problems can be represented using a network. For example, a system of roads or rail links is often thought of as a network; the task of deciding what route to take when traveling by car is

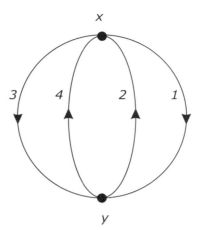

**Figure 3.39:**
Directed graph
associated with
the object in
Figure 3.38

essentially the task of finding a suitable route through a network. Several famous spatial problems have been solved by representing the problem as a network, such as the Königsberg bridge problem (see inset, on the next page). This section introduces the *graph* as a formal model of the network, starting with *abstract graphs*, and moving on to graphs embedded in the plane (*planar graphs*).

### 3.4.1   Abstract graphs

graph
node
edge

A *graph* $G$ is defined as a finite non-empty set of *nodes* together with a set of unordered pairs of distinct nodes (called *edges*). If $x$ and $y$ are nodes of $G$ and $e = xy$ is an edge of $G$, then $e$ is said to *join* $x$ to $y$, or be *incident* with $x$ and $y$. Similarly $x$ and $y$ are *incident* with edge $e$.

A graph is a highly abstracted model of spatial relationships, and represents only connectedness between elements of the space. However, in many situations such a model can be very useful, particularly if we allow some extensions, as follows:

directed graph

- A *directed graph* is a graph in which each edge is assigned a direction. Directed edges are often indicated by arrowed lines on a diagram.

labeled graph

- A *labeled graph* is a graph in which each edge is assigned a label (maybe a number or string). Such labels are usually indicated on a diagram near to the appropriate edges.

One can imagine the usefulness of a directed graph in modeling the road network in a city center, where there may be many one-way streets. A labeled graph can model a host of situations, for example, distances, travel times, or traffic usages of roads in a network.

It is usual to show the linkages of a graph by a diagram. The graph $G$ in Figure 3.40 consists of the six nodes $a$, $b$, $c$, $d$, $e$, and $f$ and the nine degree    edges $ab$, $ac$, $af$, $bd$, $be$, $cd$, $ce$, $df$, and $ef$. The *degree* of a node is the

**Königsberg bridge problem** *The Königsberg bridge problem is the question of whether it is possible to walk a circuit that crosses each of seven bridges (shown below left) in the city of Königsberg once and only once. You might try to find such a route yourself. However, don't try for too long as in 1786 the mathematician Leonard Euler succeeded in proving that the task is impossible. Euler's proof is based on a model of the topological relationships between the Königsberg bridges (shown below right). The nodes, labeled w, x, y, and z, are abstractions of the regions of dry land. The edges between nodes are abstractions of the bridges connecting regions of dry land. Euler noted that apart from the start and end nodes, the path through a node must come in along one edge and out along another edge. So, if the problem is to be solvable then the number of edges incident with each intermediate node must be even. However, in Königsberg none of the nodes is incident with an even number of edges. Thus, Euler proved that it was impossible to cross each bridge just once.*

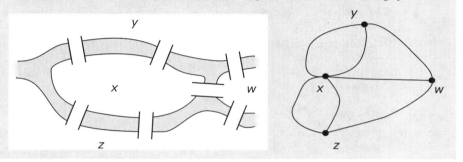

number of edges with which it is incident, and so the degree of all the nodes in $G$ is three. A *path* between two nodes is a connected sequence of edges between the nodes, and is usually denoted by the nodes that it passes through. Examples of paths between nodes $a$ and $d$ in $G$ are *afd*, *acd*, *abd*, and *abecafd*. A *connected graph* is such that there exists a path between any two of its nodes: $G$ is connected.

path

connected graph

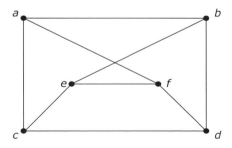

**Figure 3.40:** Graph $G$ with six nodes and nine edges

Two graphs may show exactly the same connectivity relationships, and such graphs are said to be *isomorphic*. Sometimes isomorphism can be hard to detect, because a graph can disguise itself quite well. Thus, graph $H$ in Figure 3.41 is isomorphic to graph $G$ in Figure 3.40, since it has precisely the same nodes and edges. In this case, we have made the case clearer by labeling the nodes to show the isomorphism.

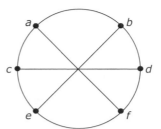

**Figure 3.41:**
The graph $H$,
isomorphic to
$G$ in Figure
3.40

cycle

acyclic graph

tree

A path from a node to itself traversing at least one edge is called a *cycle*. Examples of cycles in $G$ (or $H$) are *abeca* and *abdfeca*. A cycle that visits each node only once (like *abdfeca* in Figure 3.41) is termed a *Hamiltonian circuit*. A graph that has no cycles is called an *acyclic graph*. An especially useful class of graphs is the *tree*. A tree is a connected acyclic graph. Figure 3.42 shows the three non-isomorphic trees with five nodes.

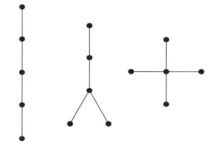

**Figure 3.42:**
The three non-
isomorphic
trees with five
nodes

rooted tree

leaf

A *rooted tree* is a tree that has one of its nodes, the *root*, distinguished from the others. Rooted trees (often we omit the word "rooted") are conventionally drawn with the root at the top and nodes occupying successive levels down, depending upon their distance (in terms of path) from the root. Nodes immediately below the root are termed *immediate descendants* of the root; they themselves may have descendants, and so on. A node with no descendants is termed a *leaf*. Figure 3.43 shows an example of a layered rooted tree. Trees provide some useful data structures for computational purposes, to which we return in later chapters.

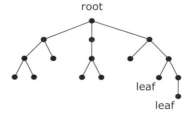

**Figure 3.43:**
Rooted tree
with five levels
and eight
leaves

directed path

It is possible to extend the notion of path and cycle to directed graphs. Thus, there is a *directed path* from node $a$ to node $b$ if there is a sequence of correctly oriented directed edges leading from $a$ to $b$. A *directed cycle* is a directed path from a node to itself. The class of directed graphs that have no cycles is particularly useful for a wide range of applications. These graphs are termed *directed acyclic graphs* or *DAGs*. A DAG defines a *partial order* on its nodes. A (strict) *partial order* is a special form of relation on a set, which is irreflexive, antisymmetric, and transitive. Transitive relations (if $x$ is related to $y$ and $y$ is related to $z$, then $x$ is related to $z$) have been introduced previously. Irreflexive and antisymmetric relations are defined as follows:

*directed cycle*

*DAG*

*partial order*

- A relation is *irreflexive* if every element of the set is not related to itself.

  *irreflexive*

- A relation is *antisymmetric* if for every pair of elements, $x$ and $y$, if $x$ related to $y$ then $y$ is not related to $x$.

  *antisymmetric*

An example of a partial ordering is the relation "greater than," between two real numbers. No number can be greater than itself (irreflexive); if $x$ is greater than $y$, then it cannot be that $y$ is greater than $x$ (antisymmetric); and $x$ greater than $y$ greater than $z$ implies $x$ greater than $z$ (transitive). For a DAG, if we define a relation $R$ where $(a, b) \in R$ means "there exists a path from $a$ to $b$" then it can be shown that this relation is a partial order.

### 3.4.2  Planar graphs

A further level of information may be added to the graph-theoretic model by considering the embedding of the graph in the Euclidean plane. A *planar graph* is a graph that can be embedded in the plane in a way that preserves its structure. In particular, a planar graph's edges are embedded as arcs that may only intersect at nodes of the graph. Figure 3.44a shows a planar graph while Figure 3.44b shows a non-planar graph. For the non-planar graph in Figure 3.44b no rearrangement of arcs in the plane will preserve the original connectivity without leading to edges crossing somewhere other than at a node.

*planar graph*

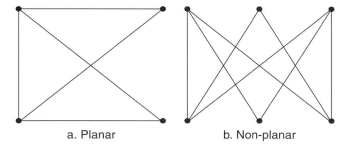

a. Planar                          b. Non-planar

**Figure 3.44:**
Planar and non-planar graphs

In general, there are many topologically inequivalent planar embeddings of a planar graph in the plane. Figure 3.45 shows three embeddings of the planar graph of Figure 3.44. The upper two embeddings are homeomorphic, but not homeomorphic to the lowest, with respect to the usual topology of the Euclidean plane. There is no way, using topological transformations (rubber-sheet geometry), to move the node of degree two inside the triangle to a position outside the triangle.

This last discussion raises an interesting consideration. Configurations that are taken to be equivalent in one model may very well be inequivalent in another. The second and third configurations are equivalent in a graph-theoretic sense, both being identically connected. However, when viewed as embeddings in the plane, they are inequivalent in the topological sense (and certainly in the metric sense). It all depends on the level of abstraction.

**Figure 3.45:** Three planar embeddings of a planar graph

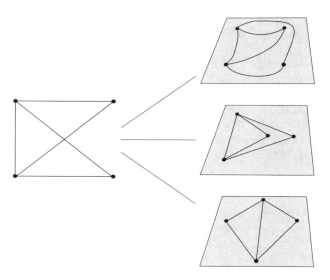

A planar embedding of a planar graph determines a subdivision of the plane into faces or regions. A simple integrity constraint upon a planar graph discussed in an earlier section is given by the Euler formula for the plane, which provides a relationship between the number of nodes $n$, arcs $a$, and faces $f$, as $f - a + n = 1$.

dual graph     A useful concept associated with planar embedded graphs is that of *duality*. The *dual* $G^*$ of a planar graph $G$ is obtained by associating a node in $G^*$ with each face in $G$. Two nodes in $G^*$ are connected by an edge if and only if their corresponding faces in $G$ are adjacent. Given an edge $e$ in $G$, the dual edge $e^*$ to $e$ joins the nodes in $G^*$ corresponding to the two faces in $G$ incident with $e$. Figure 3.46 shows a planar graph (nodes filled in black, edges marked with continuous lines) and its dual (nodes shown unfilled and edges marked with dotted lines).

When the planar graph $G$ is a diagonal triangulation of a polygon (with no Steiner points), then the dual graph $G^*$ has the properties that the degree of each node is no more than three (because triangles have three sides), and that $G^*$ is acyclic and connected: it is a tree.

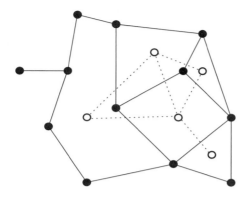

**Figure 3.46:**
A graph and its dual

## 3.5 METRIC SPACES

This section explores the kind of properties a model of space should have if it is to include the concept of distance between objects in the space. Such a space is called a *metric space*. The formal definition follows, but is should be noted that it does not always accord with our commonsense notion of distance.

A point-set $S$ is said to be a *metric space* if there exists a distance function $d$, which takes ordered pairs $(s, t)$ of elements of $S$ and returns a distance, that satisfies the following three conditions:

metric space

1. For each pair $s$, $t$ in $S$, $d(s, t) > 0$ if $s$ and $t$ are distinct points and $d(s, t) = 0$ if $s$ and $t$ are identical.

2. For each pair $s$, $t$ in $S$, the distance from $s$ to $t$ is equal to the distance from $t$ to $s$, $d(s, t) = d(t, s)$.

3. For each triple $s$, $t$, $u$ in $S$, the sum of the distances from $s$ to $t$ and from $t$ to $u$ is always at least as large as the distance from $s$ to $u$, that is: $d(s, t) + d(t, u) \geq d(s, u)$.

Put into more informal language, the first condition stipulates that the distance between points must be a positive number unless the points are the same, in which case the distance will be zero. The second condition ensures that the distance between two points is independent of which way round it is measured. The third condition states that it must always be at least as far to travel between two points via a third point rather than to travel directly.

In order to motivate this definition, we give below some possible distance functions and consider them with respect to properties 1–3 above. Let $S$ be a set of cities on the globe and distance between two cities in $S$ defined as follows (see Figure 3.47):

*Geodesic distance*: The distance "as the crow flies" is termed the geodesic distance. In our example it is the distance along the great circle of the Earth passing through the two city centers.

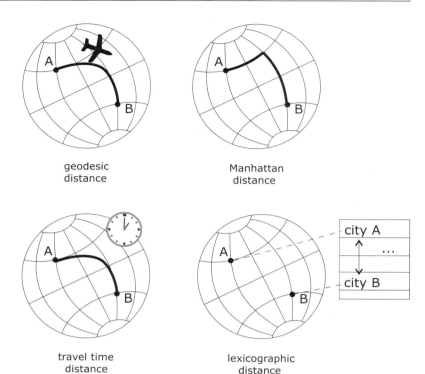

**Figure 3.47:**
Distances
defined on the
globe

*Spherical Manhattan distance*: The name "Manhattan distance" arises because in Manhattan many of the streets are arranged in a grid-like configuration. In the plane, the Manhattan distance is simply the sum of the difference in $x$ and $y$ coordinates. Similarly, the spherical Manhattan distance is the difference in latitudes plus the difference in longitudes (see "Latitude and longitude" inset, on the facing page).

*Travel time distance*: The minimum time required to travel from one city to the other (for example, using a sequence of scheduled airline flights) is the travel time distance.

*Lexicographic distance*: The absolute value of the difference between the positions of cities in a fixed list of place names (a *gazetteer*) is termed the lexicographic distance.

The first property of a metric space is quite uncontroversial, and satisfied by any self-respecting distance function. Sensible distances cannot be allowed to be negative. Also, the distance between an element and itself is always zero, whereas the distance between two distinct elements is always greater than zero. All the distance functions in our city example possess these properties.

The second property specifies that, in a metric space, distance is symmetric; that is, the distance from $a$ to $b$ is always the same as

**Latitude and longitude** *Many of the models in this chapter have been founded upon the plane. The plane is a useful approximation to the Earth's surface over small distances, but the Earth is not flat, so over larger distances other approximations are needed. The most common such approximation is the sphere. The surface of the sphere, although embedded in Euclidean 3-space, is two dimensional. Thus, any point on it may be uniquely specified by two numbers. A familiar system of coordinates for points on the surface of the sphere is latitude and longitude. In the diagram below, $O$ is origin of the coordinate system, $P$ is an arbitrary point on the sphere, and $Q$ is the projection of $P$ onto the $xy$-plane. The angle between $OP$ and $OQ$ is the latitude of the point $P$. The angle between $OQ$ and the $x$-axis is the longitude of $P$.*

*The great circle of the sphere in the $xy$-plane is the equator and the great circle in the $xz$-plane is the meridian. The usual topology of the surface of the sphere is similar to the usual topology of the Euclidean plane. Neighborhoods are sets of points with a constant distance from a fixed point, measured along a geodesic (great circle). A more detailed introduction to latitude and longitude may be found in Longley et al. (2001)*

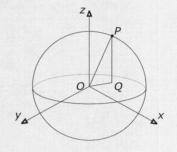

the distance from $b$ to $a$. The geodesic, Manhattan, and lexicographic distances satisfy this condition. However, the travel time distance is not symmetric. For example, it is perfectly possible (and indeed usual due to prevailing winds) for the flight times between two cities to be different in each direction. A distance function that obeys properties 1 and 3 but not 2 is called a *quasimetric*.

quasimetric

The third property is called the *triangle inequality* due to the configuration shown in Figure 3.48. In plain English, the triangle inequality implies that it is never any farther to go by a direct route rather than an indirect route. All the above examples obey the triangle inequality.

triangle inequality

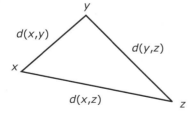

**Figure 3.48:** Triangle inequality

So, the collection of cities together with the geodesic distance function, the Manhattan distance function, and the lexicographic distance function are all metric spaces. The collection of cities with the travel time function is a quasimetric but not a metric space. The archetypal example of a metric space is Euclidean space, where the distance between two

points is defined by the Pythagorean formula given earlier (section 3.1.1).
This distance function can easily be extended to higher dimensions.

For a further example of a space that is not a metric space, consider a
plane terrain with no prevailing wind in which some areas are easy walking (maybe covered with well-cropped grass) and others more difficult
(maybe rocky). The distance function is defined as the average time it
takes to walk in a straight line from one point to another. We may assume
that conditions 1 and 2 hold, but it is unlikely that condition 3 holds. It
might well be the case that it is faster to walk round a difficult rocky
stretch than to walk straight across it (see Figure 3.49). Thus the triangle
inequality does not hold and this space is not a metric space.

**Figure 3.49:**
More haste,
less speed in a
non-metric
space

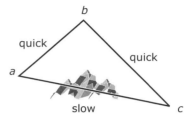

### 3.5.1   Topology of metric spaces

It turns out that a metric space has a natural topology. Let $S$ be a metric
space with distance function $d$. For each point, $x$ in $S$ and each real
number $r$, define the *open ball* $B(x, r)$ to be the set of points whose
distance from the point $x$ is less than the number $r$. Expressed formally:

$$B(x, r) = \{y | d(y, x) < r\}$$

Define the set of neighborhoods to be the set of open balls. It is
not hard to verify that this defines a topology for $S$. In the case of
the Euclidean metric, this example reduces to the usual topology of the
Euclidean plane. In the case of the Manhattan metric, then the open ball
$B(x, r)$ will contain all points for which the sum of the horizontal and
vertical distances from $x$ is less than $r$. In the case of the Manhattan metric
applied to a flat planar space, the open balls $B(x, r)$ will be squares of
diagonal $2r$, as shown in Figure 3.50.

Travel time measures have some interesting properties. We have seen
that, in general, they do not lead to metric spaces because they are not
necessarily symmetric (for example, in a one-way traffic system). Let
us for the moment make the simplifying assumption that distances are
symmetric, so a topology may be defined as above. This is the travel-time
topology introduced in section 3.3.2. Computing topological neighborhoods can be illuminating in this case. Figure 3.51 shows a travel-time
neighborhood ($t$-zone) of Liège, computed by Dussart and redrawn and
considered by Tobler (1993).

The shaded region shows the area within 1 hour's travel time of the
center of Liège in 1958 by the common means of travel available at

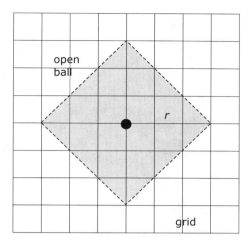

**Figure 3.50:**
An open ball
induced by the
Manhattan
metric

that time. This "neighborhood" is not even connected, being the disjoint union of a finite number of cells. This type of pattern arises because of the discontinuous nature of travel on public transport, boarding and alighting at fixed and discrete points. As Tobler points out, such travel-time configurations are not amenable to modeling within the framework of Euclidean space. Such problems force us to use network models. Later chapters further develop this theme.

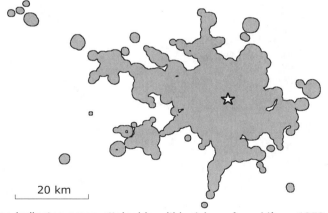

**Figure 3.51:**
A travel-time
neighborhood
of Liège,
computed by
Dussart (Tobler
1993)

20 km

Shading indicates areas attainable within 1 hour from Liège, 1958, by a combination of tramway, autobus, chemin de fer, and walking (5 km/h).

## 3.6 ENDNOTE ON FRACTAL GEOMETRY

The appearance and characteristics of many geographic and natural phenomena, like landscapes and coastlines, depend on the scale at which these phenomena are observed, termed *scale dependence*. For example, Figure 3.52 shows a satellite image of the Ganges river delta, Bangladesh,

scale dependence

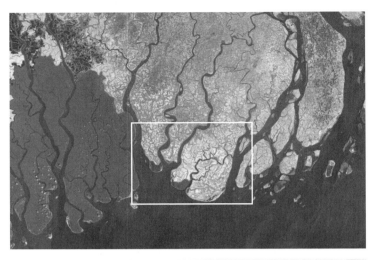

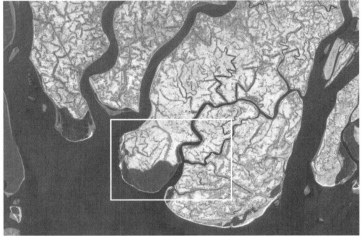

**Figure 3.52:**
Ganges River
delta at three
levels of detail
(Source: NASA
Landsat-7
image,
February 2000)

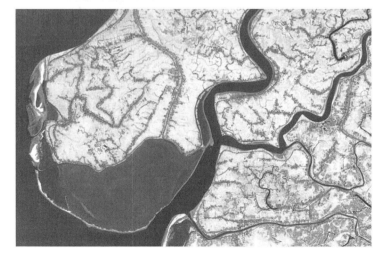

**Jordan curve theorem** *In 1887, the French mathematician Camille Jordan formulated a famous theorem about simple loops, now known as the Jordan curve theorem. The theorem states that, given any simple loop, then the complement of the simple loop is not connected, but is partitioned into two connected components, one of which is bounded (called the* inside *of the loop) and one not bounded (called the* outside *of the loop). This proposition may seem so blindingly obvious as to be the sort of thing that gives mathematicians a bad name. In fact, it is quite difficult to prove convincingly. The problem lies in the wide variety of shapes, including some fractals, that qualify as loops. For example, the Koch snowflake in Figure 3.53 is a loop, but does not have a defined slope at any point on it. Technically, it is* nowhere differentiable *(see Chapter 4). The effort involved in proving the Jordan curve theorem resulted in the development of techniques that were instrumental in the birth of topology as a major branch of mathematics. The theorem has a practical application in the field of GIS, being the foundation of the point-in-polygon operation, discussed in more detail in Chapter 5.*

at three different scales. More detail is revealed at each finer scale. Furthermore, details at finer scales tend to resemble details at coarser scales, termed *self-similarity*. In Figure 3.52 the shape of the landforms and sinuosity of the rivers are similar at each scale, making it difficult to gauge the size of landforms in Figure 3.52. The scale has been deliberately omitted from Figure 3.52; in fact, the first image is of an area about 100 miles across, the final image approximately 10 miles across. We can imagine repeating the process of zooming in over and over, perhaps until we reach the microscopic level. At every scale new and self-similar detail would be revealed. self-similar

The straight lines and smooth curves of Euclidean geometry are not well suited to modeling self-similarity and scale dependence. In a classic book, the Polish mathematician Benoit Mandelbrot (1982) argued that a fundamentally different type of geometry, which he named *fractal geometry*, provides a more faithful representation of such natural and geographic phenomena. Fractal geometry concerns the study of shapes, called *fractals*, that are self-similar across all scales. While the term "fractal" is relatively recent, fractal shapes and the problems they cause mathematicians have been known about for hundreds of years (see "Jordan curve theorem" inset, on this page). fractal geometry fractal

Fractals are self-similar because they are defined recursively, rather than by describing their shape directly. For example, Figure 3.53 shows the first four stages in the construction of a famous fractal, the Koch snowflake (named after a Swedish mathematician, Helge von Koch). Building a Koch snowflake starts with an equilateral triangle, termed the *initiator* step. Then each straight line is divided into three equal parts. The middle part of each line is replaced with a new equilateral triangle, correctly scaled and with no base, termed the *generator* step. The Koch snowflake is the result of iterating the generator step an infinite number of times. Irrespective of the scale at which the Koch snowflake is viewed, it always exhibits the same level of detail. initiator generator

**Figure 3.53:**
First four
stages in the
construction of
the Koch
snowflake

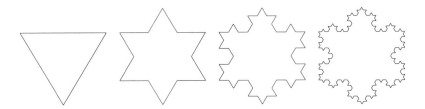

Simple fractals like the Koch snowflake form a useful analogy to natural self-similar phenomena. More complex fractals can produce more "realistic" looking shapes. Figure 3.54 shows a synthetic "fractal landscape." Like the Koch snowflake, this landscape (including the clouds and water) is the result of recursively applying a feedback generator. The generator step of the Koch snowflake is essentially a similarity transformation, introducing at each iteration scaled copies of the original shape at one-third of the size of the previous iteration. In addition to being self-similar, fractals like those in Figure 3.54 are *self-affine*. Self-affine fractals can be constructed using affine transformations within the generator, so rotations, reflections, and shears can be used in addition to scaling (recall from section 3.1.4 that all affine transformations are also similarity transformations).

*self-affine*

Aside from producing attractive pictures, fractal geometry has a number of more serious uses in GIS. One simple use of fractals in GIS is for cartographic simplification and enhancement. In order to produce maps at a range of different scales, it is often important to be able to decrease (or occasionally increase) the level of detail in the representation of geospatial data while still retaining the essential characteristics of that line (for example, its "wiggliness"). By approximating the shape of a river, for example, as a fractal we can easily generate representations of that river at arbitrary levels of detail: simplification requires less detail; enhancement requires more detail.

*fractal dimension*

Other uses of fractals are often based on the concept of *fractal dimension*. Fractal dimension is an important property of fractals, which provides a measure of the degree to which new detail is revealed at different scales. The fractal dimension of a shape lies somewhere between the Euclidean dimensions of the shape and its embedding space. For example, the fractal dimension of the Koch snowflake in Figure 3.53 lies between 1, the dimension of a line, and 2, the dimension of the Euclidean plane (in fact the Koch snowflake has a fractal dimension of 1.26). Fractal dimension is an indicator of shape complexity: a shape with a high fractal dimension is complex enough to nearly fill its embedding space (e.g., a curve with a fractal dimension of 1.8 almost fills the plane).

*space-filling*

As a result, fractals are often referred to as *space-filling*. The space-filling characteristics of fractals are useful for indexing geospatial information, as we shall see in Chapter 6.

Fractal dimension can be a useful descriptor of a geographic shape. The fractal dimension of a river may be used as an indicator of the

**Figure 3.54:** A synthetic fractal landscape

underlying geomorphological and hydrological processes involved in river formation. Similarly, fractal dimension analysis is used in landscape ecology to assess the complexity of geospatial ecological features, such as plant or animal habitats. The fractal dimension of a "true" fractal shape, like the Koch snowflake, can be determined using theoretical analysis of the fractal generator. However, the fractal dimension of natural geographic phenomena, like rivers, terrain surfaces, and animal habitats, must be determined empirically. Such empirical measurements of fractal dimension are notoriously unreliable and require careful analysis. The bibliographic notes contain further references to applications of fractal geometry in GIS.

## BIBLIOGRAPHIC NOTES

3.1 An excellent general text on geometry is Coxeter (1961), which gives a fascinating overview of the variety that geometry offers, from basic work with triangles and polygons, through tessellations and two-dimensional crystallography, to the platonic solids and golden section. Later chapters offer more advanced material, with chapters on projective, hyperbolic, differential, geodesic, and four-dimensional geometries.

3.2 Hein (2003) provides a clear introduction to elementary set theory, while Lipschutz (1997) provides a low-cost alternative.

3.3 Mathematically inclined readers can find much detailed information about metric and point-set topological spaces in Sutherland (1975) and Armstrong (1979). Algebraic or combinatorial topology can be quite inaccessible to all but the mathematically gifted. A book that has withstood the test of time as a readable elementary introduction is Giblin (1977). A further book that introduces topology in an algebraic and combinatorial framework is Henle (1979). Simplicial complexes are also discussed in chapters in Hoffmann (1989), the focus of which is computational geometry and solid modeling, but it contains in its early chapters much interesting material concerning embeddings of simplicial complexes in three dimensions.

3.4 There are now many readable elementary texts on the theory of graphs, for example, Harary (1969). Any discrete mathematics textbook, such as Hein (2003) or Lipschutz (1997), will also contain an introduction to graph theory.

3.5 The topic of metric spaces is covered in Sutherland (1975) and Armstrong (1979). Bryant (1985) is a dedicated short introduction to metric spaces.

3.6 Peitgen et al. (1992) is an excellent introductory book on every aspect of fractal geometry. Mandelbrot (1982) has become a modern classic and is well worth reading. Xia and Clarke (1997) contains an overview of fractal approaches to geospatial information. Burrough (1981) and Goodchild (1988) are two early examples of fractal geometry applied to geospatial information. Lam and De Cola (1993) is a collection of articles covering a wide range of applications of fractals in GIS, including geomorphology, hydrology, and landscape ecology. Polidori et al. (1991) and Duckham et al. (2000) are examples of fractal geometry applied to uncertain geospatial information.

# Models of geospatial information

4

A **model** defines a representation of parts of one domain in another. In GIS,    **Summary**
models are needed to define the relationship between our geographic environment
(the **source domain**) and the representation of that environment within a computer
(the **target domain**). This chapter looks at the two main classes of high-level
model for GIS: field- and object-based. The **field-based model** treats space as
an absolute framework within which attributes are measured. The **object-based
model** treats space as a relative construct defined by the objects that populate it.

The fundamental database and spatial concepts introduced in the previous
two chapters provide a necessary computational and geometric basis for
storing and manipulating geospatial information within a computer. However, GISs do more than simply store and process geometries. GISs and
geospatial information encapsulate high-level *models* of the geographic
world around us. We begin this chapter with some issues relating to what
it is that we wish to model (section 4.1). Section 4.2 goes on to define exactly what is meant by a "model," and examines the modeling process. We
then explore in detail the two high-level spatial models that have become
characteristic of GIS: the field model (section 4.3) and the object model
(section 4.4). The effects of adopting the field or object model can have
far reaching implications for spatial data structures and implementations,
in addition to how geospatial data should be understood.

## 4.1   MODELING AND ONTOLOGY

Before coming to the modeling process itself, let us first try to specify
precisely *what* it is we are trying to model. In GIS, we are primarily
interested in modeling parts of the geographic world, consisting of entities
neither at the cosmic nor the microscopic scale, but mid-scale entities

**Fiat and bona fide boundaries**  *Philosopher Barry Smith introduced the ontological distinction between fiat and bona fide boundaries (see Smith, 1995, Smith and Varzi, 2000). Fiat boundaries are boundaries that exist as the result of human cognition, while bona fide boundaries exist as the result of physical discontinuities in the world, independent of human cognition. Geopolitical boundaries, such as the border between Scotland and England, are good examples of fiat boundaries. Coastlines, rivers, and walls are all examples of bona fide geographic boundaries. Sometimes fiat and bona fide boundaries may coincide. In Roman Britain, the boundary between Scotland and England (fiat boundary) was marked by Hadrian's Wall (bona fide boundary), a wall running the width of mainland Britain. Much of Hadrian's Wall still stands today, although as a result of numerous changes in the fiat boundary between Scotland and England over the past 2000 years, the wall now lies entirely within the boundary of England. Ontological distinctions, such as that between fiat and bona fide boundaries, attempt to provide a classification of the world without reference to any particular application domain. Indeed, fiat and bona fide boundaries are not restricted to geographic entities. My skin forms the bona fide boundary to my body, while the boundary between my head and my neck is a fiat boundary. Examples of other ontological investigations of geographic boundaries and regions include Galton (2003) and Montello (2003).*

of our normal experience (cf. the discussion of "What is space?" on page 84). A precise analysis of what constitutes the world is the business
ontology of *ontology*—the study of general classifications of, and relationships between, those things that exist in the world.

For example, a basic ontological distinction is the division of the world into those entities that have identities that endure or persist through time, and those entities that occur in time. In the first category we would place a table, the USA, and you, the reader; while in the second we would have a football match, an erosion event, and your life. In a rather simple-
continuant minded way, the world is divided into objects (termed *continuants* or
occurrent *endurants*) and events or processes (termed *occurrents* or *perdurants*). The continuants may be further categorized as substances, parts of substances, aggregates of substances, locations for substances, and properties of substances. Thus, the Earth constitutes a substance; the surface of the Earth is a part of the Earth; the Solar System is an aggregate of planets; the USA is a location; and the Earth's total land area is a property of the Earth. We return to the topic of categorizing occurrents (events and processes) in Chapter 10. An example of an ontological distinction with special relevance to geographic phenomena is that between *fiat* and *bona fide* boundaries (see "Fiat and bona fide boundaries" inset, on this page).

Ontology is primarily a philosophical discipline, although in a somewhat different sense of the word it is also an important topic in artificial intelligence and computer science. While we are not suggesting that all modelers should be philosophers, some basic ontological sophistication is complementary to modeling activities. An understanding of basic ontological distinctions can help us avoid some basic modeling mistakes. Typical modeling errors arising from a lack of ontological awareness include failing to distinguish real-world entities from information system entities, and failing to distinguish substances from their properties. So, for example, a forest, the location of the forest, and the land-cover type "forested" fall in distinct ontological categories.

In the context of information systems, the study of ontology is in some ways analogous to data modeling. The key distinction between ontology and data modeling is that the former aims to develop general taxonomies of what exists, while the latter aims to develop classifications within a particular application domain. Thus, the distinction between substance and property is not a data modeling issue, although it may be useful in data modeling. Conversely, the decision whether to represent a road in a navigation system as a polyline or as an area is a data modeling question, not an ontological one. In the remainder of this chapter we are concerned primarily with the modeling process, rather than the study of ontology.

## 4.2   THE MODELING PROCESS

The word "model" has been used freely in varying contexts in the preceding chapters. A *model* is an artificial construction in which parts of one domain, termed the *source domain*, are represented in another domain, the *target domain*. The constituents of the source domain may, for example, be entities, relationships, processes, or any other phenomena of interest. The purpose of the model is to simplify and abstract away from the source domain. Constituents of the source domain are translated by the model into the target domain and viewed and analyzed in this new context. Insights, results, computations, or whatever has taken place in the target domain may then be interpreted in the source domain. A simple example of a model is a flight simulator. Objects in the real world such as an aircraft, its instrument panel, sounds, movements, views from the cockpit, and the navigation space, are simulated in an artificial environment. The pilot may manipulate the model environment, for example, by simulating a landing into Boston's Logan Airport in bad weather. This experience within the target domain may then be transferred back to experience with flying real aircraft.

> model
> source domain
> target domain

The usefulness of a particular model is determined by how closely it can simulate the source domain, and how easy it is to move between the two domains. The mathematical concept behind this is *morphism*. A morphism is a function from one domain to another that preserves some of the structure in the translation. Those readers old enough to remember schoolwork with logarithms have an excellent example of a morphism. The logarithm function translates the positive real number multiplicative structure to the real number additive structure (remember: to multiply two numbers, take the logs and add). It is then possible to return to the original domain with the result (take the anti-log).

> morphism

Cartography and wayfinding provide a second example. Suppose that the geographic world is the source domain, modeled by a map (target domain). A user needing to travel from Edinburgh to London by road consults and analyzes the map, then translates the results of the analysis back in order to navigate through the UK road network. If the map is a good model of the real road network, then the user's journey may be smooth.

**Figure 4.1:**
The modeling process as source domain, modeling function, and target domain

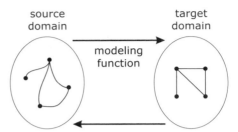

The modeling process can be shown schematically as in Figure 4.1. The left-hand oval represents the source domain to be modeled. In this example, suppose that the source domain is part of the electrical supply network. Suppose further that we wish to perform some network analysis, such as predict current flows in the case of a break at some point. The appropriate target domain in this case may be a mathematical network structure. The modeling function associates elements of the source domain with elements of the target domain. Network transformations and analyses can be made in the target domain; results can be translated back and interpreted in the source domain.

**Figure 4.2:**
Modeling as a morphism

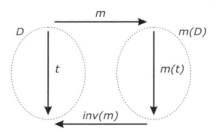

Figure 4.2 shows the modeling process more abstractly in terms of morphisms. Source domain $D$ is modeled using the modeling function $m$. A transformation $t$ in the source domain is modeled by the transformation $m(t)$ in the target domain. The result of the transformation in the target domain is then reinterpreted in $D$, using the inverse $inv(m)$ of the modeling function. The whole process "works" if the model accurately reflects the transformation $t$ in domain $D$. This is expressed by the equation:

$$inv(m) \circ m(t) \circ m = t$$

where $\circ$ indicates function composition. This relationship can be expressed more simply as:

$$m(t) \circ m = m \circ t$$

Such structural relationships are the subject of the mathematical theory of *categories*. The satisfactory way that the functions (arrows) work together in Figure 4.2 allows us to say that the diagram *commutes*. Diagrams like that in Figure 4.2 are called *commutative diagrams*. Such flights of abstraction are helpful in general, but take us rather too far from the main thread of this chapter.

category theory<br>commutes

Models for GIS operate in a wide range of different situations, from models of particular application domains (e.g., transportation models) to specific computer-based models of the physical information in the system. Figure 4.3 is an elaborated version of the more general modeling process in Figure 4.1, based on the system life-cycle introduced in Figure 1.13.

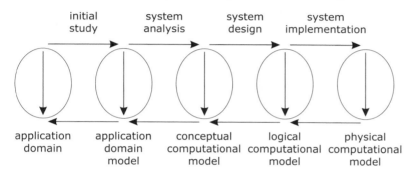

**Figure 4.3:** Modeling stages in GIS

Moving from left to right, the application domain is the subject of an application domain model, usually constructed by domain experts. An example of an application domain would be a bus transportation network. The application domain model could then be a simple network model, populated with entities, such as streets, bus stops, and bus routes. In the conceptual computational model, the elements of the transportation network may be represented, using object-oriented modeling, as object types with attributes, behavior, and relationships. For example, "bus route" might have "route number" as an attribute, "length" as an accessor behavior, and an association with the streets and stops it traverses. The conceptual computational model is usually constructed by system analysts in collaboration with domain experts. The logical computational model develops the analysis results by specifying how the conceptual computational model can be achieved within a specific paradigm, such within an OODBMS, an RDBMS, or an ORDBMS. Finally, the physical computational model is constructed by system programmers, who implement the logical computational model for a specific software system or hardware platform.

### 4.2.1   Field or object?

There are two broad and opposing classes of models of geographic information: the field and object models (see "Object versus field" inset, on the following page). A *field-based model* treats geographic information as collections of spatial distributions. In a field-based model, each

field-based model

**Object versus field** *Space may be conceptualized in two distinct ways: either as a set of locations with properties (absolute space, existent in itself) or as a set of objects with spatial properties (relative space, dependent upon other objects, see Chrisman, 1975, 1978; Peuquet, 1984). This dichotomy turns out to have far-reaching implications for spatial modeling, where absolute space is modeled as a set of fields and relative space as collections of spatially referenced objects. Couclelis (1992) draws a parallel between the object versus field distinction and a more general distinction in the philosophy of science, that of plenum versus atom and wave versus particle. It is natural to think of continuously varying geographic phenomena, like temperature or rainfall, as a field. Similarly, it is natural to think of discrete geographic phenomena, like buildings or roads, as objects. In some applications it may be appropriate to treat groups of objects as a field, or regions of a field as objects. For example, urbanization might might conceptualized as a field based on the density of houses (objects). Conversely, a zone of very low rainfall (a field) might be conceptualized as an object (a desert).*

distribution may be formalized as a mathematical function from a spatial framework (for example, a regular grid placed upon an idealized model of the Earth's surface) to an attribute domain. Patterns of topographic altitudes, rainfall, and temperature fit neatly into this view. An *object-based model* treats the space as populated by discrete, identifiable entities, each with a geospatial reference.

*object-based model*

From the viewpoint of the relational model (Chapter 2), measurable geographic phenomena may be recorded as collections of tuples. For example, Figure 4.4 shows a collection of tuples recording annual weather conditions at different locations. The tuple records the location (probably the identifier), average, maximum, and minimum temperatures, and average total precipitation over a year. While being general, this viewpoint is not necessarily the most helpful: a large collection of tuples does not provide any immediate evidence of pattern and may involve intractably large amounts of data. The field-based and object-based approaches are attempts to impose structure and pattern on such data.

**Figure 4.4:**
Relation
containing
annual climate
data
(temperature in
°C,
precipitation in
mm)

| Location | AvTemp | HiTemp | LoTemp | Precip | ... |
|----------|--------|--------|--------|--------|-----|
| London | 11 | 35 | -11 | 591 | |
| New York | 12 | 41 | -26 | 1200 | |
| Cairo | 21 | 45 | 1 | 25 | |
| Wellington | 12 | 30 | -2 | 1229 | |
| Reykjavik | 4 | 23 | -16 | 817 | |
| New Delhi | 25 | 45 | 0 | 703 | |
| Moscow | 4 | 35 | -42 | 599 | |
| ... | | | | | |

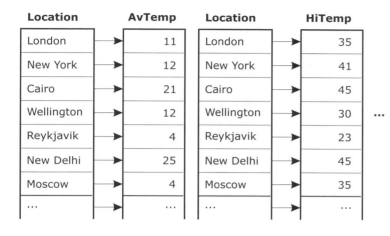

**Figure 4.5:** Field-based approach to geographic phenomena

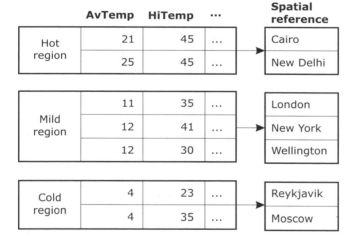

**Figure 4.6:** Object-based approach to geographic phenomena

The field-based approach treats information as a collection of fields. Each *field* defines the spatial variation of an attribute as a function from the set of locations to an attribute domain. Figure 4.5 shows the average and maximum temperature fields, constructed from the data in Figure 4.4. Note that it is the *function* that is the field, not the set of values.

The field-based approach conceptualizes the relation as divided into variations of single or multiple attributes (columns). The alternative object-based approach clumps a relation as single or groups of tuples. In the example of Figure 4.4, we may characterize certain groups of measurements of climatic variables as falling into a finite set of types. Thus (and purely fictitiously) a "Hot region" has relatively higher average temperatures, a "Cold region" relatively lower, and so on. "Hot region" is then an object, which will have some spatial references (the locations that experience higher average temperatures). This structuring of the relation is shown in Figure 4.6.

field

The field-based and object-based models are in a sense inverse to each other. With the field-based approach, the first-class entities in the model (fields) are functions from the spatial framework to the other attributes. On the other hand, the object-based approach constructs a population of entities with a spatial embedding, i.e., a function from entities to a spatial framework (see Figure 4.7).

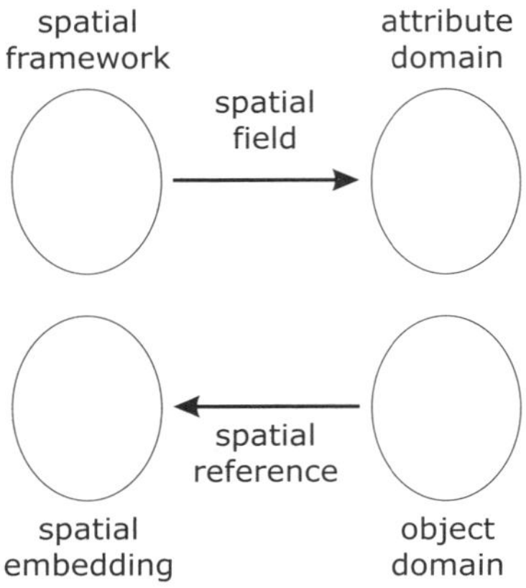

**Figure 4.7:**
Spatial field
and object are
inverse
constructs

## 4.3   FIELD-BASED MODELS

Suppose that we are set the task of providing a geographic database for a given region and we decide to use a field-based model of the information within it. One of the first tasks will be to construct a suitable spatial *spatial framework* framework for our model. A spatial framework is a partition of a region of space, forming a finite *tessellation* of spatial objects. In the plane, the elements of a spatial framework will be polygons. The spatial framework *support* for a given field-based data set is sometimes referred to as the *support* of that data set.

The tessellation used in a spatial framework can be *regular*, such as a grid of squares, or *irregular*, such as in a TIN. We return to tessellations in the next chapter. The spatial framework is a finite structure—it must be so in order to be tractable for computation. Often the application domain will not be finite, or it may be larger than can be practically accommodated in totality. This implies that the phenomena to be modeled will be *sampled* and imprecision will necessarily be introduced by the sampling process. Imprecision is one of the topics we return to in Chapter 9.

The combination of the spatial framework and the field that assigns *layer* values for each location in the framework is termed a *layer*. In general, there may be many layers in a spatial database, each with respect to the same or different underlying spatial frameworks. For example, Figure 4.8 shows two layers for the same spatial framework. The upper layer shows elevation, with lighter grays indicating higher elevation. A second layer, partially occluded by the elevation layer, shows soil moisture content,

with lighter shades indicating drier soils. The boundary of each area in
Figure 4.8 connects locations with the same attribute value (elevation or
soil moisture content). The locus of all points in a field with the same
attribute value is termed an *isoline*.

isoline

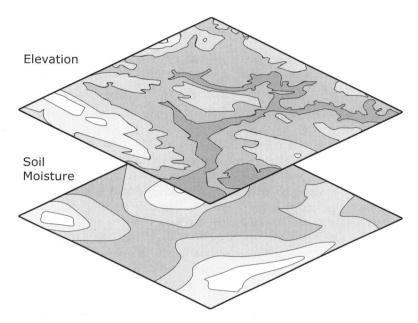

Elevation

Soil
Moisture

**Figure 4.8:**
Two layers in a
field-based
model

The field-based spatial model is now defined more precisely. For simplicity, assume that each field in the model has the same spatial framework
$F$. In general we would allow the spatial framework to be based upon any
mathematical model of space. In practice, most spatial frameworks are
based upon Euclidean space in two dimensions (the Euclidean plane),
with distance and angle defined as in Chapter 3.

- A field-based model based upon a spatial framework $F$ consists of
  a finite collection of $n$ spatial fields $\{f_i : 1 \leq i \leq n\}$.

- For $1 \leq i \leq n$, each *spatial field* $f_i$ is a computable function from    spatial field
  set $F$ to a finite attribute domain $A_i$.

In order that this model will work computationally, the number of
fields, the cardinality of the domains $A_i$, and the spatial framework $F$
are all stipulated to be finite. Also, the field functions $f_i$ are defined to
be computable (i.e., are capable of evaluation using a computer). In the
special case where the spatial framework is a Euclidean plane and the
attribute domain is a subset of the set of real numbers, then a field may
be represented as a surface in a natural way. The Euclidean plane plays
the role of the horizontal $xy$-plane and the spatial field values give the
$z$-coordinates, or "heights" above the plane.

To sum up the modeling method using a field-based approach:

- Construct or use a given suitable model of the underlying space to act as the spatial framework, $F$.

- Find suitable domains for the attribute(s) to act as the $A_i$ ($1 \leq i \leq n$).

- Sample the phenomena under consideration at the locations in the spatial framework, so as to construct the spatial field functions $f_i$ ($1 \leq i \leq n$).

- Perform analyses by computing with the spatial field functions.

An example of a simple field model would be for regional climate variations. Imagine placing a square grid over a region and measuring aspects of the climate at each node of the grid. A more realistic model might require using an irregular spatial framework to reflect where it is practical to set up weather stations to record meteorological measurements. Different fields would then associate locations with values from each of the measured attribute domains.

Another example might be regional health variations. Here the spatial framework is likely to be an irregular tessellation of the counties or districts in which people reside. For the population within each region it should be possible to measure the incidence of certain diseases. The model might then be used to analyze spatial patterns of specific diseases or relate such patterns to other spatial fields for the same spatial frame-

DEM   work. A third example would be a *digital elevation model* (DEM), where terrain elevation is measured over a region of the Earth's surface (see "Digital elevation models" inset, on the next page).

It is worth emphasizing that the conceptual models under discussion are independent of any implementation or any physical representation of the data, although certain types of spatial framework may be suited to certain types of data structure. For example, if the spatial framework is a regular tessellation, especially a regular square grid, it may be natural to represent this using raster-based data structures. In contrast, irregular tessellations, especially triangulations, often lend themselves more readily to vector-based data structures.

### 4.3.1   Properties of fields

As already highlighted, a spatial framework is a partition that may be based on any one of a range of different mathematical models of space, such as those discussed in the previous chapter. From the perspective of the field-based model, an important characteristic of a spatial framework is whether it is regular or irregular. In addition to the spatial framework, the attribute domains and the fields themselves have several important characteristics, described in this section.

**Digital elevation models** *Digital elevation models (DEMs) have been an important application of the field model since the 1950s (Miller and Laflamme, 1958). In a DEM, the spatial framework represents a set of locations within a portion of the Earth's surface, and the field values are the heights of the terrain surface at those locations. Since the field function is single-valued, overhanging terrain features, like cliffs and caves, cannot be represented in a DEM. As a consequence, DEMs are sometimes referred to as "2.5 dimensional" objects (although the term should not be confused with fractional dimensions in fractal geometry). The spatial framework for a DEM is often in the form of a regular grid, which has the advantage of simplicity. However irregular tessellations, such as a TIN, can allow the structure of the field to more closely parallel the structure of the terrain, e.g., following ridges and valleys (for a full discussion of the pros and cons of regular and irregular DEMs see Peucker, 1978; Mark, 1979). Many common GIS applications require a DEM as their basic data source, including visibility analysis (determining what can be seen from a particular location), slope and aspect analysis, and routing and shortest path calculations. DEMs are sometimes also referred to as digital terrain models (DTMs), although a DTM would normally represent other topographic features in addition to elevation (Weibel and Heller, 1991). Burrough and McDonnell (1998) contains a useful survey of principles, techniques, and applications.*

## Properties of the attribute domain

The attribute domain may contain values which are commonly classified into four levels of measurement, originally devised by Stevens (1946).

- An attribute domain that consists of simple labels is termed a *nominal* attribute. A land cover classification that consists of woodland, grassland, agricultural land, and urban land is an example of a nominal attribute. By definition, nominal attribute domains are qualitative, cannot be ordered, and arithmetic operators are not permissible for nominal attributes. *(nominal)*

- An attribute domain that consists of ordered labels is termed an *ordinal* attribute. For example, the results of a site suitability analysis to determine the best place to locate a new superstore might be expressed as values from the ordinal attribute domain: "highly suitable," "suitable," "moderately suitable," and "unsuitable." Like nominal attributes, ordinal attribute domains are qualitative and cannot be subjected to arithmetic operators, apart from ordering. *(ordinal)*

- An attribute domain that consists of quantities on a scale without any fixed point is termed an *interval* attribute. The Celsius and Fahrenheit temperature scales are examples of interval attributes. Interval attributes can be compared for size, with the magnitude of the difference being a meaningful notion. However, the ratio of two interval attribute values is not a meaningful notion. For example, if Madrid is 20°C and Edinburgh is 10°C it makes sense to say that Madrid is 10°C hotter than Edinburgh, but not that Madrid is twice as hot as Edinburgh. *(interval)*

ratio
- An attribute domain that consists of quantities on a scale with respect to a fixed point is termed a *ratio* attribute. The Kelvin temperature scale is an example of a ratio attribute, because zero Kelvin is absolute zero, the lowest physically possible temperature. Ratio measurements are capable of supporting a wide range of arithmetical operations, including addition, subtraction, multiplication, and division. There are many geographic examples of ratio measurement domains, including annual rainfall and topographical altitude above sea level. So, it makes perfect sense to say that Boston had twice as much rain in March as Washington, or that one place is three times higher than another with respect to sea level.

While Stevens' four levels of measurement are widely used, it is important to note that one has to be careful in fitting attributes to levels. An example of a common geographic measurement that needs care is population density. Population density has a fixed point (it is not possible to have less than zero population density) and can be compared like a ratio (e.g., a population density of 200 people per square kilometer is twice as dense as 100 people per square kilometer). However, population density does not behave quite like a ratio. For example, it is not possible to calculate directly the combined population density of two regions containing 100 and 200 people per square kilometer, respectively (the combined density will lie somewhere between 100 and 200 people per square kilometer depending on the relative sizes of the two regions). The bibliographic notes contain further references to this topic.

*Continuous, differentiable, and discrete fields*

continuous

differentiable
A spatial field is said to be *continuous* if small changes in location lead to small changes in the corresponding attribute value (recall that a spatial field is a *function* from a spatial framework to some attribute). Continuity is only appropriate if the notion of a "small change" is well defined in both the spatial framework and attribute domain, that is, if these domains are themselves continuous. A spatial field is *differentiable* if its rate of change (slope) is defined everywhere. As with continuity, differentiability only makes sense for a continuous spatial framework and attribute domain. Every differentiable field must also be continuous, but not every continuous field is differentiable.

Examples are shown in Figure 4.9. For simplicity, assume that the spatial framework is one dimensional, ranged along a horizontal line. Then the field may be plotted as a graph of attribute value against spatial framework. In Figure 4.9a the variation is represented as a continuously smooth curve, clearly differentiable because the slope of the curve can be defined at every point. In Figure 4.9b, although the field is continuous (the graph is connected), it is not everywhere differentiable. There is an ambiguity in the slope, with two choices at the articulation point between the two straight-line segments. In Figure 4.9c, the graph is not connected and so the field is not continuous and *a fortiori* not differentiable. Fractals, like the Koch snowflake discussed in the previous chapter, form a fourth class of functions, which are continuous but not differentiable at any point, termed *nowhere differentiable*. Figure 4.9d shows a relative of the Koch snowflake: a fractal known as the *blancmange function*.

nowhere
differentiable

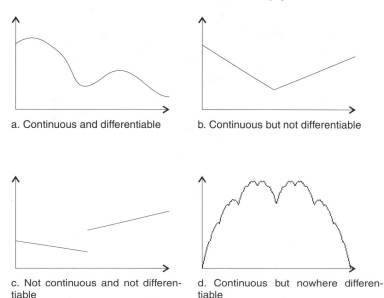

a. Continuous and differentiable       b. Continuous but not differentiable

c. Not continuous and not differen-    d. Continuous but nowhere differentiable                                  tiable

**Figure 4.9:**
Examples of
continuity and
differentiability
for fields
plotted as one-
dimensional
spatial
framework
($x$-axis) against
attribute value
($y$-axis)

In the case where the spatial framework has dimension two (or higher), the slope is dependent not only on the particular location but also on bearing at that location (see Figure 4.10).

*Isotropic and anisotropic fields*

A characteristic feature of a spatial field is whether its properties vary with direction. A field whose properties are independent of direction is called an *isotropic* field. Consider travel time in a spatial framework. Let us start with the simplest possible assumption that the time taken to travel between locations is in direct proportion to the Euclidean distance between them. Then, the locus of all points that are a constant time from an arbitrary point $X$ (an isochrone) is a circle, shown in Figure 4.11a. This field is isotropic, because the time from $X$ to any point $Y$ is dependent only upon the distance between $X$ and $Y$ and independent of the bearing of $Y$ from $X$.

isotropic

Now, let us make life more interesting. Suppose that there is a high-speed link $AB$ in the spatial framework (Figure 4.11b). For simplicity, assume that the link is so fast as to make the travel time from $A$ to $B$ negligible compared with the travel times anywhere else in the field. Then, when traveling between two points, there is a choice of whether or not to use the high-speed link. Consider points close to $X$ (say, within 14 time units) in Figure 4.11b. For many of the points near $X$ the minimum

**Figure 4.10:**
Elevation field
where the slope
at a point
depends upon
the bearing

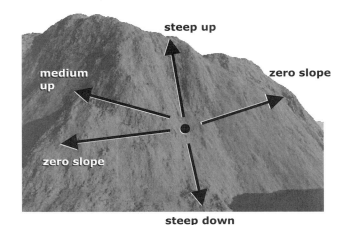

**Figure 4.11:**
Travel time
from $X$ in
isotropic and
anisotropic
fields

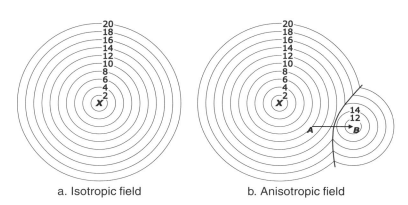

a. Isotropic field                    b. Anisotropic field

travel time is unchanged by the presence of the link. However, for points near $B$ it would be better for the traveler from $X$ to travel to $A$, take the link (zero-time), and continue on from $B$ to the destination. The hyperbola in Figure 4.11b marks the boundary between regions where it is better/worse to use the link. In the second case, to determine the travel time, it clearly matters in which direction the destination location is from

anisotropic        $X$. The field is said to be *anisotropic*.

Anisotropic fields are common in real-world situations. They are often linked to networks (in our example, the high-speed link was a very simple network). Other examples that introduce anisotropic conditions include natural and artificial barriers to direct accessibility.

*Spatial autocorrelation*

spatial           Spatial autocorrelation measures the degree of clustering of values in a
autocorrelation   spatial field. Spatial autocorrelation is a quantitative expression of Waldo Tobler's famous proposition, often referred to as the first law of geography (Tobler, 1970), that "everything is related to everything else, but near

things are more related than distant things." If a spatial field has the property that like values tend to cluster together, then the field exhibits high positive spatial autocorrelation. If there is no apparent relationship between attribute value and location neighborhoods, then there is zero spatial autocorrelation. If there is the propensity of like values to be located away from each other, then there is negative spatial autocorrelation. Spatial autocorrelation therefore measures the relationship between attribute values at a location and attribute values in the location's neighborhood. Figure 4.12 gives examples of a high positive (left-hand bull's eye), zero (random noise), and high negative spatial autocorrelation (right-hand checker).

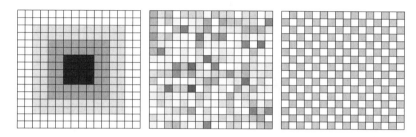

**Figure 4.12:** Patterns with high positive, zero, and high negative spatial autocorrelation (left to right)

### 4.3.2 Operations on fields

A field operation takes as input one or more fields and returns a resultant field; thus, fields form a closed structure under field operations. This section describes some of the typical operations that field-based models allow, dividing them into three main classes: *local*, *focal*, and *zonal* operations. The system of possible operations on fields in a field-based model is referred to as *map algebra*, first classified and formalized by Tomlin (1983) (see the bibliographic notes at the end of the chapter). Before the different classes of map algebra operations are described, the definition of a neighborhood function is needed.

*map algebra*

- Given a spatial framework $F$, a *neighborhood function* $n : F \to \mathcal{P}(F)$ is a function that associates with each location $x$ a set of locations that are "near" to $x$.

*neighborhood function*

Section 3.2.1 defined the power set of a set $F$, $\mathcal{P}(F)$, as the set of all subsets of $F$. Every member of $\mathcal{P}(F)$ is therefore a subset of $F$. For each location $x$ in $F$, $n(x)$ will then be a subset of $F$, called the neighborhood of $x$ (cf. neighborhood topologies in Chapter 3). Figure 4.13 illustrates the neighborhood function idea. The definition of neighborhood will depend upon the underlying spatial framework. If, as is the case with many practical applications, the space is Euclidean, then the neighborhood of $x$ might be the set of points within a specified distance and/or bearing of $x$. If the space is metric, then the neighborhood of $x$ might be the set of points within a specified distance of $x$. If the space is topological, then the neighborhood could be defined in terms of topological neighborhood. The neighborhood of $x$ will usually contain $x$ itself, although this is not a requirement of a neighborhood.

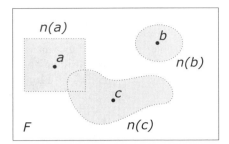

**Figure 4.13:**
Neighborhood
function $n$

*Local operations*

local operation

A local operation acts upon one or more spatial fields to produce a new field. The distinguishing feature of a local operation is that the value of the new field at any location is dependent only on the values of the input field functions at that location. Local operations may be unary (transforming a single field), binary (transforming two fields), or $n$-ary (transforming any number of fields). We give a more formal definition in the binary case. The general case may easily be extrapolated from this.

Formally, suppose we are given a framework $F$ and spatial field functions $f$ and $g$. Suppose further that $\bullet$ is a binary operation that acts on values in the attribute domains of $f$ and $g$, respectively, to produce a value in another attribute domain. Then, we may pointwise construct a new spatial field function $h$ defined as follows:

1. For each location $x$, $h(x) = f(x) \bullet g(x)$.

This binary combination of the two fields $f$ and $g$ with spatial framework $F$ is shown in Figure 4.14. The figure highlights one element of the spatial framework, $x$. Note that although the spatial framework in Figure 4.14 (and following figures) is a regular square tessellation, this is not a requirement of operations on fields: any spatial framework (partition) is admissible.

For example, suppose the attribute domains of the two fields $f$ and $g$ are real numbers describing population and incidence of cancer, respectively. For locations with non-zero population, cancer incidence may be divided by population. Then in this case, the $\bullet$ operation is division, and $f \bullet g$ is the rate of cancer incidence per unit population. The derivation of a value at any location depends only upon the values of cancer incidence and population at that location. Therefore this operation is local.

*Focal operations*

focal operation

For a focal operation the attribute value derived at a location $x$ may depend not only on the attributes of the input spatial field functions at

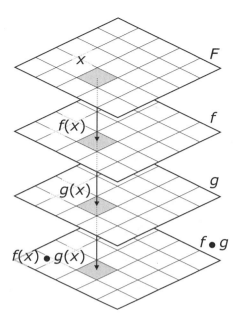

**Figure 4.14:** Local operation, •, on two fields, $f$ and $g$

$x$, but also on the attributes of these functions in the neighborhood $n(x)$ of $x$. Thus, the value of the derived field at a location may be influenced by the values of the input field nearby that location. In the unary case, suppose $F$ is a spatial framework, $n$ is a neighborhood function, $f$ is a spatial field function, and $\varphi$ (phi) is a unary focal operation. For each location $x$:

1. Compute $n(x)$ as the set of neighborhood points of $x$ (usually including $x$ itself).

2. Compute the values of the field function $f$ applied to appropriate points in $n(x)$.

3. Derive a single value $\varphi(x)$ of the derived field from the values computed in step 2, possibly taking special account of the value of the field at $x$.

A unary focal operation $\varphi$ on the field $f$ with neighborhood function $n$ is shown in Figure 4.15.

A good example of a focal operation is the operation that computes the gradient of a field of topographical altitudes over a continuous spatial framework. In this case $f$ is the topographic field, $n(x)$ will be close neighbors of $x$, and $\varphi$ will compute the differences between values of the neighbors of $x$ in different directions to produce a slope vector.

### Zonal operations

A zonal operation aggregates values of a field over each of a set of zones (arising in general from another field function) in the spatial framework.

zonal operation

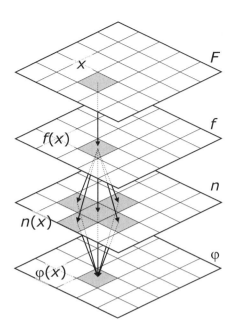

**Figure 4.15:**
Focal operation
$\varphi$ on a field

A zonal operation $\zeta$ (zeta) derives a new field based on a spatial framework $F$, a spatial field $f$, and set of $k$ zones $\{Z_1, ..., Z_k\}$ that partition $F$. For each location $x$:

1. Find the zone $Z_i$ in which $x$ is contained.

2. Compute the values of the field function $f$ applied to each point in $Z_i$.

3. Derive a single value $\zeta(x)$ of the new field from the values computed in step 2.

A unary zonal operation $\zeta$ on the field $f$ with zones $\{Z_1, ..., Z_k\}$ is shown in Figure 4.16.

For example, given a layer of precipitation and a zoning into administrative regions, a zonal operation might derive a layer of average precipitation for each region. In this case $f$ is the precipitation field, $\{Z_1, ..., Z_k\}$ will be the different administrative regions, and $\zeta$ will calculate average precipitation over all locations within each zone.

*Summary of field operations*

By way of summary, Table 4.1 contains brief notes on a sample of general field operations. For some of these, where the spatial framework is assumed to be a Euclidean plane and the field values are real numbers, it is useful to view the field as a surface. The local operations in Table 4.1 compute local arithmetical combinations of fields. The focal operations in Table 4.1 are all unary, and compute a value at each location dependent

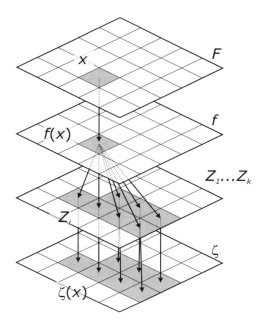

**Figure 4.16:**
Zonal
operation $\zeta$ on
a field with
zones
$\{Z_1, ..., Z_k\}$

on that location's neighborhood. The zonal operations in Table 4.1 are
also unary, and at each location compute a value dependent upon the field
values in the zone containing that location.

One of the fundamental cartographic operations is the overlay of one
layer upon another, thus producing a composite function of the underlying
space. Without using the language of overlay explicitly, it should be
clear that overlays are implicit in many of the operations performed
above. Indeed, the example of a local operation involved the overlay
of population and cancer incidence, in order to perform an arithmetic
operation (division) upon the individual pairs of elements in the layers.
Similarly, a zonal operation may be viewed as an overlay of a layer onto
a zoning layer.

## 4.4   OBJECT-BASED MODELS

Object-based models decompose an information space into *objects* or
*entities*. An entity must be:                                           entity

- identifiable
- relevant (be of interest)
- describable (have characteristics)

With respect to the last item, the description of an entity is provided by
its static properties (such as the name of a city), behavioral characteristics
(such as a method of plotting the city at a particular scale), and structural
characteristics (placing the object in the overall structure of the information space).

| Type | Name | Degree | Description |
|------|------|--------|-------------|
| Local | **lsum**, **ldiff**, **lprod**, **lquot** | binary | Pointwise sums, differences, products, and quotients |
| | **lmax**, **lmin** | $n$-ary | Pointwise maximums and minimums |
| | **lmean** | $n$-ary | Pointwise means |
| Focal | **slope** | unary | Maximum gradient at locations |
| | **aspect** | unary | Bearing of steepest slope at each location |
| | **fmean** | unary | Weighted average based on neighborhood |
| | **fsum**, **fprod** | unary | Sum and product of values in the neighborhood of each location |
| Zonal | **zmin**, **zmax** | unary | Minimum and maximum values in each zone |
| | **zsum**, **zprod** | unary | Sum and product of values in each zone |
| | **zmean** | unary | Mean values of field in each zone |

**Table 4.1:**
Sample of field
operations

The field-based model, discussed above, uses a fixed spatial framework as a reference (such as a regular grid), and then measures the variation in attribute values with respect to this reference (such as elevation at each location). In contrast, the object-based model populates the information space with spatially referenced entities (e.g., cities, towns, villages, districts) with attributes (e.g., population density, centroid, boundary). In the object-based approach the entire frame of spatial reference does not become distinguished and prescribed as it does in the field-based approach (the spatial framework), but is provided by the entities themselves (see earlier inset on "Object versus field," on page 138). In a more general context, this is precisely the object-oriented approach to data modeling discussed in Chapter 2, although the relationship between object-orientation and the object-based model is not entirely straightforward (see "Object-based or object-oriented?" inset, on the next page).

literal

At this point, it is useful to make a distinction between objects and *literals*. Literals differ from objects in that they have an immutable state that cannot be created, changed, or destroyed. For example, Figure 4.17 depicts a spatially referenced "house" object. The house has several attributes, such as registration date, address, owner, and boundary, which are themselves objects. However, the actual *values* of these attributes are literals. For example, the actual date "November 5th, 1994" cannot be created, destroyed, or changed. If in the future the house is registered to a new owner, we may change the registration attribute to a new date, say "September 5th, 2004." This alteration changes the date to which the registration attribute (an object) refers, but does not affect the date itself (a literal). Thus, "November 5th, 1994" still exists as a date, unaffected by the change. Other houses registered on the same date all refer to the

> **Object-based or object-oriented?**  *Although the object-oriented approach to systems and object-based models are superficially similar, they have some distinct differences. It is not necessary to use the object-oriented approach to implement object-based models. On the other hand, the object-oriented approach may be used as a framework for describing both object-based and field-based spatial models. For object-based models, the case is clear, but field-based models can also be set in an object-oriented context. In fact, earlier in this chapter fields were described in terms of the properties and behavior of field "objects," including a set of operations that act upon them: this is precisely the object-oriented approach. It would be perfectly reasonable to use object-oriented development methodologies and an OODBMS to develop and implement a GIS for modeling field-based spatial phenomena. Some example applications of the object-oriented approach to field-based environmental models are discussed in Kemp (1992).*

same literal. Similarly, an extension built onto the house might result in an alteration to the boundary attribute, referring to a different polygon to represent this boundary. However, the underlying points and arcs that make up a boundary are literals and cannot be created, changed, or destroyed.

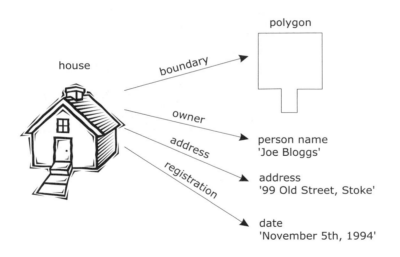

**Figure 4.17:** Spatially referenced "house" object

### 4.4.1   Spatial objects

Spatial objects are called "spatial" because they exist inside "space," called the *embedding space*. The specification of a spatial object depends upon the structure of its embedding space. For example, it would not be easy to define a circle that is embedded in a space, each point of which can only have integer coordinates. Several different types of space were introduced in Chapter 3, including Euclidean space, metric space, topological space, and set-oriented space.

embedding space

The most common situation is that the underlying space is Euclidean and each spatial object is specified by a set of coordinate tuples or computable equations. Another approach is to specify a set of primitive objects, out of which all others in the application domain can be constructed, using an agreed set of operations. Primitive spatial object classes that have been suggested include closed half-planes (although these suffer from the counterintuitive property of being infinite in extent); simplicial complexes (the disadvantage here being computational expense in building with such primitive objects); and point-line-polygon primitives (common in existing systems). Consider the following very simple analysis that may be required of a GIS.

- For Italy's capital city, Rome, calculate the total length of the River Tiber which lies within 2.5 km of the Colosseum.

**Figure 4.18:**
Satellite image of Rome centered on Colosseum (Source: NASA ASTER image, May 2003)

Figure 4.18 shows a satellite image of Rome, centered on the Colosseum (highlighted with a star). This data is not well suited to answering our question: it does not explicitly represent objects like "the Colosseum" or "the River Tiber," and contains much more detail than we need. First we need to model the relevant parts of Rome as objects. We might start by identifying a **river** object (with "Tiber" as its name), and a **building** object (named "Colosseum"). The spatial references for these objects might be **arc** and **point**, respectively. The spatial object class **circle** is also important, as it describes the region within a 2.5-km radius of the Colosseum. Operation **length** will act on **arc**, returning a real number, and **intersect** will apply to form the piece of an **arc** in common with the **disc**. The configuration of spatial objects is shown in Figure 4.19.

The analysis might proceed as with the four stages below. Each stage of the analysis uses objects defined in the model and legal operations that may be applied to those objects.

1. Find the circle that has radius 2.5 km, and is centered at the point representing the Colosseum.

2. Find the intersection of the circle obtained in stage 1 with the arc representing the River Tiber.

3. Find the length of each of the clipped arcs from stage 2.

4. Sum the lengths obtained in stage 3.

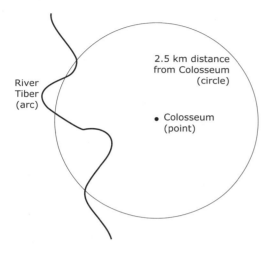

**Figure 4.19:** Continuous spatial objects in Rome

The objects and operations described above are not suitable for computation, because they are continuous and infinite. A process of discretization must convert the objects to types that are computationally tractable. For example, a circle may be represented as a discrete polygonal area, arcs by chains of line segments, and points may be embedded in some discrete space. Operations such as **intersection** and **length** may then be computed using standard algorithms from computational geometry. Figure 4.20 shows a discretized version of our Roman example.

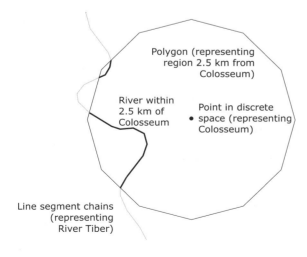

**Figure 4.20:** Discrete spatial objects in Rome

**Polynomial curves** *Polygons and polylines provide piecewise first-degree polynomial approximations to two-dimensional and one-dimensional extents, respectively. Higher-degree polynomial approximations are also computationally tractable and may be appropriate in certain circumstances, for example, when smoothly contoured lines rather than jagged edges are required. Additionally, a higher-degree curve may be more storage efficient for storing some shapes, rather than polylines with large numbers of segments. While first-degree polynomials are most common in GIS, higher-degree curves based upon cubic polynomials are in widespread use, particularly in graphics and CAD systems. A cubic polynomial $ax^3 + bx^2 + cx + d$ has four coefficients, $a$, $b$, $c$, $d$, and thus four controls (degrees of freedom) that may be varied to produce different shapes. The most widely used cubic forms are: Bézier, Hermite, and B-spline. Bézier and Hermite curves are each defined by two end-points and two controls, while a B-spline is defined by two end-points and four controls. Two examples of a Bézier cubic curve, with its controls, are shown below.*

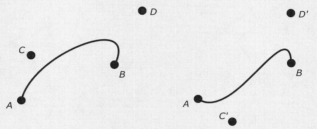

*Points $A$, $B$ are the end-points of the curve. For the left-hand diagram Points $C$, $D$ control the shape of the curve; the vectors $\underline{AC}$ and $\underline{BD}$ represent the tangent vectors at the end-points. The right-hand diagram shows how the shape of the curve can be changed by altering the tangent vectors: the end-points are the same, but the new tangent vectors are $\underline{AC'}$ and $\underline{BD'}$. An early reference for Bézier cubic curves is Bézier (1972). First-degree (polyline) and higher-degree polynomial approximations to arcs and loops are well discussed in Foley et al. (1995).*

### Spatial object types in the Euclidean plane

Figure 4.21 shows a possible inheritance hierarchy for objects in a continuous two-dimensional space, the Euclidean plane with the usual topology. The most general spatial object type **spatial** is at the top of the hierarchy. This type is the disjoint union of types **point** and **extent**, distinguishing single points and extended objects embedded in two dimensions. Class **extent** may be specialized by dimension into the types **1-extent** and **2-extent**. Two subtypes of the one-dimensional extents have been described in Chapter 3 as **arc** and **loop**, specializing to **simple arc** and **simple loop** when there are no self-crossings. The fundamental areal object was described in Chapter 3 as the regular closed set, called here by the type name **area**. A connected area we name a **region**; a region that is simply connected (no holes) is a **cell**, homeomorphic to the unit disk.

The Euclidean plane is not computable and must be discretized in order that computation can take place. In a computer, the real numbers (Euclidean 1-space) are discretized by storing digits to only a finite level of precision. For example, using 32-bit precision floating-point numbers, up to six significant decimal digits can be stored (the number 0.1234567 would be rounded up to 0.123457). Discrete forms exist for all the continuous types in Figure 4.21. Points can be discretized by storing their coordinates as finite precision numbers. Line segments can be discretized by storing their extreme points as discrete points.

A discrete polyline is a sequence of line segments connecting successive discretized points. A continuous **1-extent** may be represented using a discretized polyline. If a discretized polyline's extreme points are coincident it will be a discrete **loop**; otherwise it will discretize an **arc**. The interior and boundary of a simple looped polyline define a discrete **cell**, and so on. Although straight lines are commonly used as the building blocks of discrete spatial extents in GIS, computable curves with discrete parameters may also be used, and are especially important in design applications (see "Polynomial curves" inset, on the facing page).

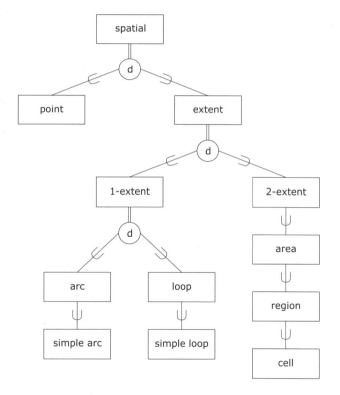

**Figure 4.21:** Inheritance hierarchy for some continuous spatial object types

## 4.4.2 Spatial operations

This section discusses some of the operations that may be applied to continuous spatial objects embedded in the Euclidean plane. The operations will need modification when acting upon discrete spatial object types. Table 4.2 catalogs operations on continuous spatial objects, embedded in the Euclidean plane, according to their group, operand type, and resulting type.

operand    The inputs to an operation are called the *operands*, while the output is called the *result*. Operations in the table may be unary (applying to a single operand) or binary (applying to two operands). Operand and result object types are given for each operation. Of course, any subtype of an operand type will also be able to inherit the operation. For example, the **union** operation may be applied to the union of an object of type **arc** with an object of type **region**.

| Group | Operation | Symbol | Operand(s) | Result |
|-------|-----------|--------|------------|--------|
| General | equals | $=$ | spatial, spatial | Boolean |
| Set-oriented | equals | $=$ | extent, extent | Boolean |
| | member of | $\in$ | point, extent | Boolean |
| | is empty | $= \varnothing$ | extent | Boolean |
| | subset of | $\subseteq$ | extent, extent | Boolean |
| | disjoint from | | extent, extent | Boolean |
| | intersection | $\cap$ | extent, extent | spatial |
| | union | $\cup$ | extent, extent | extent |
| | difference | $\setminus$ | extent, extent | spatial |
| Topo-logical | boundary | $\partial$ | area | set(loop) |
| | interior | $\circ$ | area | open area |
| | closure | $-$ | area | closed area |
| | meets | | area, area | Boolean |
| | overlaps | | area, area | Boolean |
| | is inside | | area, area | Boolean |
| | covers | | area, area | Boolean |
| | connected | | area | Boolean |
| | components | | area | set(region) |
| | extremes | | arc | set(points) |
| | is within | | point, simple loop | Boolean |
| Euclid-ean | distance | $\|\|$ | point, point | real |
| | bearing/angle | $\angle$ | point, point | $[0, 2\pi)$ |
| | length | $\|\|$ | 1-extent | real |
| | area | | area | real |
| | perimeter | | area | real |
| | centroid | | area | point |

**Table 4.2:** Catalog of spatial operations

The operations in Table 4.2 are grouped into general, set-oriented, topological, and Euclidean. In each case, an operation falls into a particular group when its definition requires the structuring of space appropriate for that group. The operation **equals** between **spatial** and **spatial** requires no particular structure, merely that it is possible to tell whether two spatial objects are the same or not. Set-oriented operations require for their specification only the structuring of space into sets of points. Spatial objects of type **extent** have an extension and may be treated as purely sets of points. Operations classified as set-oriented were defined in section 3.2.1 and will not be considered further here, except to note that they are not independent; the usual set-theoretic constraints hold. For example, for any objects $X$, $Y$, and $Z$ of type **extent**, we have the following equalities (De Morgan's laws):

$$X \backslash (Y \cap Z) = (X \backslash Y) \cup (X \backslash Z)$$
$$X \backslash (Y \cup Z) = (X \backslash Y) \cap (X \backslash Z)$$

As we saw in section 3.3, spatial object properties and relationships are considerably more complex in a topological setting than the set-oriented context above. Object types with an assumed underlying topology are **arc**, **loop**, and **area** (also sometimes **point**). Table 4.2 gives some of the possible topological operations. Operations **boundary**, **interior**, **closure**, and **connected** are defined in the usual manner (section 3.3). The operation **components** returns the set of maximal connected components of an area. Operation **extremes** acts on each object of type **arc** and returns the pair of points of the arc that constitute its end-points. Operation **is within** provides a relationship between a point and a simple loop, returning **true** if the point is enclosed by the simple loop. This relationship is the often-used point-in-polygon operation.

The Boolean topological operations **meets**, **overlaps**, **is inside**, and **covers** apply to areas. Similar topological operators can be defined for arcs and loops. Examples of these four operators are shown in Figure 4.22. Informal definitions of these operations are as follows. Let $X$ and $Y$ be objects of type **area**.

- $X$ **meets** $Y$ if $X$ and $Y$ touch externally in a common portion of their boundaries.  *meet*

- $X$ **covers** $Y$ if $Y$ is a subset of $X$ and $X$, $Y$ touch internally in a common portion of their boundaries.  *cover*

- $X$ **overlaps** $Y$ if $X$ and $Y$ impinge into each others' interiors.  *overlap*

- $X$ **is inside** $Y$ if $X$ is a subset of $Y$ and $X$, $Y$ do not share a common portion of boundary.  *inside*

Figure 4.22 also illustrates that these topological operations are more discriminating than set-oriented operations. The figure shows two pairs of objects whose relationships are indistinguishable using only set-oriented operations. From within the structure of pure sets, given two objects $X$ and $Y$ of type **area**, we cannot distinguish the relations $X$ **meets** $Y$ and $X$ **overlaps** $Y$. From the set-oriented viewpoint, the relations are both cases of the relation $X$ is not **disjoint from** $Y$. Similarly, $X$ **is inside** $Y$ and $Y$ **covers** $X$ are both instances of the Boolean set-oriented operation $X$ is a **subset of** $Y$. As for set-oriented operations, not all the given topological operations are independent. Some examples of interdependencies between the **boundary**, **interior**, and **closure** operations now follow. Let $X$ be an object of a type that supports a topology, then:

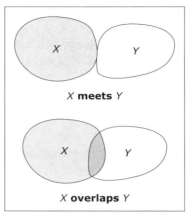

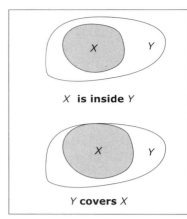

**Figure 4.22:**
Topological
and
set-oriented
operations

X **meets** Y

X **is inside** Y

X **overlaps** Y

Y **covers** X

X is not **disjoint from** Y

X is a **subset of** Y

- $\partial X$ is the set difference of $X^-$ and $X^\circ$

- $X^-$ is the set complement of the interior of the complement of $X$

- $X^\circ$ is the set complement of the closure of the complement of $X$

The operations in the topological section of Table 4.2 by no means provide a complete typology of the diversity of spatial relationships in topologically structured spaces. Figure 4.23 shows some of the infinite number of possible topological relationships that are available between objects of type **cell**. In the top left of the figure, there is a case of a meet relationship, satisfying our informal definition, but the configuration of objects $X$ and $Y$ is not homeomorphic to the objects $X$ and $Y$ shown meeting in Figure 4.22. Similarly, there is an infinite variety of cover and overlap relationships. The spatial relationship between two cells shown in the bottom right of the figure seems to be a combination of meeting, covering, and overlap. In fact, as we shall see later, it would be formally described as an overlap.

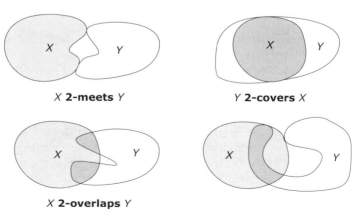

**Figure 4.23:**
Further
topological
relationships
between cells

X **2-meets** Y

Y **2-covers** X

X **2-overlaps** Y

The operations **distance**, **bearing/angle**, **length**, **area**, and **perimeter** available in metric and Euclidean spaces have already been introduced in Chapter 3. Operation **centroid** returns the center of gravity of an areal object as an object of type **point**. Distances and angles are defined between the **point** elements of the space. In practice, it is often important to measure distances and angles between objects of different dimensions. For example, we might wish to know the distance of a town from a motorway. There are several ambiguities here. Do we mean the town center or the town as an area? Are we measuring distance along roads, as the crow flies, or by some other means? If, for example, we are measuring as the crow flies, do we want the distance to the nearest point on the motorway or to the nearest intersection? These ambiguities must be resolved before the question can be answered properly. Definitions of distance and bearing/angle may be extended to operations upon objects that are not just points but are subtypes of **extent** (see bibliographic notes).

*Further topological operations*

Returning to topology, one characterization of topological relationships between spatial objects is more widely cited in the literature than any other. Here, we offer a brief overview of this characterization, originally developed by Max Egenhofer and co-workers. Readers who wish to follow this material further will find references in the bibliographic notes at the end of the chapter.

Many of the spatial operations given in Table 4.2 can be defined in terms of **boundary** and **interior** (or other combinations of two from **boundary**, **interior**, and **closure**). Although the method can be applied to any object types for which a topology can be defined, we describe it in the context of the type **cell**. The essence of the method is to ask, what can be deduced about the topological relationship between two spatial regions from the set-theoretic relationship of intersection between their interiors and boundaries? To be more precise, let $X$ and $Y$ be the spatial cells, and assume that the boundaries, $\partial X$ and $\partial Y$, and interiors, $X^\circ$ and $Y^\circ$ are known. In order to find the topological relationship between $X$ and $Y$, compute the set-oriented relationships between $\partial X$, $X^\circ$, $\partial Y$, and $Y^\circ$. In fact, for a first-pass determination of the topological relationships, all that is required is to test whether intersections are empty or non-empty. Thus, consider the following four sets:

$$\partial X \cap \partial Y \qquad\qquad \partial X \cap Y^\circ$$
$$X^\circ \cap Y^\circ \qquad\qquad X^\circ \cap \partial Y$$

Each of these sets may either be empty or non-empty. There are 16 different mutually exclusive combinations of possibilities. Each possibility is a condition on the boundaries and interiors of sets, so each possibility will lead to a relationship between the sets that is preserved under topological transformations (homeomorphisms). For general sets in a point-set topology, each of the 16 combinations can exist and lead to a distinct topological relationship between the two sets $X$ and $Y$. We are, however, concerned only with spatial cells embedded in the Euclidean plane. In this case, only eight of the sixteen combinations can occur. Table 4.3 shows the eight possibilities and their correspondence with the spatial operations given earlier in Table 4.2.

**Table 4.3:**
Eight relations between cells in the Euclidean plane

| $\partial X \cap \partial Y$ | $X^\circ \cap Y^\circ$ | $\partial X \cap Y^\circ$ | $X^\circ \cap \partial Y$ | Operation |
|---|---|---|---|---|
| $\varnothing$ | $\varnothing$ | $\varnothing$ | $\varnothing$ | $X$ **disjoint** $Y$ |
| $\neg\varnothing$ | $\varnothing$ | $\varnothing$ | $\varnothing$ | $X$ **meets** $Y$ |
| $\neg\varnothing$ | $\neg\varnothing$ | $\varnothing$ | $\varnothing$ | $X$ **equals** $Y$ |
| $\varnothing$ | $\neg\varnothing$ | $\neg\varnothing$ | $\varnothing$ | $X$ **inside** $Y$ |
| $\neg\varnothing$ | $\neg\varnothing$ | $\neg\varnothing$ | $\varnothing$ | $Y$ **covers** $X$ |
| $\varnothing$ | $\neg\varnothing$ | $\varnothing$ | $\neg\varnothing$ | $Y$ **inside** $X$ |
| $\neg\varnothing$ | $\neg\varnothing$ | $\varnothing$ | $\neg\varnothing$ | $X$ **covers** $Y$ |
| $\neg\varnothing$ | $\neg\varnothing$ | $\neg\varnothing$ | $\neg\varnothing$ | $X$ **overlaps** $Y$ |

It can now be seen that **meets** and **overlaps** are topological refinements of the set-oriented operation **intersection**, while **inside** and **covers** are refinements of the **subset of** relationship (cf. Figure 4.22). Because four intersections are tested for empty in this method, it is sometimes known as the *4-intersection* model.

*4-intersection model*

This work has been extended for higher-dimensional spaces, and where the *co-dimension* is non-zero (i.e., where the dimension of the spatial objects is less than the dimension of the space in which they are embedded). In this case, three topological operations are used: boundary, interior, and set complement. The binary spatial relation between two spatial objects $X$ and $Y$ is classified by checking for emptiness/non-emptiness of the nine combinations of the operations applied to $X$ and $Y$, known as the *9-intersection* model. For a topological space $T$ in which $X$ and $Y$ are embedded (so $T \backslash X$ is the set complement or *exterior* of $X$, written $X'$) the nine combinations are:

*co-dimension*

*9-intersection model*

$$\partial X \cap \partial Y \qquad \partial X \cap Y^\circ \qquad \partial X \cap Y'$$
$$X^\circ \cap \partial Y \qquad X^\circ \cap Y^\circ \qquad X^\circ \cap Y'$$
$$X' \cap \partial Y \qquad X' \cap Y^\circ \qquad X' \cap Y'$$

*Operations on spatial objects*

The spatial operations described in this section can be thought of as operations on spatial literals. The operands are not affected by the application of the operation. For example, calculating the length of an arc cannot affect the arc itself. Another class of spatial operation acts upon spatially referenced objects, and alters the state of those objects. The three fundamental dynamic operations are **create**, **destroy**, and **update**. We have already seen some simple creation, destruction, and update operators in the discussions of object-oriented modeling (see section 2.4.3) and of affine transformations in geometry (see section 3.1.4).

### 4.4.3   Formal theories of spatial objects

It is possible to provide formal theories (logical calculi) for spatial relationships that have very general interpretations. Such theories are set
in a logical framework, with definitions of terms, well-formed formulas,
and axioms. An example that we now summarize is Clarke's calculus of
individuals, which has been set in a many-sorted first-order logic known
as *region connection calculus*, RCC. The focus of Clarke's calculus     RCC
of individuals is a binary connection relation between regions. In this
context, the word "region" may be thought to correspond to an object
of type **area** in our earlier typology, but in fact can have much wider,
and possibly non-spatial interpretations. Write $C(X, Y)$ as shorthand for
"region $X$ is connected to region $Y$." The connection relation is reflexive
and symmetric, satisfying the following axioms:

1. For each region $X$, $C(X, X)$

2. For each pair of regions $X, Y$, if $C(X, Y)$ then $C(Y, X)$

The surprising part of Clarke's calculus is the next step. It turns out
that many of the set-oriented and topological relations between spatial
objects may be constructed using just the minimal machinery above.
Table 4.4 provides a sample of the possible constructions, where the
relationships have been named to emphasize the connection with the
spatial operation typology of Table 4.2. The **part of** relationship in Table
4.4 corresponds to the **subset of** operation in Table 4.2.

| | |
|---|---|
| $X$ **disjoint from** $Y$ | It is not the case that $C(X, Y)$ |
| $X$ is a **part of** $Y$ | For each region $Z$, if $C(Z, X)$ then $C(Z, Y)$ |
| $X$ **overlaps** $Y$ | $X$ is not a **part of** $Y$ and $Y$ is not a **part of** $X$ and there exists a region $Z$ such that $Z$ is a **part of** $X$ and $Z$ is a **part of** $Y$ |
| $X$ **meets** $Y$ | $C(X, Y)$ and it is not the case that $X$ **overlaps** $Y$ |
| $X$ **covers** $Y$ | $Y$ is a **part of** $X$, and $X$ is not a **part of** $Y$ and there exists a region $Z$ such that $Z$ **meets** $X$ and $Z$ **meets** $Y$ |
| $X$ **is inside** $Y$ | $X$ is a **part of** $Y$, and $Y$ is not a **part of** $X$ and there does not exist a region $Z$ such that $Z$ **meets** $X$ and $Z$ **meets** $Y$ |
| $X$ **equals** $Y$ | $X$ is a **part of** $Y$ and $Y$ is a **part of** $X$ |

**Table 4.4:** Defining set-oriented and topological relations based on regions and connection

References containing more details on Clarke's calculus of individuals and RCC can be found in the bibliographic notes.

# BIBLIOGRAPHIC NOTES

4.1 For an introduction to basic ontological concepts and methods, the reader could consult Smith (1998). The distinction between data modeling and ontology in information systems is explored by Spyns et al. (2002).

4.2 Herring (1991) gives a useful discussion of the application of category theory to system modeling.

4.2.1 Good discussions of the relative merits of field-based against object-based models (although sometimes not using that terminology) are contained in Peuquet (1984), Morehouse (1990), Goodchild (1991), and Couclelis (1992).

4.3.1 Chrisman (1997, 1998) provides an introduction to and critique of Stevens' four levels of measurement, from a GIS perspective. Tests for spatial autocorrelation originate from the work of Cliff and Ord (1981). An introductory and accessible account of autocorrelation is provided by Unwin (1981), with a more recent account in O'Sullivan and Unwin (2002).

4.3.2 The map algebra, the field-based approach, and local, focal, and zonal field operations, are described in detail by Tomlin (1983, 1990, 1991). A parallel development to Tomlin's algebra for image processing is the image algebra of Ritter et al. (1990). Takayama and Couclelis (1997) describes work that generalizes Tomlin's algebra, taking ideas from Ritter's image algebra and recognizing the importance of neighborhood as a foundational concept.

4.4.1 Work on the formal classification of spatial object types is a fundamental part of GIS research. Egenhofer et al. (1989), Worboys (1992), De Floriani et al. (1993), and Güting and Schneider (1995) provide frameworks in which the specification of spatial object types can be rigorously developed.

4.4.2 An early and oft-quoted paper on spatial operations is that by Freeman (1975), which declares a set of 13 operations. Peuquet and Zhan (1987) discuss an algorithm that determines a directional relationship between polygonal shapes in the plane. Egenhofer and co-workers have written a series of papers exploring the background and connections between spatial operations. The 4- and 9-intersection models are discussed in Pullar and Egenhofer (1988); Egenhofer (1989); Egenhofer and Herring (1990); Egenhofer (1991); Egenhofer and Franzosa (1991). Egenhofer and Franzosa (1995) place models of spatial relationships based upon intersection relationships in a general framework, founded on the idea of *topological similarity*.

4.4.3 Clarke's calculus of individuals is introduced in (Clarke, 1981, 1985). The relevance of Clarke's calculus of individuals to spatial databases is discussed by Cui et al. (1993). The region connection calculus is introduced in Randell et al. (1992) and discussed further in Cohn et al. (1997).

# Representation and algorithms

# 5

**Summary**

*The manner in which spatial data is represented in an information system is key to the efficiency of the computational processes that will act upon it. This chapter discusses the representations of spatial data for object- and field-based models, and some of the algorithms that are used to process the data. Common representations of discrete spatial objects include* **spaghetti** *and* **node-arc-area**. *Field-based representations involve* **tessellations** *of space. Several common* **spatial algorithms** *are described, including a discussion of the time that algorithms require to perform their functions.*

Chapter 1 introduced the idea that information system development follows a sequence of stages, beginning with high-level conceptual models through to low-level computational models and implementation. The different emphasis at each stage of the development process helps to ensure that the resulting systems are both easy to understand and computationally efficient (see section 1.3.4). In the previous chapter, the focus was on high-level conceptual models. The chapter paid special attention to the human views of applications that feature geospatial information. In this chapter we start to consider questions that are closer to low-level computational processing. Specifically, this chapter is concerned with the representation of geospatial data in a GIS, and how various spatial operations are performed on this data.

The first section of this chapter discusses some general computational issues in a GIS. In particular, the multidimensionality of geospatial data, the distinctions between field-based and object-based representations, and the question of computational efficiency are considered. Discretization emerges as a big issue for object-based representations. An approach to handling the problems resulting from discretization is discussed in section 5.2. Section 5.3 describes some computational representations of spatial objects. Section 5.4 discusses representations in field-based systems using tessellations. Fundamental geometric algorithms (section 5.5) form the engine room of a GIS, without which higher-level processing cannot take place. Conversions between field and object representations are covered in section 5.6. The final section discusses representations and algorithms for network spaces.

## 5.1   COMPUTING WITH GEOSPATIAL DATA

Much traditional computing is based upon one-dimensional data representations. The following example is intended to give a feel for the jump from one to two dimensions. Figure 5.1 shows some of the different possible relationships that can exist between two straight-line segments, each embedded in the same one-dimensional Euclidean space (the real line). To categorize the relationships, we distinguish only those that are topologically different. Thus two segments may be disjoint, touching internally or externally, overlapping, nested, or equal. These six relationships exhaust all the possibilities (see the set-oriented and topological operators of Chapter 3).

**Figure 5.1:**
Relationships
between two
line segments
embedded in a
one-
dimensional
Euclidean
space

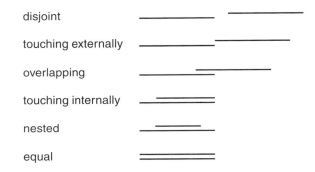

Let us now see what happens when we extend the investigation to two dimensions. In a sense, the simplest finite one-dimensional object is the line segment (1-simplex). In two dimensions, assuming discreteness, the corresponding object is the triangle (2-simplex). The extended problem is analyze the topologically distinct relationships that can exist between two triangles embedded in the same plane. Figure 5.2 shows some of the many possibilities. It is an interesting exercise to try to enumerate all the possibilities: it will probably surprise you how many there are.

As triangles are in a sense the simplest two-dimensional objects, it should be clear that the step from one to two dimensions introduces many more possibilities. It is then natural to assume that the operations underlying two-dimensional spatial data models will have a correspondingly more elaborate algorithmic support. The further jump to three dimensions introduces even greater complexities.

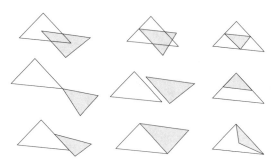

**Figure 5.2:** Nine topologically distinct relationships between two triangles embedded in the Euclidean plane

### 5.1.1 Geometric algorithms and computational geometry

An *algorithm* is a specification of the computational processes required to perform an operation. Thus, an algorithm for the addition of two numbers in a particular base will specify precisely the steps required to compute their sum. Algorithms may be straightforwardly translated into machine instructions (usually by means of a high-level programming language) that a computer is able to process.

A *geometric algorithm* is an algorithm that operates upon geometric (or spatial) objects. Geometric algorithms have some special features. Consider the problem of determining whether a point is inside a closed loop or not. Unless the loop is extremely convoluted, the human eye and brain are able to decide this instantly and without much discernible effort. But what instructions would you need to specify in order to make a machine perform this task? We shall see that this classical point-in-polygon decision procedure, while not being all that difficult, is nevertheless not trivial. This property of geometric algorithms is in stark contrast to arithmetic algorithms where many complex calculations, requiring paper, pencil, and much head-scratching from a person, can be solved with ease by a computer.

Another characteristic of geometric algorithms is that they are frequently plagued by special cases. What seems at first a simple problem often becomes more complex as special cases are considered. This feature has been vividly described by Douglas (1990), where what initially appears to be a trivial algorithm to determine whether two straight lines intersect becomes beset by a multitude of special cases.

*Computational geometry* is the study of the properties of algorithms for solving geometrical problems. Typical questions in computational geometry include:

1. Is it possible to find *any* algorithm to solve a particular geometrical problem?

2. What is the most efficient algorithm to solve a geometrical problem?

3. What is the best way of structuring the geometric data to make finding the solution to a problem most efficient?

*Margin notes:* algorithm · geometric algorithm · computational geometry

**Geometric algorithms and computation** *The notion of a geometric algorithm goes back at least as far as the geometer Euclid, living and teaching in the city of Alexandria circa 300 BC, in whose works geometric constructions are important features. In Euclid's geometry, the constructions are executed using a specified collection of instruments, straight-edge and compass, and proceeding according to a given set of basic moves. At that time, the issue was not the efficiency of the constructions but the class of objects resulting from them. A typical problem for investigation was the possibility of trisecting an angle using straight-edge and compass alone. It is only in the past century that efficiency of computation has become important. Work by Czech-born mathematician Kurt Gödel showed that a large class of logical systems, including arithmetic, must be incomplete, in the formal sense that these logics must contain true statements that cannot be proved. British mathematician Alan Turing was able to set this result within the framework of computation, and show that the same limitations apply to any computational system.*

computable

algorithmic
complexity
time complexity
space complexity

complexity
function

The first question is about computability, where a function or problem is *computable* if it is possible to find some algorithm to compute the function or solve the problem. Interest in geometric algorithms and computability dates back to ancient Greece (see "Geometric algorithms and computation" inset, on this page).

Finding an algorithm to solve a problem in theory is quite different from solving it in practice. This is the issue raised by questions 2 and 3. The efficiency of an algorithm (*algorithmic complexity*) may be measured in terms of the time that an algorithm consumes (*time complexity*) or storage space used (*space complexity*). The time taken for an algorithm to execute is the sum of the times of all the constituent operations within the algorithm.

It is almost always the case that the time and space taken to solve a problem increase with the size of the input data. For example, we would expect the time taken to compute the area of a simple polygon to increase as the number of edges of the polygon increases. The question is, "What is the relationship between the input size and time or space required by the algorithm?" This relationship may be expressed as a function, called the *complexity function*, from input size to amount of time (or space). In the remainder of this chapter we are concerned primarily with time complexity.

In the case of the area of a simple polygon, it is not too hard to show that the time taken to compute the area is directly proportional to the number of its edges, and so the complexity function is $f : n \rightarrow kn$, where $k$ is a constant of proportionality and $n$ is the number of edges. If we are interested in the general way that the computation time is related to input size, then the value of the constant $k$ does not matter; in any case it will depend upon details of the speed of the computer system and other factors. Also, the input size is not $n$ (the number of edges) but is proportional to $n$. It is the linearity of the relationship that is important.

These considerations have led to the notation $O(n)$ being used for the set of functions that are at most as large as the product of some constant and $n$. In general $O(f(n))$ is the set of functions that have a time complexity that is at most as large as the product of some constant and $f(n)$, referred to as the *order* of a function. The notation itself is termed "big-oh" notation. Big-oh notation allows us to describe the behavior of the relationship between computational time (or space) and input size, without being precise about the scalar multipliers.

*order*

*"big-oh" notation*

The most important functions that are used as yardsticks to measure the complexity of algorithms are, in increasing order of size, logarithms, fractional powers (e.g., square roots), polynomials, and exponentials. To give an idea about the rate of increase of these functions, Table 5.1 gives the approximate values for inputs between 1 and 100. Note that it is not just the static value of the function that is important, but the way that it changes as the input size $n$ increases. It is often of interest to consider the *asymptotic* nature of the function: that is, how it behaves for very large input sizes.

*asymptotic behavior*

| $n$ | $\log_e n$ | $\sqrt{n}$ | $n^2$ | $2^n$ |
|-----|-----------|-----------|-------|-------|
| 1 | 0.0 | 1.0 | 1 | 2.0 |
| 25 | 3.1 | 5.0 | 625 | $3.3 \times 10^7$ |
| 50 | 3.9 | 7.1 | 2500 | $1.1 \times 10^{15}$ |
| 75 | 4.3 | 8.6 | 5625 | $3.8 \times 10^{22}$ |
| 100 | 4.6 | 10.0 | 10000 | $1.3 \times 10^{30}$ |

**Table 5.1:** Approximate values of some common functions

A function order may then be associated with an algorithm to give some indication of how the algorithm will perform relative to the size of input. Of course, a change in the input to an algorithm will usually result in a change in performance of the algorithm, even if the size of the input remains constant. It is usual to take a pessimistic view and give functions that measure the *worst-case* performance of the algorithm: that is, the performance on the input of each size that results in the most time or space consumption. It may also be possible to determine *average-case* performance. However, it is often difficult to know how to calculate this average. For example, consider trying to generate a truly random sample of 10-sided polygons in the plane as the basis for an average-case time complexity assessment.

*worst-case performance*

*average-case performance*

| $O(1)$ | Constant time | Very fast, independent of input |
|--------|---------------|--------------------------------|
| $O(\log n)$ | Logarithmic time | Fast (e.g., binary search) |
| $O(n)$ | Linear time | Moderate (e.g., linear search) |
| $O(n \log n)$ | Sub-linear time | Moderate (e.g., optimal sorting algorithms) |
| $O(n^k)$ | Polynomial time | Moderate or slow, (e.g., shortest path algorithms) |
| $O(k^n)$ | Exponential time | Intractable (e.g., traveling salesperson algorithms) |

**Table 5.2:** Rough guide to time complexity

Returning to worst-case complexity (the usual measure), Table 5.2 gives a rough idea about the performance of algorithms with different complexity functions. Algorithms that are of order $O(k^n)$ (i.e., exponential time complexity) and greater are often called *intractable*, because they can only be used for very small input sizes.

intractable

For each geometric operation, we may now compare the performance of algorithms that implement that operation. The comparison may be experimental, generating sets of inputs and measuring the performance of a particular computer system on the input sets for each of the algorithms. Drawbacks with this approach are that:

- the results depend on the particular computer system being used as the platform for the experiments; and

- generating a truly representative sample of inputs can be difficult or impossible.

Alternatively, we may take the theoretical approach discussed above, finding and comparing the function class of each algorithm. Drawbacks here are that:

- calculating the complexity of the algorithms may be difficult;

- worst-case analysis may not be a true reflection of experience (for example, an algorithm may perform very badly in only a few cases that hardly ever occur); and

- comparing the function class may be misleading if the scalar multipliers are very different.

Overall, the theoretical approach is more general and is independent of any particular computational platform.

## 5.2   THE DISCRETE EUCLIDEAN PLANE

The issues arising from the representation of field-based and object-based data are quite distinct. For the data in a field-based model, the field functions are already discretized to the locations in a fixed spatial framework. On the other hand, with an object-based model discretization has wider ramifications, because there is no explicit spatial framework. Spatially referenced objects in a GIS are usually based on Cartesian coordinate pairs within the Euclidean plane. However, Euclidean geometry assumes an underlying continuum that can only be finitely approximated using a digital computational model. As we shall see, this approximation can lead to a large and uncontrolled accumulation of errors, which may alter the geometrical embedding of objects in the Euclidean plane and even distort topology.

### 5.2.1   Geometric domains

Assume that the number pairs chosen to represent points in discretized Euclidean space are integers. Next, set up a coordinate frame consisting of a fixed, distinguished point (origin) and a pair of orthogonal lines (axes) intersecting in the origin. A point in the plane has associated with it a unique pair of integers $(x, y)$ measuring its distance from the origin in the direction of each axis. The collection of all such points is the *discrete Euclidean plane*, $\mathbb{Z}^2$. Straight-line segments may be defined by giving a quadruple of integers, the two pairs of which represent the end-points of the line segment. $\qquad$ *discrete Euclidean plane*

We are now ready to define more precisely the notion of *geometric domain* as a triple $< G, P, S >$, where: $\qquad$ *geometric domain*

- $G$, the *domain grid*, is a finite connected portion of the discrete Euclidean plane, $\mathbb{Z}^2$ $\qquad$ *domain grid*
- $P$ is a set of points in $\mathbb{Z}^2$
- $S$ is a set of line segments in $\mathbb{Z}^2$

subject to the following closure conditions:

- Each point of $P$ is a point in the domain grid $G$.
- Any line segment in $S$ must have its end-points as members of $P$.
- Any point in $P$ that is incident with a line segment in $S$ must be one of its end-points.
- If any two line segments in $S$ intersect at a point, then that point must be a member of $P$.

The idea is that higher-level spatial objects may be constructed from objects in the geometric domain. Figure 5.3a shows a small and simple example of a geometric domain. Here the grid $G$ is 10 by 10 square, the set $P = \{a, b, c, d, e, f, g, h, i\}$ and the set $S = \{ab, ac, dh, eh, fh, gh\}$. Note that only points on the grid are allowed. Line segments must intersect at a point in $S$, so the configuration in Figure 5.3b is not a geometric domain.

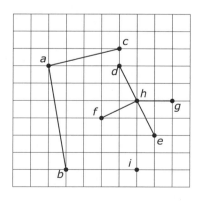

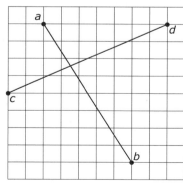

**Figure 5.3:** Grid structures

a. A structure that forms a geometric domain

b. A structure that does not form a geometric domain

### 5.2.2    Discretization and the Green-Yao algorithm

Euclidean space, based upon real numbers, is topologically dense and capable of a potentially infinite amount of precision. Discretized space, based upon the integers, is capable of only a finite precision. Thus, during the process of *discretization* (moving data from a continuous to a discrete domain) some precision will inevitably be lost. Loss of precision is unfortunate as it can lead to errors. Most people who have used a computer or calculator are aware of the rounding errors that can occur during discrete numerical applications. Similar types of errors, described below, can also occur in the geometric domain.

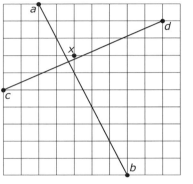

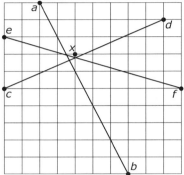

a. Adding $x$, the nearest grid point to the intersection, to the domain

b. Is the point of intersection above or below line $ef$?

**Figure 5.4:**
Shifting points and drifting lines

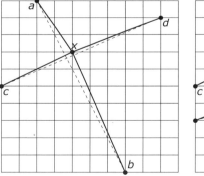

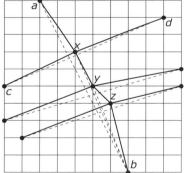

c. New split lines meet at intersection point $x$

d. Line $ab$ drifts away from its original position with further intersections

Figure 5.4a shows a configuration that is not classed as a geometric domain. The problem is that the line segments $ab$ and $cd$ intersect at a point that is not a member of the domain's point set. Indeed, it is not possible to add this point of intersection to the point set, because it is not a domain grid point. So what should we do? One solution is to introduce a grid point $x$ to the domain that is the nearest to the point of intersection (or use a convention to choose if several points are equally near). This is the two-dimensional equivalent of numerical rounding and is illustrated in Figure 5.4a. Of course, we still do not have a configuration that satisfies the domain axioms, but we could extend our notion of geometric domain to include such structures.

However, this solution on its own is unsatisfactory, because topological constraints are broken. For example, the incidence relationship that the point of intersection of two lines is on both the lines does not hold here. Also, as Figure 5.4b shows, it is possible for a point to "shift" from one side of a line to the other. The "real" point of intersection of lines $ab$ and $cd$ lies below line $ef$, but the rounded point of intersection $x$ lies above line $ef$.

A first attempt at a modification that avoids these problems is to split the line segments so as to join at the rounded intersection point, as shown in Figure 5.4c. Thus lines $ab$ and $cd$ have been split into lines $ax$, $bx$, $cx$, and $dx$. The configuration is now a geometric domain. The problem with this solution is that we lose control of the shift of the line segments as more and more intersections occur in a dynamic situation. Figure 5.4d shows some further line segments that have been added. Notice that the chain $axyzb$ has strayed well away from the original line segment $ab$.

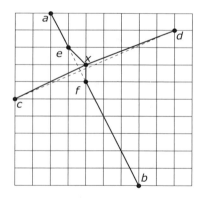

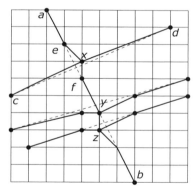

**Figure 5.5:**
Green-Yao
handling of
segment
intersections

a. Greene-Yao algorithm applied
to the $ab$ and $cd$

b. Further intersections using the
Greene-Yao algorithm

The problem of drifting lines can be solved using the *Greene-Yao algorithm*. Imagine that the grid points are pegs on a pegboard and that the lines $ab$ and $cd$ are elastic bands stretching between their end-points, shown in Figure 5.5a. When we move their point of intersection to the nearest point on the grid, we do so in such a way that the bands cannot pass over the pegs but may come to rest against some of them. In our

Greene-Yao
algorithm

example points $e$ and $f$ form such a barrier to the movement of segment $ax$ and $xb$, respectively. Thus segment $ab$ is no longer split into two segments, but in this case four, namely $ae$, $ex$, $xf$, and $fb$.

We have tried to give an appreciation of the algorithm by means of an example. In fact, the formal expression of the algorithm requires some knowledge of the mathematics of continued fractions, and is omitted from this account (but see bibliographic notes). The important result is that this algorithm places a limit on the drift of a line segment to its neighboring points (the so-called envelope of the segment), no matter how many line segment intersections are involved. In the example above, further intersections (Figure 5.5b) do not lead to the same line drift as for the earlier example (Figure 5.4d). Therefore, we now have a discretization process that is both well defined and results in a bounded error accumulation.

### 5.2.3   Discretizing arcs

So far, this section has concentrated upon problems caused by points and straight-line segments in a discretized plane. However, many spatial objects are composed of smooth curves, so here we describe approaches to the approximation of curvilinear arcs in a discrete space. The usual technique is to approximate the curve by a polyline (although as discussed in section 4.4.1 computable curves with discrete parameters may also be used).

There are of course an unlimited number of ways that a curve may be approximated by a polyline, due to the number and positioning of the segments. In general, the more segments used, the better the approximation, providing that the segments have been chosen appropriately. The problem of approximating a curve with a polyline is closely related to the problem of *line simplification*. Line simplification is the task of reducing the level of detail in the representation of a polyline, while still retaining its essential geometric character. Line simplification is commonly used in cartographic generalization, where maps need to be displayed at different scales or levels of detail.

A technique based on the *Douglas-Peucker algorithm* for arc simplification is shown in Figure 5.6. Consider the curve in Figure 5.6a. First, a point is assigned to each end of the arc. A straight-line segment $ab$ is then constructed between these points. The point $c$ on the arc farthest from the line $ab$ (in terms of length of a line perpendicular to $ab$) is added to the point set. The arc is split into two parts, $ac$ and $cb$, and the algorithm applied recursively to each of the pieces of arc. In Figure 5.6b, the arc has been split into $ad$, $dc$, $ce$, and $eb$. When the distance from a straight-line segment to the farthest point on the corresponding arc segment is below a predefined threshold, then that iteration stops. In Figure 5.6c $ce$ and $eb$ have this property, indicated with a dashed line. The smaller the threshold, the more points will be added and the better will be the approximation. The algorithm continues (Figure 5.6d) until there are no more arcs to split. In our example this stage is reached in Figure 5.6e with arcs $ah$, $hf$, $fd$,

line simplification

Douglas-Peucker
algorithm

$dg, gj, ji, ik, kc, ce, eb$. The approximating polyline is shown in Figure 5.6f.

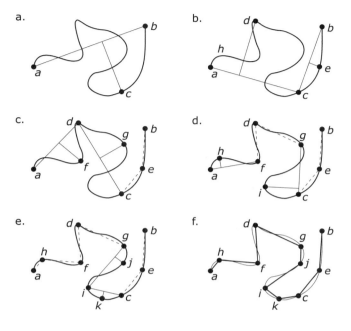

**Figure 5.6:** Douglas-Peucker algorithm for discretizing a curve

## 5.3 THE SPATIAL OBJECT DOMAIN

This section considers some different approaches to representing the structure of spatial objects, often called "capturing the topology." The first representation—spaghetti—provides only a minimal level of topological information. We then progress to representations that handle topology more explicitly. Some of these representations mix topology with the spatial embedding.

### 5.3.1 Spaghetti

The *spaghetti* data structure represents a planar configuration of points, arcs, and areas. Geometry is represented as a set of lists of straight-line segments. Each such list is the discretization of an arc that might exist independently, or as part of the boundary of an area. There is no explicit representation of the topological interrelationships of the configuration, such as adjacency relationships between constituent areas. The term "spaghetti" is an expressive metaphor for this representation: lists of straight line segments are akin to strands of spaghetti on a plate. (Actually, as the details below show, "spaghetti rings" would be a more accurate metaphor.)

An example is shown below, where the planar configuration of areas shown in Figure 5.7 is represented as spaghetti. Each of the polygonal

spaghetti

areas is represented by its boundary loop. Each loop is discretized as a closed polyline (for example, using the Douglas-Peucker algorithm of section 5.2.3). Each such polyline is represented as a list of points, each point being an extreme of a line segment in the polyline. For the example in Figure 5.7, the point lists are:

$A$: $[1, 2, 3, 4, 21, 22, 23, 26, 27, 28, 20, 19, 18, 17]$

$B$: $[4, 5, 6, 7, 8, 25, 24, 23, 22, 21]$

$C$: $[8, 9, 10, 11, 12, 13, 29, 28, 27, 26, 23, 24, 25]$

$D$: $[17, 18, 19, 20, 28, 29, 13, 14, 15, 16]$

To represent the geometric embedding in the plane, each point would be specified by its coordinate pair. These sequences are structured as cycles, so, for example, $[w, x, y, z] = [x, y, z, w] = [y, z, w, x] = [z, w, x, y]$. The spatial structure has thus been translated into a set of lists.

**Figure 5.7:**
A planar configuration (continuous and discrete) that partitions a subset of the plane into areas $A, B, C, D$

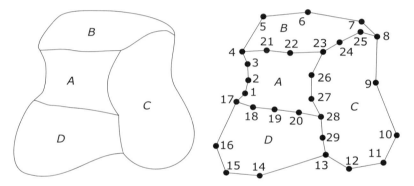

Any geometrical configuration based upon points and lines embedded in the Euclidean plane may be represented in this way. Spaghetti provides the basic connectivity within a spatial configuration, but few other spatial relationships are explicitly represented. For example, the fact that areas $A$ and $B$ are adjacent must be calculated by noting a common point sequence in their point lists. Thus the representation is inefficient for classes of spatial analysis where spatial relationships are important. Also, the representation is inefficient in space utilization because there is redundant duplication of data (point sequences are recorded in lists for both areas that they bound).

Spaghetti representations are useful in situations involving large numbers of simple geometric operations where structural relationships are not required. For example, a spaghetti representation is sufficient when all that is required is the graphical presentation of geospatial data on a computer screen or other output device. The spaghetti representation stems from a conceptually simple data model and has been popular in GIS. It is possible to take the basic spaghetti structure and enhance it to provide a richer representation. One example of an enhancement is to associate additional attributes to the lists (maybe giving feature type or line display style).

### 5.3.2 Representing more topology

This and the following sections introduce representations that capture explicitly some of the spatial relationships not inherent in the spaghetti representation. Such representations are often called "topological," to show that they are able to represent topological relationships, like adjacency. However, this terminology is not very helpful as even spaghetti is capable of expressing the topological connectivity relationship between edges.

The next representation, called NAA for *node-arc-area*, represents explicitly the adjacency relationships between areas in a subdivision of a surface. NAA shows clearly in its symmetry the duality between nodes and areas in a surficial subdivision. It is convenient to describe NAA in the form of a set of relation schemes for a relational database. In fact, this representation was used as an example for E-R modeling in section 2.3.1. The primary constituent entities are **directed arc**, **node**, and **area** (here, **area** is a face enclosed by arcs homeomorphic to a cell, except for the external area). The system rules for NAA are:

NAA

- Each directed arc has exactly one start and one end node.
- Each node must be the start node or end node (maybe both) of at least one directed arc.
- Each area is bounded by one or more directed arcs.
- Directed arcs may intersect only at their end nodes.
- Each directed arc has exactly one area on its right and one area on its left.
- Each area must be the left area or right area (maybe both) of at least one directed arc.

These rules may be expressed as an EER diagram, as shown earlier in Figure 2.16. Entity type **area** has overlapping subtypes **left area** and **right area**. Each instance of these subtypes enters into a many-to-one relationship with the directed arcs that bound it on the left and right, respectively. Entity type **node** has overlapping subtypes **begin node** and **end node**. Each instance of these subtypes enters into a many-to-one relationship with the arcs for which it is the begin or end node, respectively.

Figure 5.8 shows a decomposition of the configuration in Figure 5.7 into nodes, directed arcs, and areas ready for NAA representation. The constituent areas are labeled $A$, $B$, $C$, $D$. The external area, $X$, is necessary so that the system rules are satisfied. Arcs are labeled $a, ..., i$ and nodes are labeled $1, ..., 6$. Area $A$ is bounded by four arcs $a$, $e$, $h$, and $g$. Arcs $a$ and $e$ have $A$ as a left area, while arcs $g$ and $h$ have $A$ as a right area. Node 1 has incident arcs $a$, $b$, and $e$. Arcs $e$ and $b$ have node 1 as an end node, while arc $a$ has node 1 as a begin node.

The duality between node and area leads to a pleasing symmetry in the NAA representation. The symmetry is only marred by the existence of the specially distinguished area that is the external ($X$ in Figure 5.8): there is

**Figure 5.8:**
Decomposition
of the planar
configuration
in Figure 5.7
into nodes,
directed arcs,
and areas

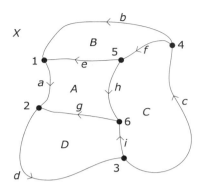

no such distinguished node. However, if we change our embedding space to be the surface of the sphere (the perfect 2-space?), then the need for an external area no longer exists and we have perfect symmetry! So much for mysticism, back to computation.

Following the usual database design procedures, the EER diagram forms the basis for a normalized relational database scheme as a single relation scheme:

ARC(<u>ARC ID</u>, BEGIN_NODE, END_NODE, LEFT_AREA, RIGHT_AREA)

The corresponding relation for the example in Figure 5.8 is shown in Table 5.3.

**Table 5.3:**
Relation
corresponding
to areal
configuration
in Figure 5.8

| ARC ID | BEGIN NODE | END NODE | LEFT AREA | RIGHT AREA |
|--------|-----------|----------|-----------|------------|
| $a$ | 1 | 2 | $A$ | $X$ |
| $b$ | 4 | 1 | $B$ | $X$ |
| $c$ | 3 | 4 | $C$ | $X$ |
| $d$ | 2 | 3 | $D$ | $X$ |
| $e$ | 5 | 1 | $A$ | $B$ |
| $f$ | 4 | 5 | $C$ | $B$ |
| $g$ | 6 | 2 | $D$ | $A$ |
| $h$ | 5 | 6 | $C$ | $A$ |
| $i$ | 3 | 6 | $D$ | $C$ |

The NAA representation up to now has given only the connectivity and adjacency relationships between planar objects. It can be extended to include details of the embedding. New entity types **point**, **polyline**, and **polygon** are required for the discretized embedded spatial objects. We also require a coordinate entity to pin down positions to the surface. An EER diagram for the extended NAA representation is given in Figure 5.9. Note that every **polygon** is an embedding of an **area** and is constituted as a sequence (cycle) of **polyline** entities. Each **polyline** is an embedding of a (directed) **arc** and is constituted as a sequence of **point** entities. Every **node** is a **point**, but not every **point** is a **node**, because points trace the embedding of arcs between nodes.

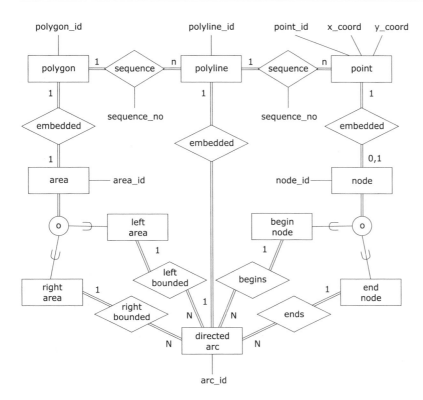

**Figure 5.9:** EER diagram for the extended NAA representation

As before, this diagram may be used as the basis for the design of a normalized relational database scheme. To simplify matters, since some entities are in one-to-one correspondence with no added attributes, we may make identifications between **area** and **polygon**, and between **arc** and **polyline**. In this case, the tables are:

ARC(<u>ARC ID</u>, BEGIN_NODE, END_NODE, LEFT_AREA, RIGHT_AREA)
POLYGON(<u>AREA ID</u>, <u>ARC ID</u>, SEQUENCE_NO)
POLYLINE(<u>ARC ID</u>, <u>POINT ID</u>, SEQUENCE_NO)
POINT(<u>POINT ID</u>, X_COORD,Y_COORD)
NODE(<u>NODE ID</u>, POINT_ID)

The relation ARC is unchanged from the discussion of the basic NAA relation scheme. Relation POLYGON gives the embedding of each area as a sequence of arcs (polylines): each sequence of arcs should be interpreted as a cycle. Relation POLYLINE gives the embedding of each arc as a sequence of points. Relation POINT gives the embedding of each point and relation NODE gives the point corresponding to each node.

### 5.3.3    Doubly connected edge list (DCEL)

The *doubly connected edge list* (DCEL) allows further capture of topology in surficial configurations. The DCEL provides a complete represen-                    DCEL tation of the topology of a connected planar graph. It omits the details of the actual embedding (i.e., polygons, chains, and coordinates of points) but focuses upon the topological relationships embodied in the entities node, arc (edge), and area (face). The DCEL provides in a single table the information necessary to construct:

- the sequence (cycle) of arcs around the node for each node in the configuration; and

- the sequence (cycle) of arcs around the area for each node in the configuration.

The word "sequence" is important here. The earlier NAA representation will give the sets of arcs that bound an area or meet at a vertex, but the sequencing cannot be determined unless the embedding is also given.

The EER diagram for the DCEL, shown in Figure 5.10, is an extension of the NAA representation. Two new relationships have been added: every arc has a unique next arc and a unique previous arc. Given an arc $a$, to find the previous arc to $a$, select the begin node (say, $n$) of $a$ and starting at $a$ proceed around $n$ in an counterclockwise direction until the first arc is found. To find the next arc to $a$, proceed around the end node of $a$ in an counterclockwise direction starting at $a$, until the first arc is found. The relationships **previous** and **next** are shown in Figure 5.11. An example of a DCEL representation is given in Table 5.4, representing the configuration shown in Figure 5.8.

**Figure 5.10:**
EER diagram showing the data structure of the DCEL

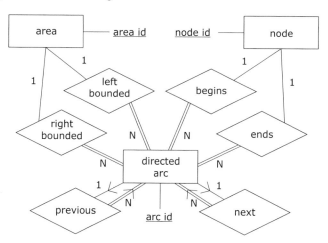

Algorithm 5.1 gives the procedure for computing the counterclockwise sequence (cycle) of arcs surrounding node $n$ for a DCEL. The input to the algorithm is the NODE_ID $n$. The algorithm then arbitrarily selects an arc $x$ which is incident with $n$ (line 1). For the relation DCEL in Table 5.4 we might specify this step using SQL or as an expression in the relational algebra. For example:

$$\pi_{\text{ARC\_ID}}\left(\sigma_{\text{BEGIN\_NODE}=n \text{ OR END\_NODE}=n}\left(\text{DCEL}\right)\right)$$

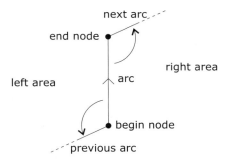

**Figure 5.11:** Relationships to a single arc in the DCEL

| ARC ID | BEGIN NODE | END NODE | LEFT AREA | RIGHT AREA | PREVIOUS ARC | NEXT ARC |
|--------|-----------|----------|-----------|------------|--------------|----------|
| $a$ | 1 | 2 | $A$ | $X$ | $e$ | $d$ |
| $b$ | 4 | 1 | $B$ | $X$ | $f$ | $a$ |
| $c$ | 3 | 4 | $C$ | $X$ | $i$ | $b$ |
| $d$ | 2 | 3 | $D$ | $X$ | $g$ | $c$ |
| $e$ | 5 | 1 | $A$ | $B$ | $h$ | $b$ |
| $f$ | 4 | 5 | $C$ | $B$ | $c$ | $e$ |
| $g$ | 6 | 2 | $D$ | $A$ | $i$ | $a$ |
| $h$ | 5 | 6 | $C$ | $A$ | $f$ | $g$ |
| $i$ | 3 | 6 | $D$ | $C$ | $d$ | $h$ |

**Table 5.4:** DCEL relation representing the configuration in Figure 5.8

returns the relation containing all arcs incident with node $n$ (see Chapter 2 for an explanation of relational algebra). The algorithm stores this initial arc $x$ as the first in a sequence (indicated with the assignment operator $\leftarrow$ in line 2 of Algorithm 5.1), then finds the arc that is counterclockwise from $x$ (lines 5–8). The relational algebra expression:

$$\pi_{\text{NEXT\_ARC}}\left(\sigma_{\text{ARC\_ID}=x}(\text{DCEL})\right)$$

might be used to retrieve the relation containing the next arc to $x$ (a similar operation retrieves the previous arc). The new arc is stored in sequence (line 4), and the process reiterates until we return to the original arc $x$.

**Input:** Node $n$
1: find some arc $x$ which is incident with $n$
2: $arc \leftarrow x$
3: **repeat**
4:     store $arc$ in sequence $s$
5:     **if** $begin\_node(arc) = n$ **then**
6:         $arc \leftarrow previous\_arc(arc)$
7:     **else**
8:         $arc \leftarrow next\_arc(arc)$
9: **until** $arc = x$
**Output:** Counterclockwise sequence of arcs $s$

**Algorithm 5.1:** Computing the counterclockwise sequence of arcs surrounding node $n$

To illustrate, consider applying Algorithm 5.1 to node 6 in Figure 5.8 (Table 5.4). The algorithm might select arc $i$ as the initial arc (node 6 is the end node for $i$). The first iteration of the algorithm would then store $i$ and find $h$ ($h$ is the next arc for arc $i$). The algorithm stores $h$ and reiterates to find next arc $g$. Finally, $g$ is stored, and the algorithm terminates because the previous arc for $g$ is $i$, which completes the sequence.

In a similar way, the clockwise sequence (cycle) of arcs around an area $X$ may be computed by modifying Algorithm 5.1 as in Algorithm 5.2. For example, area $C$ is the left area for arc $c$. On the first iteration of Algorithm 5.2, arc $c$ is stored, and arc $i$ is selected as the previous arc from $c$. The next iteration stores $i$ and selects $h$ as the next arc, which in turn is stored and used to select the arc $f$. Finally, $f$ is stored. The algorithm then terminates because $c$ is the previous arc from $f$, which completes the sequence $cihf$. Again, we might specify SQL or relational algebra expressions to actually perform some of the steps in the algorithm. For example, the SQL statement:

> **SELECT** ARC_ID **FROM** DCEL
> **WHERE** LEFT_AREA $= A$ OR RIGHT_AREA $= X$

retrieves those arcs that bound area $X$ from the relation DCEL.

---

**Algorithm 5.2:**
Computing the
clockwise
sequence of
arcs
surrounding
area $X$

**Input:** Area $X$
1: find some arc $x$ which bounds $X$
2: $arc \leftarrow x$
3: **repeat**
4:     store $arc$ in sequence $s$
5:     **if** $left\_area(arc) = X$ **then**
6:         $arc \leftarrow previous\_arc(arc)$
7:     **else**
8:         $arc \leftarrow next\_arc(arc)$
9: **until** $arc = x$
**Output:** Clockwise sequence of arcs $s$

---

### 5.3.4   Object-DCEL

object-DCEL

The same kind of DCEL structure may be used to describe aggregations of strongly connected areal objects. Assume that the objects under consideration in this section are regular closed in the Euclidean plane: that is, they are "pure area." The formal concept of the combinatorial map was introduced in section 3.3.5. The representation now described, which we call the *object-DCEL*, is based upon the combinatorial map, and has the important characteristic that it is *faithful*, up to homeomorphism and cyclic reordering of the arcs around a polygon. That is, two such areal aggregations that are not homeomorphic must have different representations, subject to cyclic reordering of arcs around the polygon.

The areal object $A$ shown shaded in Figure 3.38 seems to have two equally natural representations, as either the union of two lunes $B$ and $C$

or the difference of circular and elliptical disks $D$ and $E$. It does not really matter which one is chosen, as long as we are consistent; in fact, we will adopt the former. The object-DCEL representation relies on the notions of strong and weak connectivity, discussed in section 3.3.5. For Figure 3.38, object $A$ is not itself strongly connected, there being two articulation points at each end of its vertical diameter. $A$ is weakly connected and has the lunes $B$ and $C$ as strongly connected components. The object-DECL representation has the effect of decomposing an areal object into its strongly connected components.

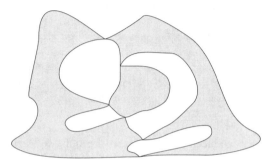

**Figure 5.12:**
Complex weakly connected areal object

The requirement is to provide a method that will allow the specification of unique and faithful representations of complex weakly connected areal objects of the kind shown in Figure 5.12. Such a method is now presented. First, overlay upon the object a collection of arcs and nodes, as shown in Figure 5.13. As usual in this section, the arcs are directed. Construct the direction of the arcs so that the object's area is always to the right of each arc. As for the DCEL representation, the weakly connected areal object may be represented in object-DCEL form as a table (see Table 5.5). In this case, no area identifiers need be added, because it is assumed that the whole structure is the spatial reference of a single areal object. Also, we have been sparse with the node information, because an arc's end node may be retrieved as the begin node of the next arc.

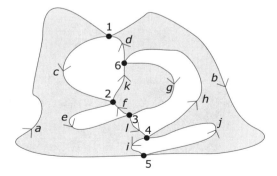

**Figure 5.13:**
The object in Figure 5.12 with arcs and nodes defined

The arc boundaries of the strongly connected cells that are components of the weakly connected object may now easily be retrieved by

**Alternative spatial object representations**  *Spaghetti, NAA, DCEL, and object-DCEL are four of the most common representations of planar spatial objects. However, several other representations exist, particularly in specialized domains. The winged-edge representation (Baumgart, 1975) is a variant of the DCEL. It may be used to represent subdivisions of orientable surfaces (i.e., surfaces for which the notion of "clockwise" is well defined). Unlike the DCEL, where edge orientation is explicitly stored as begin and end nodes, in a winged-edge representation orientation is implicit within the data structure. For a winged-edge representation, each edge is associated with four other edges: the next edge clockwise, the previous edge clockwise, the next edge counterclockwise, and the previous edge counterclockwise. The basic winged-edge representation does not allow loops and isthmuses, but can be extended to handle such features (Weiler, 1985). The quad-edge representation (Guibas and Stolfi, 1985) allows orientable or non-orientable surfaces to be represented (i.e., the notion of "clockwise," invariant over the surface, is not required). The quad-edge representation is generally more complicated than the winged-edge representation. The main advantage of quad-edge over winged-edge is that the dual graph (see Chapter 3) is automatically embedded within the quad-edge representation.*

following the sequences of arc to next arc until we arrive back at the starting arcs. For the above example these arc boundaries are:

$$[a, c, e, l, i]$$

$$[f, k, g]$$

$$[b, j, h, d]$$

In summary, the object-DCEL representation comprises the lists of arcs bounding the strongly connected components of the object. The representation provides a complete and faithful description of the topology of the object. Two related representations of planar areal objects are introduced in the "Alternative spatial object representations" inset, on this page.

**Table 5.5:**
Object-DCEL
relation
representing
the weakly
connected areal
object in Figure
5.13

| ARC ID | BEGIN NODE | NEXT ARC |
|--------|------------|----------|
| $a$ | 5 | $c$ |
| $b$ | 1 | $j$ |
| $c$ | 1 | $e$ |
| $d$ | 6 | $b$ |
| $e$ | 2 | $l$ |
| $f$ | 3 | $k$ |
| $g$ | 6 | $f$ |
| $h$ | 4 | $d$ |
| $i$ | 4 | $a$ |
| $j$ | 5 | $h$ |
| $k$ | 2 | $g$ |
| $l$ | 3 | $i$ |

Before leaving this section on the representations of planar areal objects, it is worth dwelling for a short time on the problem of holes and islands. In the context of regular closed areal objects, a hole (or island) is defined as a subregion contained by the main region, with a boundary that is disjoint from the main region's boundary. Figure 5.14 gives examples of combinations of holes and islands. Note that a hole may not touch the boundary of the main region or another hole. Areal objects with holes can be represented using extensions of the structures above, although care must be taken if faithful representations are still required. The bibliographic notes contain some references to extensions capable of representing holes and islands.

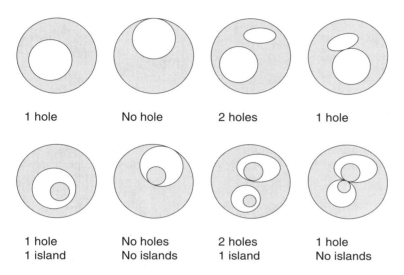

| 1 hole | No hole | 2 holes | 1 hole |
|---|---|---|---|

| 1 hole | No holes | 2 holes | 1 hole |
| 1 island | No islands | 1 island | No islands |

**Figure 5.14:** Areal objects with varying numbers of holes and islands

## 5.4 REPRESENTATIONS OF FIELD-BASED MODELS

The overall development of the last section steered a path in the direction of ever more comprehensive representations of spatial objects embedded in the plane. These representations explicitly hold information about topological relationships between constituent parts of spatial objects. We now change course in order to describe an important class of representations based upon tessellated structures, for which topological relationships are often implicit. Tessellations have already been introduced in the context of the spatial framework described in the preceding chapter. A *tessellation* is a partition of the plane, or a portion of the plane, as the union of a set of disjoint areal objects. If these constituent objects are all exact copies of the same regular polygon and each of the vertices is also regular (in a sense shortly to be explained) then the tessellation is *regular*; otherwise the tessellation is *irregular*. By far the most important class of regular tessellated representations is the grid or *raster* representation, based upon a tessellation of squares. The irregular tessellated representations of most interest are TINs (triangulated irregular networks). Triangles are the sim-

tessellation

**Regular tessellations**  *Tessellations have been used for hundreds of years in the decorative tiling of the floors, walls, and ceilings of buildings. In fact, tiling is an alternative term for tessellation. In the 17th century, the mathematician and astronomer Johannes Kepler was the first person to systematically study the geometry of tessellations. Kepler observed that in the case of the Euclidean plane, only three regular tessellations are possible, based upon equilateral triangles, squares, and regular hexagons (illustrated below). We can show that no other regular tessellations are possible by noticing that in order to tessellate, the interior angles of the regular polygon at each vertex of a tessellation must be a factor of 360°. Triangles (interior angle 60°), squares (interior angle 90°), and hexagons (interior angle 120°) are the only regular polygons that satisfy this criterion.*

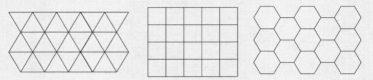

plest discrete areal objects that populate the Euclidean plane, and so it is interesting to use them as the primitive areal elements in representations.

### 5.4.1  Regular tessellated representations

In terms of discrete areal objects, a tessellation of a surface is a covering of the surface with an arrangement of non-overlapping polygons. A *regular polygon* is a polygon for which all edges have the same length and all internal angles are equal. At each vertex of a tessellation, the *vertex figure* is the polygon formed by joining in order the mid-points of all edges incident with the vertex. A tessellation of a surface for which all the participating polygons and vertex figures are regular and equal is termed a *regular tessellation*. For example, a regular tessellation of equilateral triangles has a regular hexagon for its vertex figure.

*regular polygon*

*vertex figure*

*regular tessellation*

Of the regular tessellations (see "Regular tessellations" inset, on this page), by far the most commonly used for spatial representations is the square grid. This provides the raster representation of spatial data, whereby planar spatial configurations are decomposed into a pattern of squares (*pixels*) on a grid. Regular triangular and hexagonal tessellations are rarely used for planar data representation. Nested regular triangular tessellations have been suggested for representing spherical data (see Chapter 6). The regular tessellation representation based upon squares fits well with standard and well-supported programming data types, such as two-dimensional arrays. However, this representation does not necessarily accord with the way that the data has been collected, often with measurement locations arranged irregularly through the region. In this case irregular tessellations are more appropriate, and such representations are now considered.

### 5.4.2 Irregular tessellated representations

An *irregular tessellation* is a tessellation for which the participating polygons are not all regular and equal. The most commonly used irregular tessellation is the TIN. A TIN may represent the variation of a field function over a spatial framework. Its irregularity allows the resolution to vary over the surface, capturing finer details where required. A useful concept in what follows is again the notion of duality of planar graphs, discussed in section 3.4.2, where faces become nodes and nodes become faces. If we label each triangular face of the TIN with a node and join with edges those faces that are adjacent in the TIN, then the planar mesh that results is the *dual* of the original mesh. If $T^*$ is the dual of the TIN $T$, then the degree of each node (the number of edges incident with that node) in $T^*$ must be three.

irregular tessellation

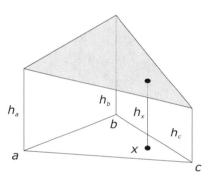

**Figure 5.15:** Interpolation is required to find the height of point $X$

The height at any point on a TIN can be determined using linear interpolation. Figure 5.15 shows the linear interpolation process working for a single general triangular facet of the surface. The triangular facet $abc$ is part of a TIN. The coordinates (in vector form) $a$, $b$, and $c$ of the vertices are assumed known, as are the heights $h_a$, $h_b$, and $h_c$, at $a$, $b$, and $c$, respectively. The problem is to estimate the height $h_x$ of the surface at point $x$, the coordinates of which are known to be $x$. Point $x$ is inside or on the boundary of the triangle $abc$, so we have the following relationship:

$$x = \alpha a + \beta b + \gamma c$$

where $\alpha$, $\beta$, and $\gamma$ are scalar coefficients that can be uniquely determined, such that:

$$\alpha + \beta + \gamma = 1$$

The height $h_x$ can now be found by using the equation:

$$h_x = \alpha h_a + \beta h_b + \gamma h_c$$

The preceding equation gives an estimate of the height on an intermediate point between measured points, based on the assumption that the discretization is into planar facets, that is, first-order (linear) surfaces. It is also possible to base the interpolation upon higher-order surfaces (see, for example, Akima, 1978).

### 5.4.3   Delaunay triangulations and Voronoi diagrams

Given a set of irregularly spaced points within a region, there remains the problem of constructing the tessellation of irregular triangles, the vertices of which are the given set. In general, there will be many candidate tessellations and some may be judged better than others. Figure 5.16 shows two possible triangulations based upon the same set of vertices. The right-hand triangulation has a higher proportion of long, thin triangles and so may be judged inferior for some purposes.

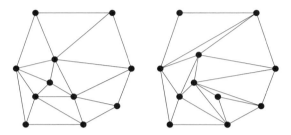

**Figure 5.16:**
Two
triangulations
based upon the
same vertex set

Delaunay
triangulation

Voronoi diagram

proximal polygon

A triangulation that has many desirable properties, both for itself and for its dual, is the *Delaunay triangulation*. Loosely, it can be said that the constituent triangles in a Delaunay triangulation are "as nearly equilateral as possible." The dual of a Delaunay triangulation is a *Voronoi diagram* (or *Thiessen polygons*). Delaunay triangulations can be introduced by considering their dual, Voronoi diagrams.

Imagine that we are given the point locations of fire emergency response units in an area. Suppose that we wish to surround each response point $p$ by a region $R_p$ that has the property that every location in $R_p$ is nearer to $p$ than to any other response unit. Suppose for simplicity that distance is measured "as the crow flies" using the usual Euclidean metric. Then it turns out that these areas of closest proximity, $R_p$, are polygons, termed *proximal polygons*. The set of proximal polygons constitutes a Voronoi diagram. Given a set of points with the property that no three are collinear (to avoid degenerate cases), a Voronoi diagram surrounds each point with an area that contains all the locations that have that "seed" point as their nearest.

The Delaunay triangulation is now simply formed as the dual of the Voronoi diagram. Points in the Voronoi diagram become vertices in the Delaunay triangulation. Edges in the Delaunay triangulation connect adjacent proximal polygons in the Voronoi diagram (Figure 5.17). A Delaunay triangulation has the property that each circumcircle of a constituent triangle does not include any other triangulation point within it. This property is illustrated in Figure 5.18.

To understand the intricate details of Delaunay triangulations and Voronoi diagrams, a considerable amount of geometry outside the context of this book would be required (see bibliographic notes). Below is a list of some of the more striking properties of the Delaunay construction. Given

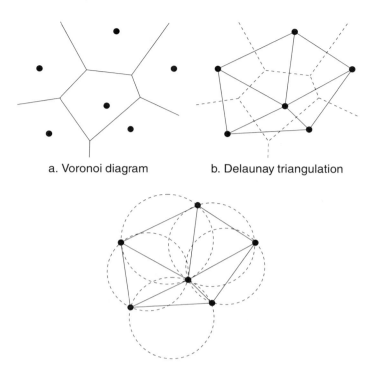

**Figure 5.17:** Voronoi diagram of proximal polygons and its dual Delaunay triangulation

a. Voronoi diagram       b. Delaunay triangulation

**Figure 5.18:** Circumcircles of a Delaunay triangulation

an initial point set $P$ for which no sets of three points are collinear (again to avoid degenerate cases):

1. The Delaunay triangulation is unique.

2. The external edges of the triangulation form the convex hull of $P$ (i.e., the smallest convex set containing $P$).

3. The circumcircles of the triangles contain no members of $P$ in their interior. Note that this is a defining property, in that if it holds, then the triangulation must be a Delaunay triangulation.

4. The triangles in a Delaunay triangulation are best-possible with respect to regularity (closest to equilateral).

Property 4 is vague and needs some explanation. Suppose that we have a triangulation (not necessarily Delaunay) and make a list of the minimum angles in each of the constituent triangles in such a way that the list starts with the smallest angle and moves along in sequence to the largest angle. Call this list the *ordered minimum angle vector* for the triangulation. We may compare two such vectors *lexicographically* as if they were ordered in a telephone directory. Now Property 4 can be made more precise as follows:

*ordered minimum angle vector*

4′. The triangles in a Delaunay triangulation have the greatest ordered minimum angle vector of any triangulation of $P$.

A Delaunay triangulation has the greatest possible ordered minimum angle vector (property 4′), so it will contain the smallest collection of long, thin triangles. Informally, this implies that the Delaunay triangulation is "as regular as possible" (i.e., property 4).

### 5.4.4  Triangulations of polygons

In general, a triangulation tessellates an unbounded region with triangles based upon a prescribed set of vertices. This section describes several approaches to the triangulation of a bounded polygonal region. It would be nice to modify the Delaunay triangulation for an unbounded region, but if we take the collection of points comprising the vertices of a polygon and form the Delaunay triangulation based upon them, this will not necessarily be the triangulation of the polygon, because some of the edges of the polygon may not be edges in the triangulation. However, it is possible to use the *constrained Delaunay triangulation*, which is constrained to follow a given set of edges.

constrained
Delaunay
triangulation

Figure 5.19 shows a simple example. The Delaunay triangulation is as in Figure 5.19a. The triangulation in Figure 5.19b has been constrained by a polygon (boundary shown as a thick line). The constrained triangulation includes the edge *ab*, which is not part of the Delaunay triangulation in Figure 5.19a.

**Figure 5.19:**
Unconstrained
and constrained
Delaunay
triangulations

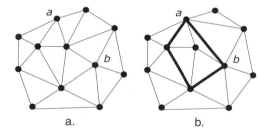

a.                              b.

Apart from Delaunay methods, there are many other methods for triangulations of sets of points and polygons (see bibliographic notes). For example, the *greedy triangulation* has the objective of minimizing the total edge length in the triangulation. This is achieved by introducing the shortest possible internal diagonal at each stage. Figure 5.20 shows the greedy and Delaunay triangulations of a simple quadrilateral. Algorithms for triangulations are considered later in the chapter.

**Figure 5.20:**
"Greedy" (left)
and Delaunay
(right)
triangulations
of a simple
polygon

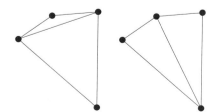

### 5.4.5  Medial axis of a polygon

The *medial axis* of a polygon is the Voronoi diagram computed for the line segments that make up the boundary of that polygon. For example, Figure 5.21 shows the medial axis (thick lines) of the boundary of the polygon (thin lines). The medial axis is used within a variety of algorithms, some of which are mentioned later on, and is sometimes referred to as the *skeleton* of a polygon. The medial axis for a convex polygon is composed of straight-line segments, but in general may comprise arcs of a parabola, as in Figure 5.21. The medial axis of a polygon can be computed in linear time using a *medial axis transform* (MAT), based on an extension of the algorithm for computing the Voronoi diagram.

*medial axis*

*skeleton*

*medial axis transform*

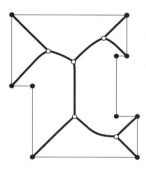

**Figure 5.21:** Medial axis of a polygon

### 5.4.6  Tessellations of the sphere

The representations described so far have all been referenced to planar objects. Of course, the Earth is topologically a sphere, and so it is natural to consider representations appropriate for spherically referenced data. The reasons that so much emphasis has been placed upon the plane are:

- Planar data is computationally simpler than the sphere.
- Small areas of the Earth's surface may be approximated by planes.

Regular tessellations of the sphere correspond to the five Platonic solids of antiquity: tetrahedron, cube, octahedron, dodecahedron, and icosahedron. For example, the tetrahedral tessellation comprises four spherical triangles (called *triangular facets*) bounded by parts of great circles. Each spherical triangle has internal angle of 120°. Three triangles meet at each vertex. In the case of the octahedral tessellation, there are initially eight triangular facets, each having an internal angle of 90°. Four triangles meet at each vertex. Figure 5.22 shows an octahedral tessellation on the sphere, having triangular facets *nab, nbc, ncd, nda, sab, sbc, scd, sda*. Unfortunately, unlike the plane, regular tessellations of arbitrary fine resolution are not possible on the sphere. However, it is possible to recursively *nest* regular polygons within the facets of the Platonic solids to produce the resolutions required (see "Nested tessellations" inset, on the following page).

*triangular facets*

*nested tessellation*

**Nested tessellations**  A *nested tessellation* is a tessellation the cells of which are themselves partitioned using a further and finer tessellation. This process can be recursive, providing a graded granularity to which spatial phenomena may be referenced. *Regular nested tessellations* occur where the regular constituents are themselves partitioned using the same regular figure. Examples of planar regular and irregular nested tessellations are shown below. The first two (triangle and square) are regular, but the third (hexagon) is not, because hexagons cannot fit together to make a larger hexagon (see Peuquet, 1984). The nested square tessellation of the plane leads to the quadtree data structure, discussed in Chapter 6. The nested triangular tessellation may be used to provide the nested octahedral tessellation of the sphere (Dutton, 1984, 1989), also considered in Chapter 6. Returning to the plane, there is a distinction between *hierarchical* models, based upon recursive regular or irregular tessellations and *stratified* models in which each stratum is defined to have particular properties. A survey of hierarchical models is given by De Floriani et al. (1994) with particular emphasis on hierarchical TIN in De Floriani and Puppo (1992). A stratified Delaunay triangulation, where the circumcircle property is satisfied at each level, is defined in the *Delaunay pyramid* of De Floriani (1989). Stratified models are discussed in general by Bertolotto et al. (1994).

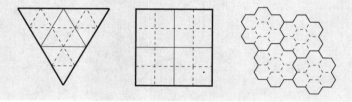

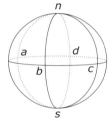

**Figure 5.22:** Octahedral tessellation of the sphere

## 5.5   FUNDAMENTAL GEOMETRIC ALGORITHMS

It would be impossible to describe in this single section all the spatial algorithms that have been devised. The aim of this section is to provide a representative sample, giving preference to commonly used algorithms and those that show an example of a generic approach. The representations introduced in earlier sections will be used in the algorithms.

All of the algorithms described below assume an embedding in the Euclidean plane. Thus point objects have a position that may be given by a coordinate pair $(x, y)$. The set of points on a straight line in the Euclidean plane may be constrained by a linear equation to give $\{(x, y)|ax + by + c = 0\}$, where $a$, $b$, $c$, are constants. The set of points on a straight-line segment between two distinct points $p$ and $q$ may also be represented in parametric form as $\{\lambda p + (1 - \lambda)q|\lambda \in [0, 1]\}$ (see Chapter 3).

### 5.5.1 Metric and Euclidean algorithms

*Distance and angle between points*

For ease of reference, we re-state these formulas from Chapter 3. The (Euclidean) distance $|pq|$ between points $p = (x_p, y_p)$ and $q = (x_q, y_q)$ is given by the formula:

$$|pq| = \sqrt{(x_q - x_p)^2 + (y_q - y_p)^2}$$

The bearing, $\theta$, of $q$ from $p$ is given by the unique solution in the interval $[0, 360[$ of the simultaneous equations:

$$\sin \theta = \frac{x_q - x_p}{|pq|}$$

$$\cos \theta = \frac{y_q - y_p}{|pq|}$$

*Distance from point to line*

Distance is meant here as the *minimum* distance between spatial objects. To compute the distance between a point and a straight line, the most compact formula arises when the straight-line $l$ is given in the form $\{(x, y) | ax + by + c = 0\}$. Suppose that the point $p$ is given by the coordinate pair $(x_p, y_p)$. Then the distance from $p$ to $l$, measured by the length of the line segment through $p$ and orthogonal to $l$, is given by the formula:

$$\textbf{distance}(p, l) = \frac{|ax_p + by_p + c|}{\sqrt{a^2 + b^2}}$$

For the distance between a point $p$ and a straight-line *segment* $l$, the calculation is no longer quite so straightforward. Figure 5.23 shows that the distance will be measured as the distance from $p$ to one of the end-points of $l$ or from $p$ to the line incident with the end-points of $l$, depending on the relative positions of $p$ and $l$. The line segment $l$ defines a partition of the plane into two point sets, a connected set that we have called **middle**$(l)$ and a disconnected set called **end**$(l)$. The calculation to determine the distance of $p$ from $l$ depends upon whether $p$ is in **middle**$(l)$ or **end**$(l)$.

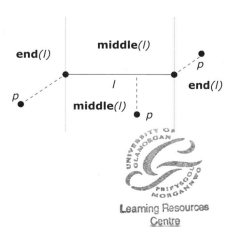

**Figure 5.23:** Distance between a point $p$ and line segment $l$

The problem becomes more complex for a polyline, as can be seen from Figure 5.24. First, the distance to each line segment in the polyline must be calculated, as above. The distance from $p$ to the polyline will then be the minimum of all these distances. If the number of segments in the chain is $n$, then the time complexity of the computation is $O(n)$. An approximation to the distance may be achieved by considering only the distances to vertices of the chain. This approximation will be good in general if the segment lengths are small compared with the distance of the point from the polyline.

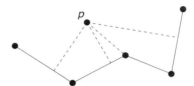

**Figure 5.24:** Distance between a point and a polyline

Distances from a point to a polygon or between polygons may be required in terms of their boundaries. Thus the distance between two polygons may be interpreted as the distance between their closest points. In that case, a calculation involving boundary polylines may be necessary. More usually, the distance between two polygons is interpreted as the distance between their centroids (discussed in a later section).

*Area of a simple polygon*

Let $P$ be a simple polygon (no boundary self-intersections) with vertex vectors $(x_1, y_1), (x_2, y_2), ..., (x_n, y_n)$ where $(x_1, y_1) = (x_n, y_n)$ (i.e., the polygon is closed and has the same start and end vertices). Then a formula for the area is:

$$\mathbf{area}(P) = \frac{1}{2} \sum_{i=1}^{n-1} x_i y_{i+1} - x_{i+1} y_i$$

In the case of a triangle $pqr$, where $p = (x_p, y_p)$, $q = (x_q, y_q)$, $r = (x_r, y_r)$, the formula gives:

$$\mathbf{area}(pqr) = \frac{x_p y_q - x_q y_p + x_q y_r - x_r y_q + x_r y_p - x_p y_r}{2}$$

Notice that the area calculated has a sign (positive or negative). Also, according to this formula $\mathbf{area}(pqr) = -\mathbf{area}(qpr)$. If point $p$ is to the left-hand side of the directed segment $qr$, then $\mathbf{area}(pqr)$ will be positive; if point $p$ is to the right-hand side of the directed segment $qr$, then $\mathbf{area}(pqr)$ will be negative; and if $p$, $q$, $r$ are collinear, then $\mathbf{area}(pqr)$ will be zero. Thus a useful spin-off from our work on area is an algorithm for the **side** operation (Figure 5.25):

$$\textbf{side}(p, q, r) = \begin{cases} 1 & \text{if } \textbf{area}(pqr) > 0 & (p \text{ is left of } qr) \\ 0 & \text{if } \textbf{area}(pqr) = 0 & (pqr \text{ are collinear}) \\ -1 & \text{if } \textbf{area}(pqr) < 0 & (p \text{ is right of } qr) \end{cases}$$

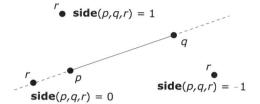

**Figure 5.25:** The **side** operation

### Centroid of a polygon

The *centroid* or *center of gravity* of an areal object is the point at which it would balance if it were cut out of a sheet of material of uniform density. For a regular polygon or a set of points the centroid is simply the mean of the vertex vectors. Computing the centroid of an irregular polygon requires a more complex calculation, based on the weighted sum of vertex vectors. For a (possibly irregular) polygon $P$ with $n$ vertex vectors $(x_1, y_1), (x_2, y_2), ..., (x_n, y_n)$ where again $(x_1, y_1) = (x_n, y_n)$ the $x$ and $y$ coordinates of the centroid are given by:

centroid

$$\textbf{centroid}_x(P) = \frac{1}{6.\textbf{area}(P)} \sum_{i=1}^{n-1} (x_i + x_{i+1})(x_i y_{i+1} - x_{i+1} y_i)$$

$$\textbf{centroid}_y(P) = \frac{1}{6.\textbf{area}(P)} \sum_{i=1}^{n-1} (y_i + y_{i+1})(x_i y_{i+1} - x_{i+1} y_i)$$

Notice that the center of gravity of a polygon is not guaranteed to lie within its boundary (consider the center of gravity of a "horseshoe" shape, for example).

### 5.5.2 Topological algorithms

### Point-in-polygon

Point-in-polygon is one of the most common operations in GIS. Given a point $p$ and a polygon $P$, **point_in_polygon**$(p, P) = $ **true** if and only if $p$ is in the interior of $P$.

point-in-polygon

If the polygon is convex, then the operation **side** discussed in the previous section may be used. Assume that the vertices of the polygon are $p_1, p_2, ..., p_n$, ordered counterclockwise around the polygon (again $p_1 = p_n$). Then **point_in_polygon**$(p, P) = $ **true** if and only if:

$$\textbf{side}(p, p_1, p_2) = \textbf{side}(p, p_2, p_3) = ... = \textbf{side}(p, p_{n-1}, p_n) = 1$$

The more usual and interesting case is where the polygon is not necessarily convex. Two algorithms are considered here: the *semi-line algorithm* and the *winding number algorithm*.

semi-line
algorithm

The underlying principle behind the semi-line algorithm is simple. Given a point $p$ and polygon $P$, assume that $p$ is not on the boundary of $P$ (to avoid degenerate cases). Draw an infinite semi-line (or *ray*) out from $p$. Count the number of times that the ray intersects with the boundary of $P$. If the number is even, then $p$ is outside $P$. If the number is odd, then $p$ is inside $P$ (see Figure 5.26).

ray

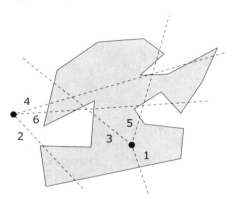

**Figure 5.26:**
Semi-line
algorithm for
determining
whether a point
is inside a
polygon

However, as with many of these algorithms, several special cases can arise. For example, if a ray intersects with a vertex of the polygon, how should it be counted? Figure 5.27 shows two such cases. At vertex $a$, the intersection should not be counted, while at vertex $b$, the intersection should be counted. It is important that the polygon actually crosses the ray for the intersection to be counted, and this does not happen at $a$. Figure 5.27 also shows the intersection of the entire segment $cd$ with the ray. The same principle may be applied: it is only the *crossing* that is counted.

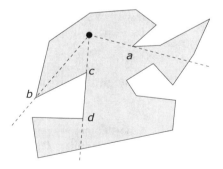

**Figure 5.27:**
Special cases in
the semi-line
algorithm

With regard to implementation, it simplifies matters to choose the semi-line to be horizontal and directed to the right, then to translate the polygon so that the semi-line is the positive $x$-axis with the point $p$ as the origin. The polygon is then traversed and, for each edge, a simple calculation with coordinates determines whether the edge crosses the positive $x$-axis. The algorithm has time complexity $O(n)$, where $n$ is the number of edges of the polygon.

The winding number algorithm provides an alternative and quite distinct approach. It has the same complexity as the semi-line algorithm, but in practice is much less efficient due to the trigonometrical calculations. Imagine that an observer is moving counterclockwise along the boundary of $P$, always facing the point $p$. If, after having completed the journey round $P$, the observer has turned through one complete circle, then $p$ is within $P$. Otherwise, $p$ is outside $P$. Figure 5.28 shows examples of this. In each case a path is walked visiting the vertices in the order 1-2-3-4-5-6-1. In the left-hand configuration, where the point is inside the polygon, the observer will rotate through a full turn. In the right-hand configuration, where the point is outside the polygon, the net turning effect will be zero.

winding number algorithm

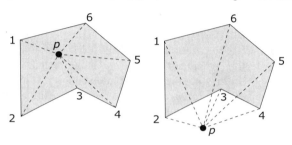

**Figure 5.28:** Traversing the boundary of the polygon 123456 always facing point $p$

Given the coordinates of the polygon vertices, the angle formed by each pair of consecutive vertices and point $p$ may be calculated using elementary trigonometry, as in Chapter 3. The angles are then summed over the entire walk round the polygon to find the total turn.

### 5.5.3  Set-based algorithms

Throughout this section spatial objects are treated as sets of points. It is assumed, unless the contrary is indicated, that all the sets are topologically closed; that is, they contain their boundary points and edges.

*Collinearity*

Given three points $a$, $b$, and $c$, the Boolean operation **collinear**$(a, b, c)$ determines whether $a$, $b$, and $c$ are collinear (lie on the same straight line). The algorithm for this operation needs no new ideas. From section 5.5.1:

> **collinear**$(a, b, c) = $ **true** if and only if **side**$(a, b, c) = 0$

*Point on segment*

Let straight-line segment $l$ have end-points $q = (x_q, y_q)$ and $r = (x_r, y_r)$. For any point $p = (x_p, y_p)$, the operation **point_on_segment**$(p, l)$ returns the Boolean value **true** if $p \in l$.

The first step in calculating **point_on_segment**$(p, l)$ is to use the previous algorithm to decide whether $p$, $q$, and $r$ are collinear. If not, the determination is negative and there is no need to proceed further.

minimum
bounding box
Otherwise, consider the *minimum bounding box* (MBB) of $l$, $\mathbf{mbb}(l)$, which is the smallest rectangle with sides parallel to the coordinate axes that contains $l$. Then, $p \in l$ if and only if $p \in \mathbf{mbb}(l)$. Finally $p \in \mathbf{mbb}(l)$ if and only if both the following inequations hold:

$$\mathbf{min}(x_q, x_r) \leq x_p \leq \mathbf{max}(x_q, x_r)$$
$$\mathbf{min}(y_q, y_r) \leq y_p \leq \mathbf{max}(y_q, y_r)$$

where $\mathbf{min}$ and $\mathbf{max}$ return the minimum and maximum of the arguments, respectively.

*Segment intersection*

There are two related operations for straight-line segment intersection: one Boolean, testing whether the two lines have at least one point in common; the other returning the point of intersection if it exists. Clearly, the result of the latter operation immediately gives a result for the former. Assume, as usual, that the segments are closed and therefore contain their end-points.

The intersection detection operation is a decision method, returning **true** if and only if the segments have at least one point in common. The operation may be defined in terms of previous operations, in particular the **side** operation. Figure 5.29 shows an example of how the **side** operation is used. Segments $ab$ and $cd$ will intersect if points $a$ and $b$ are on opposite sides of the line through $cd$ and points $c$ and $d$ are on opposite sides of the line through $ab$. More formally, $ab$ and $cd$ intersect if both the inequations below hold:

$$\mathbf{side}(a, b, c) \neq \mathbf{side}(a, b, d)$$
$$\mathbf{side}(c, d, a) \neq \mathbf{side}(c, d, b)$$

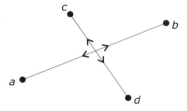

**Figure 5.29:**
Using the **side**
operation for
intersection
detection

The converse of this statement is not true, however. It is possible for $ab$, $cd$ to intersect when one or both of these inequations is false. This might occur when one of segments partially lies on the other. Thus, a robust intersection detection algorithm must also take account of the special cases where **side** takes the value zero. This may be accomplished by also checking the following possibilities:

$$\textbf{point\_on\_segment}(a, cd) = \textbf{true}$$
$$\textbf{point\_on\_segment}(b, cd) = \textbf{true}$$
$$\textbf{point\_on\_segment}(c, ab) = \textbf{true}$$
$$\textbf{point\_on\_segment}(d, ab) = \textbf{true}$$

Our work is not yet done. Even if two segments are shown to intersect by the above decision procedure, it still remains to determine their point of intersection. The calculation of a point of intersection of two line segments, if it exists, is at first sight a trivial exercise in coordinate geometry, reducing to the simultaneous solution of two simple linear equations. However, as with many geometric algorithms, special cases (e.g., vertical segments) and errors due to the discretization process provide several degenerate cases for which a robust algorithm must allow (see Douglas, 1990).

Suppose that we are given segments $l$ and $l'$, with respective parametric forms for their point sets:

$$l = \{\lambda p + (1 - \lambda)q | \lambda \in [0, 1]\}$$
$$l' = \{\lambda' p' + (1 - \lambda')q' | \lambda' \in [0, 1]\}$$

Let $r$ be a point of intersection of these sets. Then there are $\alpha, \beta \in [0, 1]$ such that:

$$\alpha p + (1 - \alpha)q = r = \beta p' + (1 - \beta)q'$$

which may be rearranged to give $\alpha(p - q) + \beta(q' - p') = q' - q$. This vector equation may be decomposed into two simultaneous equations for the two unknowns $\alpha$ and $\beta$, based on the $x$ and $y$ coordinates of the endpoints of $l$ and $l'$. Solving for $\alpha$ and $\beta$ yields:

$$\alpha = \frac{(x_q - x_{q'})(y_{p'} - y_{q'}) - (x_{p'} - x_{q'})(y_q - y_{q'})}{(x_{p'} - x_{q'})(y_p - y_q) - (x_p - x_q)(y_{p'} - y_{q'})}$$
$$\beta = \frac{(x_p - x_q)(y_{q'} - y_q) - (x_{q'} - x_q)(y_p - y_q)}{(x_{p'} - x_{q'})(y_p - y_q) - (x_p - x_q)(y_{p'} - y_{q'})}$$

Provided that $l$ and $l'$ are not parallel, the solutions for these two equations must then be checked to ensure that the line segments do intersect (i.e., $\alpha, \beta \in [0, 1]$). If so, the coordinate of the intersection point can be recovered by substituting $\alpha$ or $\beta$ into the corresponding original parametric line equation.

*Intersection, union, and overlay of polygons*

Algorithms for the intersection, union, and overlay of polygons are based on iterative application of the line segment intersection procedure, described in the previous section. For two simple polygons $P = [l_1, ..., l_m]$

and $P' = [l'_1, ..., l'_n]$, each given as a cycle of edge segments, the key is to find the points of intersection of line segments from $P$ with line segments from $P'$. In the general case, we must examine each pair $(l_i, l'_j)$ where $1 \leq i \leq m, 1 \leq j \leq n$ and determine the point of intersection, if it exists. (If the segments overlap or meet at their end-points, then special cases must also be constructed, not considered here.) There are $mn$ such pairs to examine and so the time complexity of intersection, union, and overlay algorithms is usually $O(mn)$.

Examples of individual algorithms are not described in detail here (but see the bibliographic notes for references to such algorithms). Figure 5.30 illustrates the core features of intersection, union, and overlay for a pair of intersecting line segments $ab$ and $cd$ within two polygons, $P$ and $P'$. Note that the topology of the polygons is important in deciding which line segments to discard and which to keep in the case of intersection and union. For intersection, the points $a$ and $d$ along with the point of intersection $x$ form part of the resulting structure, while $xc$ and $xb$ are discarded. For union, the points $c$, $b$, and $x$ are part of the resulting polygon, and $ax$ and $dx$ are discarded. For overlay, all the points are retained.

**Figure 5.30:** Intersection, union, and overlay of two polygons

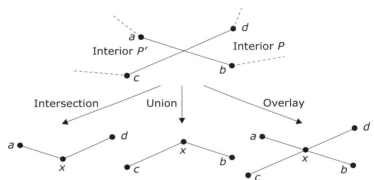

### 5.5.4   Triangulation algorithms

*Polygon triangulation algorithms*

This section describes two of the simpler non-Delaunay algorithms for triangulation, taking triangulation of a polygon as an exemplar. The triangulations will not introduce any extra interior vertices (Steiner points). The objective is to arrive at reasonably regular triangulations: that is, with few long, thin triangles.

greedy
triangulation

greedy algorithm

*Greedy triangulation algorithm*   Greedy triangulation has been briefly considered in section 5.4.4 and illustrated in Figure 5.20. In general, a *greedy algorithm* is a localized processing procedure that never revisits data that has already been processed. Algorithm 5.3 provides a procedure for the greedy triangulation of a simple polygon. The algorithm first sorts the diagonals of the polygon vertices into ascending order of length (line 2). For each diagonal $d_i$, beginning with the shortest, the algorithm tests whether $d_i$ properly intersects any diagonals already within the triangulation $T$ (i.e., intersects at a point in the interior of a diagonal) and whether $d_i$ is an internal diagonal (line 5). If not, $d_i$ is added to the triangulation $T$ (line 6).

---

**Input:** Simple polygon $P$ with $n$ vertices

1:   form the set $D$ of $m = \frac{n(n-3)}{2}$ diagonals of $P$

2:   sort $D$ into ascending order of length $d_1, ..., d_m$

3:   triangulation $T \leftarrow P$

4:   **for** $i \leftarrow 1$ to $m$ **do**

5:     **if** $d_i$ does not properly intersect any line segment in $T$ and $d_i$ is an internal diagonal of $P$ **then**

6:       $T \leftarrow T \cup d_i$

**Output:** Triangulation $T$ of $P$

**Algorithm 5.3:** Greedy polygon triangulation algorithm

---

The algorithm is termed "greedy" because it processes at each step the next available diagonal, never revisiting diagonals that have already been processed. However, it is also "greedy" for computational time. In Algorithm 5.3 all diagonals $m = n(n-3)/2$ will be considered. Each of these edges must be tested for intersection with the other edges in the triangulation. The triangulation will contain $n$ edges. Assuming a naive test for intersection of two edges, running in constant time, the algorithm is of worst-case complexity $O(n^3)$.

We can make some immediate improvements to the efficiency of this algorithm by keeping a count of the edges added to the triangulation. The number of such edges must be $n-3$, so when we arrive at this point the algorithm may terminate. This change will not improve the worst-case complexity of $O(n^3)$, although it will improve the average-case complexity. Consequently, while the greedy algorithm is useful for illustration purposes, as it is simple to understand, it is not a good choice for practical implementations due to its high computational complexity.

*The triangulation algorithm of Garey et al.*   The algorithm of Garey et al. (1978) is more suitable for the triangulation of simple polygons. With respect to time complexity, it improves on the greedy method above. The algorithm proceeds by first decomposing the polygon into monotone polygons and then triangulating each of the monotone polygons. Monotonicity was introduced in section 3.1.3. Monotone polygons have the property that their vertices are sorted with respect to a particular orientation, the *line of monotonicity*. As a result of this property, monotone polygons can be triangulated in linear time, $O(n)$, where $n$ is the number of vertices.

line of monotonicity

A procedure for triangulating a monotone polygon is outlined in Algorithm 5.4. Let $P$ be a monotone polygon with respect to the $y$-axis.

**Input:** Monotone polygon $P$ with $n$ vertices
1: merge the two monotone chains of $P$ into a list of vertices $V = [v_1, ..., v_n]$, sorted according to descending $y$-coordinate
2: define triangulation $T \leftarrow P$ and list $X = [x_1, ..., x_m]$ of nodes to be processed
3: initialize list $X \leftarrow [v_2, v_1]$
4: **for** $i \leftarrow 3$ to $n - 1$ **do**
5:     **if** $v_i$ and $x_1$ are in different monotone chains **then**
6:         add diagonals $v_i x_1, ..., v_i x_{m-1}$ to triangulation $T$
7:         update $X \leftarrow [v_i, v_{i-1}]$
8:     **else**
9:         **for** $j = 2$ to $m$ **do**
10:             **if** diagonal $v_i x_j$ lies entirely within $P$ **then**
11:                 add diagonal $v_i x_j$ to $T$
12:                 remove $x_{j-1}$ from $X$
13:         add $v_i$ to the beginning of $X$
14: add diagonals $v_n x_k$ to $T$ for $x_k \in X$ where $1 < k < m$
**Output:** Triangulation $T$ of monotone polygon $P$

**Algorithm 5.4:**
Triangulation
of a monotone
polygon

Merging the two monotone chains of $P$ to produce a list $V = [v_1, ..., v_n]$ can be achieved in linear time, because the two chains are already ordered. Processing each element of $V$ in turn, diagonals are added to the triangulation when either a pair of vertices are on opposite monotone chains or a pair of vertices are intervisible, in the sense that their diagonal lies entirely within the polygon $P$.

To illustrate Algorithm 5.4, consider the monotone polygon shown in Figure 5.31. The two monotone chains are $[a, b, c, d, e, f, g]$ and $[a, m, l, j, k, i, h, g]$. Sort-merging the chains gives the vertex list $V = [a, m, b, c, d, l, k, j, e, i, f, h, g]$. List $X$ is initialized to $[m, a]$. The first vertex to be processed is $b$. Since $b$ and $m$ are members of different monotone chains in $P$, we add the edge $mb$ to the triangulation. The following two vertices to be processed, $c$ and $d$, do not result in any new edges in the triangulation. Subsequently processing vertex $l$ results in three new edges $bl$, $cl$, and $dl$. Processing vertex $k$ results in no new edges, but subsequently processing $j$ leads to the edge $jl$ being added to the triangulation. Although $j$ and $l$ are members of the same monotone chain, they satisfy the condition that edge $jl$ lies completely within $P$ (line 10). The algorithm continues in this way until eventually 10 new edges are added and the algorithm terminates.

Having established that it is possible to triangulate a monotone polygon in linear time, the next question is how to partition a simple polygon into monotone polygons. The details of this procedure are beyond the scope of this text, but can be found in the references cited in the bibliographic notes. Briefly, let us assume that monotonicity is with respect to the $y$-axis and that no two different vertices of the polygon have the same

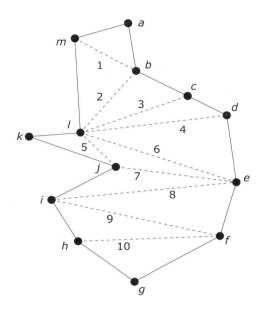

**Figure 5.31:** Triangulation of monotone polygon (edges numbered in order of processing)

$y$-coordinate. Then the initial polygon $P$ must be *regularized* to ensure that:

1. Edges intersect only at polygon vertices.

2. Apart from the highest vertex (with respect to the $y$-axis), every vertex is joined by at least one edge to a higher vertex.

3. Apart from the lowest vertex (with respect to the $y$-axis), every vertex is joined by at least one edge to a lower vertex.

It can be proved that this regularization process results in the polygon $P$ being partitioned into regions, each of which is monotone. Note that this is a different sense of the term "regularization" from that used in Chapter 3. After regularization, any new edges that lie outside the polygon must be removed. The time complexity of the regularization stage is $O(n \log n)$, where $n$ is the number of vertices of the polygon. As a result, the time complexity of this procedure dominates the algorithm, and the entire triangulation algorithm can be completed in $O(n \log n)$ time. Figure 5.32 illustrates the different stages of the Garey et al. triangulation algorithm.

*Delaunay triangulation algorithms*

Algorithms for the Delaunay triangulation and its dual Voronoi diagram are abundant in the literature. The worst-case time complexity of any Delaunay triangulation algorithm is at least $O(n \log n)$ (see "Faster triangulation algorithms" inset, on page 209). A useful technique common in many algorithms is the *divide-and-conquer* strategy, whereby a complex problem is recursively broken up into smaller and simpler problems,

divide-and-conquer

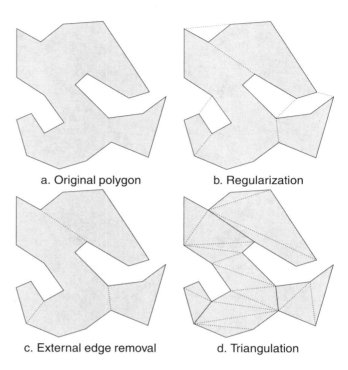

**Figure 5.32:**
Stages of
Garey et al.
algorithm for
$O(n \log n)$
simple polygon
triangulation

a. Original polygon                    b. Regularization

c. External edge removal               d. Triangulation

which are solved individually and then combined to give a complete
solution. Algorithm 5.5 provides an overview of the divide-and-conquer
method for Delaunay triangulation.

This description glosses over the crucial details of how the merger
is conducted (see bibliographic notes for more information). Briefly,
to merge two triangulations we form their respective convex hulls and
construct upper and lower common tangents. Figure 5.33 shows the merging process by means of an example. Starting from the lower common
tangent, the merging process combines the triangulations, adding and
deleting edges as appropriate. The merging is a little like "stitching" the
two triangulations together. In Figure 5.33b, one edge from the right-
hand partial triangulation is deleted and eight edges are added. The time
complexity of the merger operation is linear, $O(n)$. The basic partial triangulations are constructed only from sets of two or three edges (Algorithm
5.5, lines 6–9) so this stage can be completed in constant time, $O(1)$. As
a result, the sorting stage of the algorithm dominates the time complexity,
so the whole algorithm runs in $O(n \log n)$ time.

As noted in section 5.4.3, the Voronoi diagram is the dual of the
Delaunay triangulation. Consequently, algorithms for computing the Delaunay triangulation can also be used to compute Voronoi diagrams, and
vice versa. Figure 5.34 illustrates one method for computing a Voronoi
diagram incrementally. The general idea is to construct the Voronoi diagram by modifying it as new vertices from the vertex set are considered.
At each stage, the newly introduced vertex is shown unfilled in Figure

**Input:** $n$ points $p_1, ..., p_n$ in the Euclidean plane
1: sort the $n$ input points into ascending order of $x$-coordinates, with ties sorted by $y$-coordinate
2: divide the points into two roughly equal halves $L$ and $R$
3: **if** $|L| > 3$ **then**
4:   recursively apply triangulation algorithm on $L$ to create $T(L)$; similarly if $|R| > 3$ recursively create $T(R)$
5: **else**
6:   **if** $|L| = 2$ (i.e., $L = \{l_1, l_2\}$) **then**
7:     create $T(L)$ containing edge $l_1 l_2$; similarly for $T(R)$
8:   **if** $|L| = 3$ (i.e., $L = \{l_1, l_2, l_3\}$) **then**
9:     create $T(L)$ containing triangle $l_1 l_2 l_3$; similarly for $T(R)$
10: merge $T(L)$ and $T(R)$ by triangulating $T(L) \cup T(R)$
**Output:** Delaunay triangulation $T$ of points $p_1, ..., p_n$

**Algorithm 5.5:**
Divide-and-conquer Delaunay triangulation

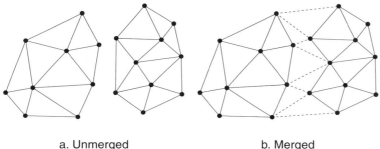

a. Unmerged          b. Merged

**Figure 5.33:**
Merging two Delaunay triangulations

5.34. The polygon in which the new vertex lies is calculated, and a trail (shown as a dotted line) is made around the new vertex, creating a new proximal polygon. The other edges are then modified appropriately and the process continues until no more new vertices are available.

*Conclusions*

Computational geometry has traditionally been concerned with measures of efficiency derived from theory—especially complexity theory. As usual, theory and practice are not always in accord. Methods provided by computational geometers for computing triangulations that are theoretically efficient may not be practical to implement. However, computational geometry provides a basis from which practical approaches may be constructed and evaluated.

### 5.6   VECTORIZATION AND RASTERIZATION

Manually converting data from raster to vector format (vectorization), perhaps using digitization, is an expensive and labor-intensive process. The conversion process is also difficult to fully automate because the

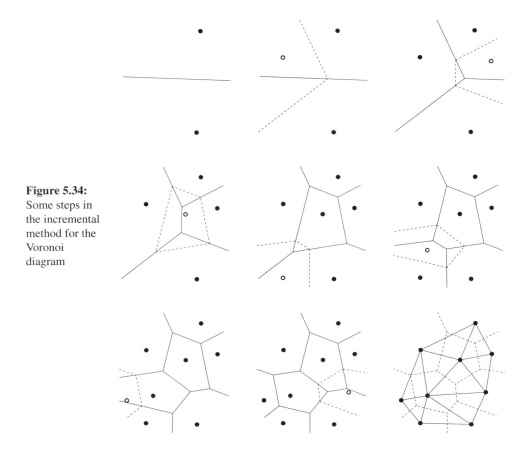

**Figure 5.34:**
Some steps in
the incremental
method for the
Voronoi
diagram

structure of spatial objects is not explicitly represented in the raster data format. Consequently, most vectorization systems are semi-automatic, requiring a human operator to guide an automatic system, checking the reliability of the vectorization and providing assistance when the automatic system gets stuck. The series of steps in vectorization may be summarized as:

1. Form a binary image from a raster, a process known as *thresholding*. For *simple thresholding*, a fixed value is chosen for the threshold value throughout the raster. In *adaptive thresholding*, the value varies according to the local conditions in part of the image. Thus in a generally dark area, it might be necessary to choose a different value to a generally light area.

2. Remove random noise, usually in the form of speckles, termed *smoothing*.

3. Thin lines so that they are one pixel in width. An example *thinning* algorithm is given below. Other thinning algorithms make use of the medial axis of a region.

thresholding

smoothing

thinning

**Faster triangulation algorithms** *The history of triangulations is interesting. We have seen that a greedy triangulation has a time complexity of $O(n^3)$. Triangulation algorithms of time complexity $O(n^2)$ date back to Lennes (1911). Garey et al. published their $O(n \log n)$-time algorithm in 1978, an achievement of the newborn field of computational geometry. Although a lower bound on the time complexity of a general planar triangulation was known to be $O(n \log n)$-time, it was not known whether this still held for triangulations of simple polygons or whether an even lower bound was possible. It was not until nearly 10 years later that Tarjan and Wyk (1988) published an $O(n \log \log n)$-time algorithm. Of course, the best that can be hoped for is a linear time algorithm, and this was actually achieved by Chazelle (1991), thus providing a climax to years of intense research effort. A concise introduction to Chazelle's remarkable algorithm can be found in O'Rourke (1998, p56). The worst-case Delaunay triangulations algorithm is limited by the vertex sorting algorithm, which requires at least $O(n \log n)$-time (Aho et al., 1974). In fact, optimal $O(n \log n)$ Delaunay algorithms have been achieved (Shamos and Hoey, 1975; Lee and Schachter, 1980). Despite a worst-case time complexity of $O(n^2)$, an incremental method enhanced by an efficient data structure is often more useful in practice (Ohya et al., 1984a,b), because it is conceptually simple and has linear average time complexity. Dwyer (1987) published a variant of the divide-and-conquer approach that led to an algorithm with optimal worst-case time complexity and average time complexity of $O(n \log \log n)$.*

4. Transform the thinned raster image into a collection of chains of pixels each representing an arc, termed *chain coding*.

chain coding

5. Transform each chain of pixels into a sequence of vectors. In this final stage there is a balance to be struck between line smoothness and conversion accuracy. Vectors that follow the arc chains accurately will often be jagged, exhibiting a "staircase" effect. Smoothing vectors will lead to greater deviations from the original raster.

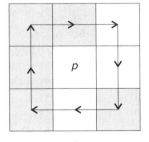

**Figure 5.35:** Pixel neighborhood

A commonly used erosion algorithm for thinning originated with Zhang and Suen (1984). Assume that the binary raster has 0 to represent white and 1 to represent black and that we want to thin the black shapes (gray, in the case of our diagrams). For each pixel position $p$, let $N(p)$ be an integer that is the sum of the values of the eight neighbors of $p$. Also, $p_N, p_S, p_E, p_W$ denote the values at pixels positioned above, below, right, and left, respectively, of $p$. Finally, let $T(p)$ be a count of the number of transitions from 0 to 1, visiting the immediate neighbors of $p$ in rotation. In Figure 5.35, $N(p) = 5$, $p_N = p_W = 1$, $p_S = p_E = 0$, and $T(p) = 2$. The Zhang-Suen algorithm is described in Algorithm 5.6, and an example showing the successive stages of an erosion is given in Figure 5.36.

---

**Input:** $m \times n$ binary raster (0 = white, 1 = black)

1: **repeat**
2:     **for all** points $p$ in the raster **do**
3:         **if** $2 \leq N(p) \leq 6$ and $T(p) = 1$ and $p_N.p_S.p_E = 0$ and $p_W.p_E.p_S = 0$ **then** mark $p$
4:     **if** there are no marked points **then** halt
5:     **else** set all marked points to value 0
6:     **for all** points $p$ in the raster **do**
7:         **if** $2 \leq N(p) \leq 6$ and $T(p) = 1$ and $p_N.p_S.p_W = 0$ and $p_W.p_E.p_N = 0$ **then** mark $p$
8:     set all marked points to value 0
9: **until** there are no marked points

**Output:** $m \times n$ thinned binary raster (black is thinned)

---

**Algorithm 5.6:** Zhang-Suen erosion algorithm for raster thinning

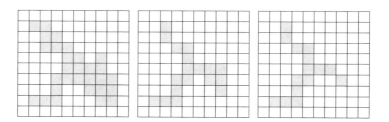

**Figure 5.36:** Zhang-Suen erosion example

Chains are formed from sequences of pixels in a thinned raster image. For each pixel, we need to determine whether it is on a line segment, the end-point of a line segment, or forms a point of intersection between segments. A chain-coding algorithm will find a pixel that constitutes the end-point of a strand, and then follow the pixels along the line, stopping at an end-point or an intersection point. Thus a sequence of pixels, the *chain*, is generated. The algorithm continues until all pixels in the raster image have been processed. The algorithm can be refined a little to handle loops (with no end-points).

chain

The final stage in the vectorization process is vector reduction, the conversion of each chain to a set of vectors. A long, "wiggly" line will of course need more vectors than a straighter line, depending upon the level of accuracy required. The Douglas-Peucker algorithm (section 5.2.3) is often used for the vector-reduction process.

The converse process of vector-raster conversion (rasterization) is a basic graphics operation in most computer systems. Raster data is often required for display purposes, for example, in the pixels of a display screen (VDU). This topic is known in the graphics literature as *scan conversion*. However, aside from display, rasterization is infrequently

scan conversion

used in a GIS, and so not considered further here (but see bibliographic notes).

## 5.7  NETWORK REPRESENTATION AND ALGORITHMS

Network models are fundamental to many GIS applications, for example, in transport applications. As discussed in Chapter 3, the fundamental structure for modeling a network is the graph. This section explores the representation of networks in a computer and goes on to consider algorithms for some of the most common operations on networks.

### 5.7.1  Network representation

A straightforward approach to representing a network for computational purposes using a graph is to represent the network as a set of node pairs, each pair representing an edge of the network. If multiple edges between two nodes are allowed, then a multiset will be required. If the network is directed, then the order of the nodes in the node pair representing an edge will be significant.

Another approach is to represent the graph as a square $n$ by $n$ *adjacency matrix*, in which each row and column is labeled by a node. A "1" is placed in a matrix cell if and only if the nodes corresponding to the appropriate row and column are connected by an edge, otherwise a "0" is placed in the cell. If multiple edges between two nodes are allowed, then the number in the cell can represent the number of edges. If the network is directed, then the matrix may not be symmetric.

adjacency matrix

Often it is important to label the edges. For example, in a traffic flow application where the nodes represent cities and the edges represent flows between cities, the edges may be labeled with distances or flow rates. Edge labels may be accommodated by adding an extra field to the node pair in the edge set representation, and for the adjacency matrix representation, placing the label in the matrix. Figure 5.37 shows a hypothetical example of a network of average travel times (in minutes) for tram routes between places of interest in the Potteries.

The network in Figure 5.37 may be represented as the set of labeled edges:

$$\{(ab, 20), (ag, 15), (bc, 8), (bd, 9), (cd, 6),$$
$$(ce, 15), (ch, 10), (de, 7), (ef, 22), (eg, 18)\}$$

In this case, each edge has associated with it an appropriate travel time. The adjacency matrix representing the Potteries network is given in Table 5.6. A non-zero entry in the matrix indicates the existence of an edge between the appropriate row and column nodes and gives the average tram travel time. A zero entry indicates no edge between the appropriate row and column nodes.

Using the set of labeled edges to represent the graph is highly efficient in terms of storage, but much less efficient than the adjacency matrix

**Figure 5.37:**
Network of
hypothetical
tram routes
between some
Potteries
locations,
labeled by
average travel
times

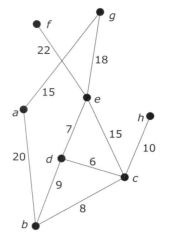

$a$. Newcastle Museum
$b$. Trentham Gardens
$c$. Beswick Pottery
$d$. Minton Pottery
$e$. City Museum
$f$. Westport Lake
$g$. Ford Green Hall
$h$. Park Hall Country Park

**Table 5.6:**
Adjacency
matrix for the
network in
Figure 5.37

|   | $a$ | $b$ | $c$ | $d$ | $e$ | $f$ | $g$ | $h$ |
|---|---|---|---|---|---|---|---|---|
| $a$ | 0 | 20 | 0 | 0 | 0 | 0 | 15 | 0 |
| $b$ | 20 | 0 | 8 | 9 | 0 | 0 | 0 | 0 |
| $c$ | 0 | 8 | 0 | 6 | 15 | 0 | 0 | 10 |
| $d$ | 0 | 9 | 6 | 0 | 7 | 0 | 0 | 0 |
| $e$ | 0 | 0 | 15 | 7 | 0 | 22 | 18 | 0 |
| $f$ | 0 | 0 | 0 | 0 | 22 | 0 | 0 | 0 |
| $g$ | 15 | 0 | 0 | 0 | 18 | 0 | 0 | 0 |
| $h$ | 0 | 0 | 10 | 0 | 0 | 0 | 0 | 0 |

for computing most basic network operations. The adjacency matrix in
Table 5.6 contains repeated and redundant information that is not present
in the labeled-edge representation. For example, in the adjacency matrix
in Table 5.6 both edges $ab$ and $ba$ are stored even though the graph is
undirected. Further, a zero must be stored in the matrix where there exists
no edge between two vertices (for example, between vertices $a$ and $c$). On
the other hand, deciding whether a particular edge exists within the graph
using the adjacency matrix can be computed in constant time (by simply
checking if there is a non-zero entry in the corresponding matrix cell). The
same operation requires linear time for the labeled-edge representation
(because we may need to search through the entire set of labeled edges).

adjacency list

In fact, computational geometry algorithms often use a close relation
to the adjacency matrix, the *adjacency list*. For each vertex, the adjacency
list stores only those edges that are incident with that vertex, effectively
eliminating all the 0's from the adjacency matrix. Consequently, the
adjacency list achieves a good balance between storage efficiency and
computational efficiency, especially for relatively sparse graphs (where
the corresponding adjacency matrix contains many zeros). Table 5.7
shows the adjacency list for the network in Figure 5.37.

| | |
|---|---|
| $a$ | $(b, 20), (g, 15)$ |
| $b$ | $(a, 20), (c, 8), (d, 9)$ |
| $c$ | $(b, 8), (d, 6), (e, 15), (h, 10)$ |
| $d$ | $(b, 9), (c, 6), (e, 7)$ |
| $e$ | $(c, 15), (d, 7), (f, 22), (g, 18)$ |
| $f$ | $(e, 22)$ |
| $g$ | $(a, 15), (e, 18)$ |
| $h$ | $(c, 10)$ |

**Table 5.7:**
Adjacency list
for the network
in Figure 5.37

### 5.7.2  Depth-first and breadth-first traversals

A fundamental operation on any connected network is the systematic traversal of the nodes within it. Such traversals have many applications, for example, searching a problem space (future possible positions in a chess game) or traversing a cable network to locate the source of a fault. There are two main alternative approaches: *depth-first* or *breadth-first* traversal. We may either traverse the network by probing ever deeper and only coming up when all alternatives below are exhausted (depth-first), or traverse the network by examining all nodes connected to a given node before going deeper (breadth-first). A single breadth- or depth-first traversal will visit every node in the network if and only if the underlying graph is connected (there exists a path from every node to every other node in the network).

depth-first traversal

breadth-first traversal

Algorithm 5.7 gives the procedure for a breadth-first traversal of a graph. To illustrate the breadth-first traversal, Figure 5.38 gives the series of iterations for a breadth-first traversal of the tram network in Figure 5.37. Starting with node $b$, each of the adjacent nodes $a$, $c$, $d$ is visited before the search proceeds to the next level. The set $V$ in Algorithm 5.7 is used to keep track of the nodes that have been visited by the algorithm. The tree on the right-hand side of Figure 5.38 shows the traversal order diagrammatically.

**Input:** Adjacency matrix $M$, starting node $s$
1: queue $Q \leftarrow [s]$, visited set $V \leftarrow \varnothing$
2: **while** $Q$ is not empty **do**
3:     remove the first node $x$ from $Q$
4:     add $x$ to $V$
5:     **for** each node $y \in M$ adjacent to $x$ **do**
6:         **if** $y \notin V$ and $y \notin Q$ **then** add $y$ to the *end* of $Q$

**Algorithm 5.7:**
Breadth-first
network
traversal

Algorithm 5.7 uses a special type of list to store the active nodes waiting to be processed, called a *queue* ($Q$, line 1). In a queue, the first elements to be added to the list are also first to be removed. A queue is often referred to as a *FIFO* structure, meaning *first-in-first-out*. The complementary structure used in the depth-first traversal is the *stack*, in which the last elements to be added are removed first, termed *LIFO* or *last-in-first-out*.

queue

FIFO

stack

LIFO

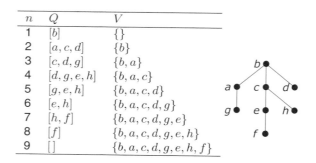

**Figure 5.38:**
Iterations for
breadth-first
traversal of the
tram network
in Figure 5.37

| $n$ | $Q$ | $V$ |
|-----|-----|-----|
| 1 | $[b]$ | $\{\}$ |
| 2 | $[a, c, d]$ | $\{b\}$ |
| 3 | $[c, d, g]$ | $\{b, a\}$ |
| 4 | $[d, g, e, h]$ | $\{b, a, c\}$ |
| 5 | $[g, e, h]$ | $\{b, a, c, d\}$ |
| 6 | $[e, h]$ | $\{b, a, c, d, g\}$ |
| 7 | $[h, f]$ | $\{b, a, c, d, g, e\}$ |
| 8 | $[f]$ | $\{b, a, c, d, g, e, h\}$ |
| 9 | $[]$ | $\{b, a, c, d, g, e, h, f\}$ |

Algorithm 5.8 gives the procedure for a depth-first traversal. Algorithm 5.8 is almost identical to Algorithm 5.7. The key difference is that Algorithm 5.8 uses a stack where Algorithm 5.7 uses a queue. As a consequence, in line 6 of Algorithm 5.8 the adjacent nodes are added to the beginning of the stack, ensuring the last nodes added to the list are the first to be processed.

**Algorithm 5.8:**
Depth-first
network
traversal

**Input:** Adjacency matrix $M$, starting node $s$
1: stack $S \leftarrow [s]$, visited set $V \leftarrow \varnothing$
2: **while** $S$ is not empty **do**
3:     remove the first node $x$ from $S$
4:     add $x$ to $V$
5:     **for** each node $y \in M$ adjacent to $x$ **do**
6:         **if** $y \notin V$ and $y \notin S$ **then** add $y$ to the *beginning* of $S$

Figure 5.39 illustrates the series of iterations for the depth-first traversal of the same tram network in Figure 5.37, again starting with node $b$. The tree on the right-hand side of Figure 5.39 illustrates the way the depth-first traversal follows each path until exhausted, before backtracking to search the next path.

**Figure 5.39:**
Iterations for
depth-first
traversal of the
tram network
in Figure 5.37

| $n$ | $Q$ | $V$ |
|-----|-----|-----|
| 1 | $[b]$ | $\{\}$ |
| 2 | $[a, c, d]$ | $\{b\}$ |
| 3 | $[g, c, d]$ | $\{b, a\}$ |
| 4 | $[e, c, d]$ | $\{b, a, g\}$ |
| 5 | $[f, c, d]$ | $\{b, a, g, e\}$ |
| 6 | $[c, d]$ | $\{b, a, g, e, f\}$ |
| 7 | $[h, d]$ | $\{b, a, g, e, f, c\}$ |
| 8 | $[d]$ | $\{b, a, g, e, f, c, h\}$ |
| 9 | $[]$ | $\{b, a, g, e, f, c, h, d\}$ |

### 5.7.3  Shortest path

shortest path

Another fundamental network operation is the computation of the *shortest path* between nodes in a network. Shortest path algorithms are needed, for example, within in-car navigation systems to determine the most direct route, in terms of Euclidean distance, through a street network to some destination. With the appropriate data, a shortest path algorithm might may also be used to find, among other things, the shortest route in terms of travel time.

A well-known shortest path algorithm is *Dijkstra's algorithm* (Dijkstra, 1959), devised by Edsger Dijkstra, an influential pioneer in computer science. Dijkstra's algorithm, outlined in Algorithm 5.9, requires a weighted graph in which the edge weights are non-negative. This restriction is not normally problematic, because negative distances or travel times are not sensible in most applications (cf. discussion of metric spaces in section 3.5). In Algorithm 5.9, distances are represented as a weighting function $w$ from the set of edges $E$ to the positive reals $\mathbb{R}^+$. An additional target weighting function $t : N \rightarrow \mathbb{R}^+$ is used to store the minimum distances from the start node $s$ to each node in the graph.

Dijkstra's algorithm

---

**Input:** Undirected simple connected graph $G = (N, E)$, starting node $s \in N$, weighting function $w : E \rightarrow \mathbb{R}^+$, target weighting function $t : N \rightarrow \mathbb{R}^+$
1: initialize $t(n) \leftarrow \infty$ for all $n \in N$, visited node set $V \leftarrow \{s\}$
2: set $t(s) \leftarrow 0$
3: **for all** $n \in N$ such that edge $sn \in E$ **do**
4:     set $t(n) \leftarrow w(sn)$
5: **while** $N \neq V$ **do**
6:     find, by sorting, $n \in N\backslash V$ such that $t(n)$ is minimized
7:     add $n$ to $V$
8:     **for all** $m \in N\backslash V$ such that edge $nm \in E$ **do**
9:         $t(m) \leftarrow \min(t(m), t(n) + w(nm))$
**Output:** Graph weights $t : N \rightarrow \mathbb{R}^+$

**Algorithm 5.9:** Dijkstra's shortest path algorithm

---

Dijkstra's algorithm initializes the target weights $t$ to infinity (line 1), except for the starting node ($t(s) \leftarrow 0$, line 2) and the nodes adjacent to $s$ (lines 3–4). The algorithm then proceeds to traverse the entire graph from the start node, at each step sorting the unvisited nodes $N\backslash V$ into ascending order of target weight (line 6) and recalculating the minimum target weights $t$ (line 9).

The progress of Dijkstra's shortest path algorithm for the tram network in Figure 5.37 is shown in Figure 5.40. At the initial iteration (Figure 5.40a), all the target weights are set to infinity, except for the start node, $b$, and the nodes adjacent to $b$: $a$, $c$, and $d$. Since $c$ has the lowest weight of the unvisited nodes, $t(c) = 8$, $c$ is the next node to be visited (Figure 5.40b). The target weights for the adjacent nodes to $c$ ($h$ and $e$) are updated, $t(h) = 18$ and $t(e) = 23$. At the third iteration (Figure 5.40c) node $d$ is visited. At this iteration, the distance from $b$ to $e$ via $d$ is found to be less than $b$ to $e$ via $c$. As a result $t(e)$ is updated to 16. The next iteration (Figure 5.40d) visits $e$, generating new target weights for $f$ and $g$. The algorithm continues until all nodes are visited, although in this case no further changes to the target weights would occur. Algorithm 5.9 only stores target weights $t$, which give the length of the shortest path, but does not explicitly store the actual paths themselves. It requires only a simple modification of Algorithm 5.9 to additionally store this information.

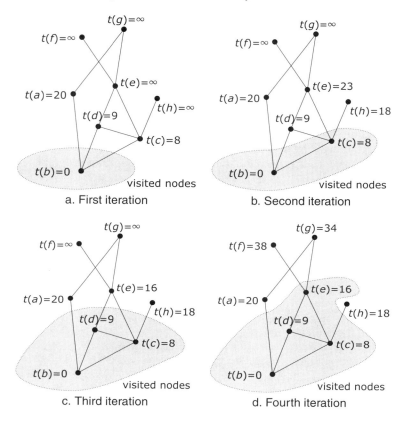

**Figure 5.40:** First four iterations of Dijkstra's algorithm for the tram network in Figure 5.37

Note that Dijkstra's algorithm is able to calculate the shortest path from a single node to *all* the other nodes in the network, termed a *single-source shortest path* algorithm. Dijkstra's algorithm is able to achieve this with a time complexity of $O(n^2)$, where $n$ is the number of nodes (because it visits each node once, and must recalculate target weights each time it visits a node). If we are interested in the shortest path from a single starting node to a single destination node, we could improve the average-case time complexity by halting the algorithm when the destination node is added to the visited list. However, this change does not affect the worst-case time complexity, as the destination node might be the very last node to be visited.

A close relative of Dijkstra's algorithm is the A* (pronounced "ay-star") algorithm. The A* algorithm is a *goal-directed* version of Dijkstra's algorithm, meaning that it aims to find the shortest path from the starting node to a particular destination (goal). A* uses a *heuristic* to achieve this: at each iteration it preferentially visits those nodes that are closest

single-source shortest path

A* algorithm

goal directed

**NP-completeness** *Many common search problems, like the traveling salesperson problem, belong to a class of problems known as NP-complete. NP-complete problems have the property that it is relatively easy to check whether a solution is correct, but very difficult to actually find the solution. For example, the NP-complete subset sum problem aims to find whether a subset of a finite set of integers sums to a given number. For example, even for the small set $S = \{-15, -3, 7, 11, 29, 34, 38\}$ deciding whether any subset of $S$ sums to 55 is not immediately obvious. It is, however, trivial to check that the subset $\{-15, 7, 29, 34\}$ does indeed sum to 55. Similarly, we may rephrase the traveling salesperson problem as a decision problem (a problem with a "yes" or "no" answer) by asking "For an edge-weighted graph is there a Hamiltonian circuit with total weight less than $k$?" (see section 3.4 for the definition of a Hamiltonian circuit). In general, the answer to this question requires exponential time to compute; checking whether a specific route is indeed a Hamiltonian circuit with total weight less that $k$ can be computed very rapidly. All known NP-complete problems can be solved in exponential time (they are intractable). However, no one has ever managed to prove that NP-complete problems cannot be solved in polynomial time. What makes NP-complete problems even more interesting is that all NP-complete problems are interchangeable, in the sense that if an algorithm were ever found for solving one NP-complete problem in polynomial time, the same solution could be applied to every other NP-complete problem. As a result, NP-completeness has been the subject of extensive research by the theoretical computer science community (e.g., Brookshear, 1989).*

to the destination node. To do this requires an *evaluation function* that gives a consistent underestimate of the remaining distance from each node to the destination. A* can be particularly useful for geospatial information, since the Euclidean distance between two points often forms a convenient evaluation function. For example, when traveling through a road network from node $n_1$ to $n_2$, the shortest possible distance is the Euclidean straight-line distance between $n_1$ and $n_2$ (assuming the Earth is locally planar). The actual distance we need to travel will always be at least as far as the Euclidean distance, because the twists and turns of a road network usually prevent us from traveling directly to our destination. In the example of our travel time network, in Figure 5.37, there is no such obvious evaluation function, so A* could not be used.

*evaluation function*

The A* algorithm is essentially the same as Dijkstra's algorithm, so the worst-case time complexity for A* is still $O(n^2)$. However, in practice A* offers considerable improvements on Dijkstra's algorithm for average-case time complexity when computing goal-directed shortest paths where a suitable evaluation function exists. There are occasions when the shortest path from *all* possible starting locations to *all* possible destinations is required, termed an *all-pairs shortest path* algorithm. Clearly, a simple way of achieving this is to iterate Dijkstra's algorithm using every node in the network in turn as the start node. As a result, the time complexity of an all pairs shortest path algorithm is usually $O(n^3)$, although some algorithms exist that are able to improve on this slightly.

*all-pairs shortest path*

### 5.7.4  Other network operations

In addition to the operations described so far in this section, there are
many other useful network operations. In this section we briefly introduce
just two. As mentioned above, computing the shortest path between two
nodes assumes these nodes are connected (i.e., there exists a path between
them). If the nodes are not connected, Dijkstra's algorithm will calculate
the shortest path between the two nodes as having an infinite distance, $\infty$.
It is, therefore, possible to determine connectivity between two nodes $n_1$
and $n_2$ by computing Dijkstra's algorithm for $n_1$, and checking whether
the shortest path distance associated with $n_2$ is infinite.

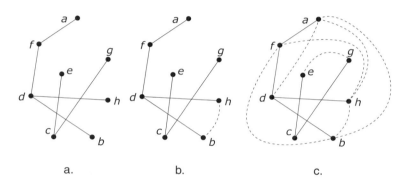

**Figure 5.41:**
Transitive
closure of a
disconnected
network

       a.                    b.                   c.

transitive closure      However, shortest path algorithms are not an efficient method of
testing whether two nodes are connected within a network. The *transitive
closure* operation augments the edge set of a network by placing an edge
between two nodes if they are connected by some path. Figure 5.41a
shows a disconnected network. Figure 5.41b shows the completion of the
triangle $bdh$. Figure 5.41c shows the full transitive closure of the network.
Deciding whether the two nodes $f$ and $h$, for example, are connected
is then simply a matter of searching through the edges of the transitive
closure of the network for edge $fh$.

traveling
salesperson
algorithm
     Finally, the *traveling salesperson algorithm* (originally mentioned
in section 1.2.2) computes the round-trip traversal of an edge-weighted
network, visiting all the nodes in such a way that the total weight for
the traversal is minimized. A naive solution, calculating all possible
routes and choosing the best, has exponential time complexity and so is
impractical for any but the smallest examples. Heuristic methods, such
as at each stage visiting the nearest unvisited node, allow good approximations in reasonable time. In fact, the traveling salesperson operation
NP-complete    is a member of a large class of problems termed *NP-complete* (see "NP-
Completeness" inset, on the preceding page).

## BIBLIOGRAPHIC NOTES

5.1 Computational geometry is sometimes said to have begun with the publication of Shamos (1975). The classic book in the field is Preparata and Shamos (1985). O'Rourke (1998) is a readable elementary text that introduces computational geometry in a practical way, using many examples and implementations in the C programming language. A beautifully presented alternative introductory text on computational geometry is de Berg et al. (2000).

5.2.1 Work on the discrete specification of the geometric domain is described by Güting and Schneider (1995).

5.2.2 A detailed description of the Greene-Yao algorithm may be found in Greene and Yao (1986).

5.3 More details on DCEL and other representations may be found in the computational geometry literature cited above, for example, O'Rourke (1998). Bryant and Singerman (1985) and David et al. (1993) describe extensions to DCEL representations and combinatorial maps capable of representing holes and islands.

5.4.3 The literature on triangulated irregular networks (TIN) is extensive. A seminal work is Peucker et al. (1978). Example GIS applications of TIN are discussed in Burrough and McDonnell (1998), Gold and Cormack (1987), Gold (1994), and Lee (1991). A summary and overview of polygonal triangulations may be found in Jayawardena and Worboys (1995). Delaunay triangulations and their dual Voronoi diagrams also have a very large literature. Voronoi diagrams are used in the work of the mathematicians Dirichlet (1850) and Voronoi (1908), although similar ideas appear even earlier. Delaunay triangulations appear in Voronoi (1908) as the dual of Voronoi diagrams. An extensive and scholarly work in this area is Okabe et al. (1992).

5.5.4 An ingenious plane sweep method for constructing a Voronoi diagram is given in Fortune (1987), summarized in O'Rourke (1998, p166). Incremental methods for Delaunay triangulations may be found in Lee and Schachter (1980) and Bowyer (1981). Tsai (1993) contains a survey of Delaunay triangulations.

5.6 Raster algorithms and algorithms related to rasterization and vectorization have a large literature within graphics and image processing that does not generally form part of the GIS corpus. A highly recommended and comprehensive introduction to graphics is provided by Foley et al. (1995). For more specialist research material on the automatic vectorization process, the reader could begin by consulting Musavi et al. (1988) and Flanagan et al. (1994).

5.7 Graph traversal and shortest path algorithms are covered in introductory texts on artificial intelligence, for example, Russell and Norvig (2002) and Luger and Stubblefield (1998).

# Structures and access methods

6

**Summary**

*This chapter is about the organization of data in computer storage to facilitate efficient retrieval. The chapter begins by surveying some basic data structures and* **index** *methods for general-purpose databases, and then focuses of efficient spatial data retrieval. Some of the more important spatial data structures introduced include the* **region quadtree**, **point quadtree**, *and* **2D-tree**. *The chapter concludes with issues related to storing data referenced to the sphere rather than the plane.*

In this chapter there is a further movement from high-level conceptual models toward the machine level. In the last chapter, the emphasis was upon computationally appropriate ways of representing and manipulating different kinds of spatial data. We now move on to consider storage questions, and in particular storage structures that allow acceptable *performance*. In this chapter, the performance of an information system is taken to be measured in terms of database size and response times to queries. Other aspects of performance, such as the suitability and usability of the interface, are considered in Chapter 8. The discussion here begins with an introduction to the main issues of performance for general-purpose databases and then takes up these issues in the special context of geospatial data. We will see that while the fundamental principles of good database indexing practice still apply, spatial data access presents its own special problems.

performance

## 6.1 GENERAL DATABASE STRUCTURES AND ACCESS METHODS

The typical organization of a database is as a collection of *files*, each containing a collection of *records*, stored on a set of disks. Throughout this chapter we assume disks, either magnetic or optical, are being used as secondary storage media. We noted the main physical characteristics of a computer disk in Chapter 1. The atomic unit of data held on a disk is *block* referred to as the disk *block*. The time taken to transfer a disk block to or from a disk has three components:

- the time taken for the mechanical movement of the disk heads across the disk to the correct track, termed *seek time*;

- the time taken for the disk to rotate to the correct position, termed *latency*; and

- the time required to transfer the block into the CPU, *CPU transfer time*.

*block*

*latency*

*CPU transfer time*

As noted in Chapter 1, seek time is the dominant factor determining performance in data retrieval from disks. Data structures in secondary storage that lessen the mechanical movement of the disk heads will therefore result in improved database performance. To minimize disk head movement, data is ideally placed in such a way that blocks which are often accessed together are close together on the disk. Lessening disk head movement will also be a consequence of the construction of appropriate indexes, so that unnecessary searches, maybe of entire files, are avoided.

Databases, by their very nature, are usually designed to be flexible and to respond well to a wide variety of queries. In general, queries are unknown in advance, although where predictable query patterns do exist they should be identified during the system development process. Another difficulty is caused if the database is highly dynamic, with its content changing considerably in the course of its lifetime. Therefore, the physical placement of data on the disk can provide only a partial solution to performance issues. Efficient indexes and access methods are also needed to achieve good performance.

### 6.1.1   File organization and access methods

Before presenting the main discussion, we introduce the fundamental concepts of file, record, and field.

field
- A (named) place for a data item in a record is termed a *field*. Note that this sense of "field" is unrelated to "field" in the sense of field-based models.

- A sequence of fields related to a single logical entity is termed a *record*.

record

file
key field
- A sequence of records, usually all of the same type, is organized into a *file*. A field that serves to identify each record within a file unambiguously is called a *key field*.

The terms field, record, and file in data organization are clearly analogous to attribute, tuple, and relation in database terminology. In file organization terms, a database is a collection of related files. Figure 6.1 illustrates the organization of fields, records, and files within an example weather database. Fields, such as the temperature field for Bangor, provide a place for data items. Individual data items, such as '82°F', are stored together in records, such as the weather report record for Bangor, Maine, on July 7, 2003. Sequences of related records, such as the weather for Bangor for 2003, are organized into files. Collections of related files, such as the weather report for all cities in Maine, form a database.

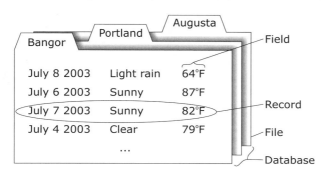

**Figure 6.1:** Data organization into field, record, file, and database

Files are physically placed upon a disk by assigning disk blocks to hold records. If the disk block is smaller than the record size, then records will be spread across blocks. Otherwise (and more usually) each block contains several records. The term *file organization* is used to refer to the physical organization of the records on the secondary storage, the manner in which blocks of records are linked, and the way new records are inserted into storage.

file organization

### 6.1.2   Unordered files and linear searches

In an *unordered* file organization, new records are inserted into the file in the next physical position on the disk, either in the last used disk block or in a new disk block. Insertion of a new record is therefore very efficient. However, the unordered file has no structure beyond entry order, so retrievals will require a search through each record of the file in sequence. If all that is required is that the file is accessed sequentially, for example, to print out lists of weather reports for a particular location, then an unordered file organization is acceptable. However, problems arise with direct access, where specific information is required from targeted records. For example, retrieving those days on which the temperature rose above 80°F requires that the value of the temperature field of each record be examined in turn. Such a search is called a *linear search*, or a *brute-force* approach, and is to be avoided if possible. For $n$ records a linear search may need to retrieve every record. A linear search therefore has linear time complexity, $O(n)$. We will see shortly how indexes can help to solve this problem.

unordered file organization

linear search

Another difficulty arises for a highly dynamic file, where as records are deleted "holes" are created. It is possible to modify unordered file organization to mark holes and insert new records into these positions, but this will slow the insertion operation.

### 6.1.3   Ordered (sequential) files and binary searches

In an ordered file organization, each record is inserted into the file in the order of values in one or more of its fields. For example, we may decide to order the Bangor weather report file by date. In this case the

ordering field

date field is called the *ordering field*. The values in the ordering field must of course be capable of being totally ordered (e.g., integers with the usual ordering, character strings with lexicographic ordering, dates with temporal ordering). The great advantage of an ordered file organization is

binary search

that *binary searches* on ordered fields become possible. The binary search algorithm is given in Algorithm 6.1. Note that the operation **div** forms the integer part of the quotient of two integers (thus 15 **div** 2 = 7).

---

**Input:** An ordered file with an ordering field, placed on $n$ disk blocks (labeled 1 to $n$), and a search value $V$

1:   $low \leftarrow 1$;

2:   $high \leftarrow n$;

3:   **while** $high \geq low$ **do**

4:     $mid \leftarrow (low + high)$ **div** 2

5:     read block $mid$ into memory

6:     **if** $V <$ value of ordering field in first record of block $mid$ **then**

**Algorithm 6.1:**     7:      $high \leftarrow mid\text{-}1$

Binary search    8:     **else**

9:      **if** $V >$ value of ordering field in last record of block $mid$ **then**

10:       $low \leftarrow mid\text{+}1$

11:      **else**

12:       linear search block $mid$ for records with value $V$ in their ordering field, possibly proceeding to next block(s), then halt

**Output:** Records from the file with value $V$ in their ordering field

---

The binary search algorithm takes advantage of the ordering in the file to successively chop the file in half until the targeted records are found and retrieved or no matching records are found. With a file utilizing $n$ disk blocks, the number of chops can be at most $\log_2(n)$. Thus the time complexity of the binary search method is logarithmic for retrieval of a single record—a huge improvement on linear search.

For our previous example, retrieving the weather record for a given date, suppose that the file is placed upon 1000 blocks. Then linear search would require on average 500 block accesses. If the records were in order of date, binary search would require approximately $\log_2(1000) = 10$ accesses. Note that if multiple records match the search condition, then the time complexity may no longer be logarithmic. In the worst case, where all records match the search condition, the time complexity is clearly linear. However, in all those cases where the number of records

> **Hashing** *Hashing is a form of file organization that has much in common with a simple index. A hash file is organized using a hash function. For each record, a hash function transforms the values of a field into the addresses of the disk blocks where each record is stored. The field upon which a hash function operates is referred to as a hash field. To search for records on a hash field, the hash function is applied to the search value, and the address of the disk block is calculated. As a result, only one disk block access is required. Taking a highly simplified example, suppose that we have 1000 blocks for a weather report file. We could hash on the date field by using a label for each block based on the day, month, and final two digits of the year. Thus the weather report for August 21st 2003 would be placed in a block labeled 210803. To retrieve the record for August 21st 2003 then simply involves using the hash function to find the block address for label 210803. Hashing is in principle a simple technique for file organization. However, complexities arise when we consider what happens when disk blocks overflow, or how to ensure an even distribution of records through the available disk blocks. Also, in order to choose the most suitable hash function, it helps to know in advance the number of disk blocks to be allocated.*

retrieved is small compared with the size of the data file, the complexity is approximately logarithmic.

Many retrievals are based on range search conditions, rather than equality conditions. For example, the search might involve finding all weather records for dates in June. Records close in the range will be close in the ordering, so a binary search will also effect significant performance improvement for range queries. It is straightforward to modify Algorithm 6.1 to handle range searching.

Although ordering a file makes a large difference to performance with searches on the ordering fields, retrievals on other fields are once again reduced to linear search. For a regularly updated file, insertion of a record can be expensive, as all succeeding records must be shunted along. The same problem arises with deletion, although we could just mark records that are deleted and clean the file up periodically.

### 6.1.4 Indexes

The preceding discussion introduced some simple file organization techniques. However, the purely physical organization of files on disks cannot solve all the performance problems that arise. Other more sophisticated file organization techniques, such as hashing (see "Hashing" inset, on this page), are able to offer some improvements. Most databases rely on *indexes* as the primary mechanism for efficient search and retrieval. An index is an auxiliary structure that is specifically designed to speed the retrieval of records. Indexes speed performance of retrieval at the expense of the extra space required for the index itself.

index

The concept of an index to a file is similar to the index to a book. The index for this book, for example, contains an ordered sequence of important terms used to reference pieces of text. Without an index, the book would need to be searched from cover to cover every time you needed to refer to a particular topic. An index is ordered, so it can be searched quickly (perhaps with a binary search). To find a topic in the main text, you may search for the item in the index (for example, "GIS," listed alphabetically) and retrieve a reference to the positions (addresses) in the main text where the item occurs (for example, page 2).

single-level index     A *single-level index* acts in just the same fashion as a book index, described above. It is an ordered file, each of whose records contain two fields:

index field     • An *index field* contains the ordered values of the indexing field in the data file.

pointer field     • A *pointer field* contains the addresses of disk blocks that have a particular index value.

In the case that the indexing field is a key field, then the pointer field of each index record will contain just a single pointer; otherwise an index field may require multiple pointers within a pointer field. Figure 6.2 shows a schematic for an index to student last names in a student data file physically ordered by student identification number.

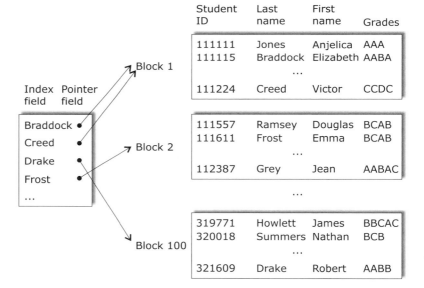

**Figure 6.2:** Student file indexed by last name

To retrieve a record from an indexed file, based upon a search condition on the indexing field, the index is searched. The index file is ordered, so the search can be binary. Having located the record in the index, the pointer indicates the location of the record in the data file. Thus, like an ordered file, an index provides an improvement from linear to logarithmic time complexity for searches. Unlike an ordered file, an index may operate on multiple fields. As we shall see in later sections, some indexes are also simple to update as records are inserted and deleted, unlike an ordered file.

A factor so far neglected in the discussion is the blocking of the index. For a large data file, the index file will itself be substantial and may need to be placed on the disk in many blocks. The search time on the index will be proportional to the logarithm of the number of these blocks. We can cut down the search time even further by allowing the index to be itself indexed, and so on recursively. This is the principle of the *multi-level index*.

*multi-level index*

### B-trees

As index structures become more complex, maintaining the index structure through changes and updates becomes more difficult. When a record is deleted or added to a data file the index must also be modified. If the index is multi-level, then a modification may have ramifications throughout its levels. A *B-tree* is an index structure that handles these modifications in a particularly elegant fashion. B-trees remain *balanced* in a dynamic setting, in the sense that the various branches of a balanced tree remain of equal length as the tree grows and shrinks.

B-tree

Suppose that we have a set of data records, indexed by an integer field (any type that is capable of being linearly ordered can be used as a B-tree index domain). Like any tree, a B-tree contains two sorts of nodes: leaf nodes and internal (non-leaf) nodes. Each index field contains a pointer to the record that is indexed; for simplicity these pointers are not shown in the figures below. Each node of a B-tree contains a list of index fields (with pointers to data). Within each node, the index fields are in strictly increasing order. Additionally, internal nodes contain a list of pointers to immediate descendants. Figure 6.3 shows an example of an internal node.

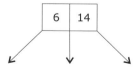

**Figure 6.3:** Internal node of B-tree

In our example, the B-tree is said to be of *fan-out ratio* 3, meaning that each non-leaf node may contain at most two index fields and have at most three immediate descendants. The basic property of a B-tree is that the value of the index field for all descendants is within the range set by the index fields of the ancestor node. In our example, the left-most pointer will point to nodes containing index values less than 6; the second pointer will point to nodes containing index values between 6 and 14; and the right-most pointer will point to nodes greater than 14. Figure 6.4 shows a B-tree with fan-out ratio 3.

fan-out ratio

The B-tree operations of search, insert, and delete are described below. Searching a B-tree, based upon a value of the indexing field, is the most straightforward operation, because no restructuring of the tree is required. The other operations restructure the tree to keep it balanced.

**Figure 6.4:**
Two-level
B-tree with
fan-out ratio 3

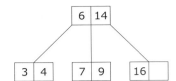

*Search*: The search begins at the root node. Depending upon the value of the indexing field, a route will be traced to one of the root's descendants, and so on down the tree until an exact match is found or the bottom of the tree is reached. Upon exact match, the value of the pointer field enables the required records to be retrieved from the data file.

*Insert*: When a record is inserted into the data file, the B-tree is searched to determine where its index record is placed. If the record index can be inserted into an empty space in the B-tree, no restructuring is required. For example, a new record with index value 19 can be inserted into the B-tree in Figure 6.4 simply by adding 19 to the empty space in the existing 16 node. If inserting the record index leads to an *overflow* for a non-root node, then that node is split into two equal parts, with the middle value promoted to the parent level. If there is an overflow in the root node, that node is split into two equals parts, each demoted one level, with the middle value retained as the new root node. For example, in Figure 6.5 inserting index value 10 into the B-tree leads to the 7-9 node being split, with 9 promoted to the next level. This in turn leads to the root node being split, with 6 and 14 being demoted one level.

overflow

*Delete*: A similar algorithm to insertion ensures a balanced tree on restructuring the tree following a deletion. If the deletion of the appropriate index value from a node $n$ leaves that node less than half-full (an *underflow*), then a process of merging nodes will take place, propagated from $n$. This process is inverse to the insertion restructuring and will not be described in detail. Of course, if $n$ is still at least half-full after deletion, then no merging is required.

underflow

The B-tree structure has the following properties, which are maintained as the structure is dynamically modified with insertion and deletion of records from the data file. Let $n$ be the number of records in the data file to be indexed.

- A B-tree is completely balanced: at any stage in its evolution, the path from root to leaf is of constant depth regardless of the leaf. Search time is bounded by the length of this path and therefore has time complexity $O(\log n)$.

- Insertion and deletion of a record from the file (and possible restructuring of the index) have time complexity $O(\log n)$.

- Each node is guaranteed to be at least half-full (or almost half-full for B-trees with an even fan-out ratio) at all stages of the tree's evolution.

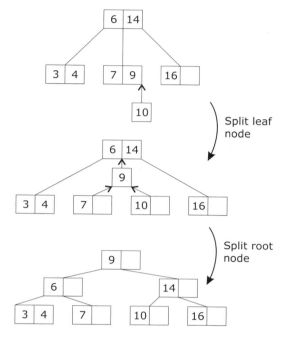

Split leaf node

Split root node

**Figure 6.5:**
The B-tree of Figure 6.4 during and following an insertion

The B-tree provides a balanced multi-level index in a dynamic setting and constitutes one of the first substantive achievements for research into general-purpose data structures. B-trees allow indexes to be placed in a dynamic environment and achieve significant performance improvements. B-trees, or their close relations, are at the core of most modern database management systems. In fact, the $B^+$-*tree* is the refinement most often used in actual implementations. In a $B^+$-tree, pointers to records are stored only at the leaf nodes. Thus the leaf nodes must contain every index value and will be more numerous than for the B-tree; on the other hand, the non-leaf nodes will have a simpler structure and will occupy less storage space.

$B^+$-tree

## 6.2   FROM ONE TO TWO DIMENSIONS

The discussion so far has applied to any general-purpose database. Files in such a database are multidimensional, each entity having several independent attributes. The student file used as an example has dimensions corresponding to at least student identification number, first name, last name, and grades. While there may be dependencies between the attributes in some cases, usually attributes are independent. When dependencies do occur, they are ideally eliminated by normalization or are of a more statistical nature (e.g., correlations between job descriptions and salaries in an employee database).

With spatial data, the spatial dimensions are orthogonal, but there is a dependency between them expressed in Euclidean space by the Euclidean metric. To illustrate the point, consider the data on the positions of key points in the Potteries, as shown in tabular form in Table 6.1 and on a grid in Figure 6.6.

**Table 6.1:**
Potteries places
of interest

| ID | Site | East | North |
|---|---|---|---|
| 1 | Newcastle Museum | 14 | 58 |
| 2 | Waterworld | 31 | 65 |
| 3 | Gladstone Pottery Museum | 74 | 23 |
| 4 | Trentham Gardens | 20 | 00 |
| 5 | New Victoria Theater | 18 | 55 |
| 6 | Beswick Pottery | 66 | 25 |
| 7 | Coalport Pottery | 54 | 36 |
| 8 | Spode Pottery | 37 | 43 |
| 9 | Minton Pottery | 36 | 39 |
| 10 | Royal Doulton Pottery | 31 | 87 |
| 11 | City Museum | 41 | 62 |
| 12 | Westport Lake | 17 | 92 |
| 13 | Ford Green Hall | 53 | 99 |
| 14 | Park Hall Country Park | 86 | 44 |

**Figure 6.6:**
Location of
some places of
interest in the
Potteries
(labels refer to
Table 6.1)

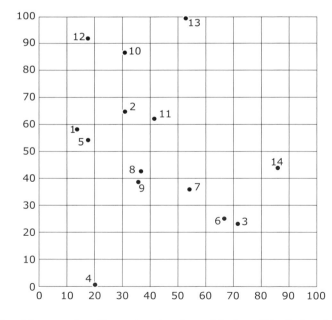

Consider now the effect of retrievals on this data. We distinguish two types of queries on the planar points: point and range queries.

- A *point query* retrieves all records with spatial references located at a given point coordinate.

  point query

- A *range query* retrieves all records with spatial references located within a given range. The range may be of any shape, although typically a rectangular area will be specified by the coordinates of two opposite vertices or a disk identified by center and radius.

  range query

Consider the following example queries:

*Query One (non-spatial query)*:  Retrieve the point location of Trentham Gardens.

*Query Two (spatial point query)*:  Retrieve any site at location $(37, 43)$.

*Query Three (spatial range query)*:  Retrieve any site in the rectangle with southwest and northeast vertices $(20, 20)$, $(40, 50)$, respectively.

The first query, although involving spatial location in the result, is concerned only with searching the file using a non-spatial field. Consequently, the general-purpose access methods discussed above would be appropriate for satisfying this query. The second query is a retrieval using a spatial search condition. Without any index or file ordering, a linear time search would be necessary, such as in Algorithm 6.2.

---

1:  open Potteries file
2:  **while** there are records to examine **do**
3:      get the next record
4:      **if** the value of the first coordinate field is 37 **then**
5:          **if** the value of the second coordinate field is 43 **then**
6:              Retrieve site name from record

**Algorithm 6.2:**
Point query
linear search

---

For the third range query, a similar linear-time search may be adopted, as in Algorithm 6.3.

---

1:  open Potteries file
2:  **while** there are records to examine **do**
3:      get the next record
4:      **if** the value of the first coordinate field is in the range $[20, 40]$ **then**
5:          **if** the value of the second coordinate field is in the range $[20, 50]$ **then**
6:              retrieve site name from record

**Algorithm 6.3:**
Range query
linear search

---

To improve on these linear time queries, we might build two indexes for the two spatial coordinate fields, using the general-purpose database approach. The indexes are shown (with the site names indicating the pointer fields) in Table 6.2, and the two sequences of points are shown as paths in Figure 6.7. For the second query, we conduct a binary search of the east index to locate records that have a first coordinate value of

37. We then go to the full records to check if the second coordinates have value 43, and retrieve those records for which this is the case. For the third query, we could do a range [20, 40] search on the east index, resulting in a list of pointers to the data file. For each pointer in the list, the data file record may be accessed and the north value checked to be in the range [20, 50], in which case the site name is retrieved.

**Table 6.2:**
East and North indexes to the Potteries places of interest

| East | Site | North | Site |
|------|------|-------|------|
| 14 | Newcastle Museum | 00 | Trentham Gardens |
| 17 | Westport Lake | 23 | Gladstone Pottery Msm |
| 18 | New Victoria Theater | 25 | Beswick Pottery |
| 20 | Trentham Gardens | 36 | Coalport Pottery |
| 31 | Waterworld | 39 | Minton Pottery |
| 31 | Royal Doulton Pottery | 43 | Spode Pottery |
| 36 | Minton Pottery | 44 | Park Hall Country Park |
| 37 | Spode Pottery | 55 | New Victoria Theater |
| 41 | City Museum | 58 | Newcastle Museum |
| 53 | Ford Green Hall | 62 | City Museum |
| 54 | Coalport Pottery | 65 | Waterworld |
| 66 | Beswick Pottery | 87 | Royal Doulton Pottery |
| 74 | Gladstone Pottery Msm | 92 | Westport Lake |
| 86 | Park Hall Country Park | 99 | Ford Green Hall |

**Figure 6.7:**
The two indexes shown as paths through the places of interest in the Potteries (labels refer to Table 6.1)

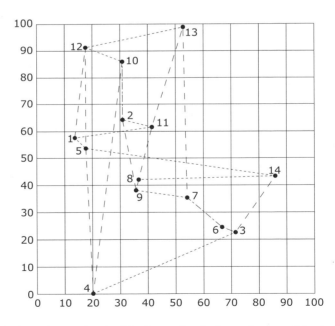

The key observation in this discussion is that only *one* of the indexes is used in these retrievals. What we would like to do is to construct an index that takes advantage of ordering in two dimensions. Particularly for range searches, we would like to order the records in such a way that records that are near each other in space are near each other in the index.

### 6.2.1 Two-dimensional orderings

The principal problem confronting designers of data structures for two-dimensional and multidimensional Euclidean space is that computer storage is essentially one dimensional. Even though a disk of course physically exists in three-dimensional space, the computational model of a disk is one dimensional (as can be seen by noting its one-dimensional addressing scheme). Thus we have to unravel the higher-dimensional space so as to minimize the distortion, in a similar way to unraveling a piece of two-dimensional knitting into its one-dimensional yarn. This section examines some of the ingenious techniques that have been proposed for performing the unraveling process in the case of two dimensions.

More formally, let us take a finite portion of the Euclidean plane $P$. Without loss of generality, let $S$ be the real interval $[0, 1]$ where the ordering is the usual ordering of real numbers, and let $P$ be the unit square $[0, 1] \times [0, 1]$. We are looking for a bijective function $f : P \to S$ that has the property that, for all $x, y \in P$, $x$ is near $y$ in the region if and only if $f(x)$ is near $f(y)$ in the interval $[0, 1]$.

In the discrete case, assume that $S$ is regularly tessellated into squares. Then the function $f$ can be represented as a path through the grid. Suitable functions, called *tile indexes*, will have the property that points close in $P$ will be close on the path, and vice versa. In practice, it is not possible to find tile indexes that have this nearness property for all pairs of points, but functions have been found that work well in most cases. *(tile index)*

Figure 6.8 shows six common tile indexes. The *row ordering* shown in the top-left of the figure is the first and most obvious possibility. It corresponds to the scan of a raster image across the rows of pixels. Although simple, it has the unfortunate property that close points at row ends are often quite distant on the path, particularly for tessellations with a large number of cells. The *row prime* order provides a simple modification that allows some improvement, but there is still a problem with half of the end-points. Somewhat better at preserving nearness are the Cantor-diagonal order (row-prime rotated, top-right) and the *spiral* order (bottom-left). *(row order, row-prime order, Cantor-diagonal order, spiral order)*

The *Morton* (bottom-center) and *Peano-Hilbert* (bottom-right) orderings are rather more ingenious. The Morton ordering can be seen as a recursive "Z," with the letter "Z" at various scales, or if rotated through a right-angle, recursive "N." Morton orderings may be constructed by interleaving the bits of the row and column numbers. The Peano-Hilbert ordering relates to the Morton ordering as the row-prime to the row ordering. Both Morton and Peano-Hilbert orderings are discretized fractals, generated using a finite number of recursive iterations of a fractal generator (see Chapter 3). Morton and Peano-Hilbert orderings are therefore space-filling and can be constructed at arbitrary levels of detail (section *(Morton order, Peano-Hilbert order)*

3.6). They are also better at preserving nearness than the other orderings in Figure 6.8. The bibliographic notes contain further references on the properties of these orderings.

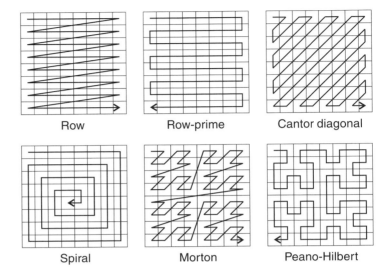

**Figure 6.8:**
Six common
tile indexes

Row    Row-prime    Cantor diagonal

Spiral    Morton    Peano-Hilbert

## 6.3   RASTER STRUCTURES

This section reviews some of the basic methods for storing and compressing raster data. Each cell in a raster is addressed by its position in the array (row and column number). For example, in Figure 6.9 (reproduced from Figure 1.12), the raster may be represented as a $32 \times 32$ array, each cell occupied by 0 or 1. Rasters are able to represent a large range of computable spatial objects: a point may be represented by a single cell; a strand by a sequence of neighboring cells; and a connected area by a collection of contiguous cells. Rasters are managed naturally in computers, because most commonly used programming languages support array handling. However, a raster when stored in a raw state with no compression can be extremely inefficient in terms of computer storage. For example, a large uniform area with no special characteristics must be stored as a large collection of cells, each holding the same value. In this section, we examine ways of improving raster space efficiency.

### 6.3.1   Chain codes, run-length codes, and block codes

Freeman chain
coding

Chain codes, introduced in the previous chapter, may be used to represent the raster boundary of a region. Based on a starting cell, a chain code known as *Freeman chain coding* uses the numbers 0 to 7, arranged clockwise around the 8 directions N = 0, NE = 1, E = 2, SE = 3, S = 4, SW = 5, W = 6, NW = 7. Starting at the northeast corner and

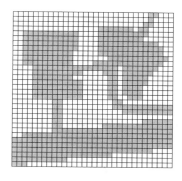

**Figure 6.9:**
An example
raster structure

proceeding clockwise, the Freeman chain code for Figure 6.10a would
then be:

$$[2, 2, 2, 2, 2, 2, 2, 2, 2, 2, 4, 4, 4, 6, 4, 4, 4, 4, 4, 4, 4, 2, 2, 4, 4, 4,$$
$$6, 6, 6, 6, 6, 6, 6, 6, 6, 0, 0, 0, 0, 0, 0, 0, 6, 6, 0, 0, 0, 0, 0, 0, 0]$$

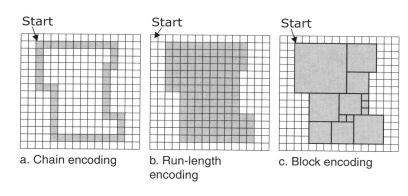

**Figure 6.10:**
Chain code and
block code for
a portion of the
raster in Figure
6.9

a. Chain encoding       b. Run-length         c. Block encoding
                           encoding

*Run-length encoding* (RLE) is an alternative representation, which
counts the length of "runs" of consecutive cells with the same value, 1
or 0. For example, counting in row order Figure 6.10b can be run-length
encoded as:

run-length
encoding

$$[18, 11, 5, 11, 5, 11, 5, 11, 5, 10, 6, 10, 6, 10, 8, 8, 8, 8, 8, 8, 8,$$
$$10, 6, 10, 6, 10, 6, 10, 18]$$

Using a different tile index results in a different run-length encoding.
For example, using Morton ordering the same figure may be encoded as:

$$[6, 2, 4, 4, 2, 2, 2, 10, 4, 4, 4, 20, 2, 2, 2, 10, 2, 1, 5, 1, 1, 1, 5,$$
$$17, 31, 16, 16, 10, 2, 2, 2, 16, 10, 2, 4, 10, 2, 2, 2, 4, 4, 2, 2, 4]$$

Freeman chain coding and RLE can be used together to produce a
compact boundary representation. For example, the Freeman chain code

representation of Figure 6.10a, given above, could be run-length encoded as:

$$[2, 10, 4, 3, 6, 1, 4, 7, 2, 2, 4, 3, 6, 9, 0, 7, 6, 2, 0, 6]$$

Direction codes and the number of consecutive occurrences of each direction code are interlaced in the encoding above (i.e., $[2, 10, ...]$ means 10 consecutive occurrences of the Freeman code for "east," 2).

*block encoding*     *Block encoding* is a generalization of run-length encoding to two dimensions. Instead of sequences (runs) of 0's or 1's, square blocks are counted. Each block is defined in terms of a distinguished point (center or southwest vertex, for example) and the length of a side. The medial axis transform (section 5.4.5) is used to generate efficient packing of the shape with square blocks, as in Figure 6.10c.

### 6.3.2  Region quadtrees

*quadtree*     A widely used structure for holding planar areal raster data is the *region quadtree*. The term "quadtree" arises because the underlying data structure is a tree where all non-leaf nodes have exactly four descendants. *region quadtree*     The principle of the region quadtree is a recursive subdivision of a non-homogeneous square array of cells into four equal-sized quadrants. The decomposition is applied to each sub-array until all the sub-arrays bound homogeneous regions. This process has been applied to the raster in Figure 6.9, and the result shown in Figure 6.11. To see the process of subdivision in detail, Figure 6.12 shows successive subdivisions of part of the northwest quadrant of the raster.

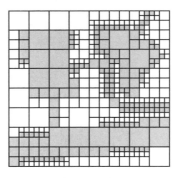

**Figure 6.11:**
Quadtree
subdivision for
the raster in
Figure 6.9

**Figure 6.12:**
Successive
subdivisions of
a small portion
of the raster

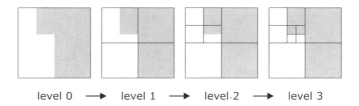

level 0  ⟶  level 1  ⟶  level 2  ⟶  level 3

Quadtrees are stored in a leveled tree data structure, with the root at the top level (level 0). For each non-leaf node, its four constituent quadrants are represented by four descendant nodes. A homogeneous quadrant, in which no further subdivision is required, is stored as a leaf node. Leaf nodes may have attributes associated with them, such as a color code, or they may point to records of more detailed information in a data file. We need a convention for the order of the descendants, say, NW, NE, SW, and SE at each level of the tree. Figure 6.13, shows the tree corresponding to the small portion of the raster in Figure 6.12.

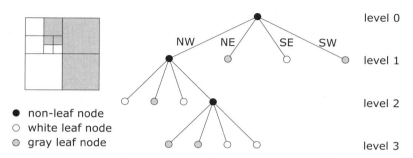

**Figure 6.13:** The quadtree for part of the raster in Figure 6.9

The region quadtree structure takes full advantage of the two-dimensional nature of the data; this is not the case for run-length encoding. Further, region quadtrees have variable adaptive resolution, devoting more depth of the tree in places as needed to capture more detail. There is a body of results on space complexity of the quadtree structure, for example:

> The quadtree representing a polygon with perimeter $m$ embedded in a [raster] image of size $2^n \times 2^n$ contains $O(m+n)$ nodes in the worst case. (Rosenfeld et al. quoted by Samet, 1990a.)

Algorithms for Boolean operations such as union, intersection, and difference of rasters based on the region quadtree structure can be elegantly coded and perform well. These algorithms are able to take advantage of the recursion and variable resolution inherent in the quadtree structure. Algorithms for other operations, such as labeling and counting the number of connected components in an areal spatial object, also perform well. Figure 6.14 shows the complement, intersection, union, and difference operations applied to two planar spatial regions $Q$ and $R$.

For example, the algorithm for quadtree complement involves a simple recursive traversal of the tree. Algorithm 6.4 shows the complement algorithm, based on a breadth-first traversal. Algorithm 6.4 is simpler than the breadth-first traversal algorithm given in the last chapter because the quadtree structure means there is no need to keep a list of visited nodes (see section 5.7.2).

An algorithm for constructing a quadtree $S$ from the Boolean intersection of two binary quadtrees $Q$ and $R$ is given in Algorithm 6.5, similar to the breadth-first complement algorithm given in Algorithm 6.4. When constructing a new node in $S$, the Algorithm 6.5 must deal with four cases:

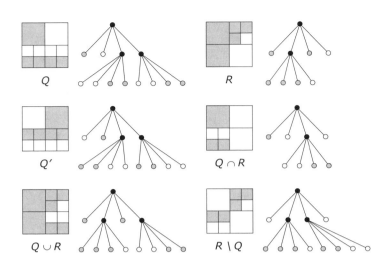

**Figure 6.14:**
Quadtree
complement,
intersection,
union, and
difference

**Algorithm 6.4:**
Quadtree
complement
algorithm
(breadth-first
traversal)

**Input:** A non-null binary quadtree $Q$
1: $r \leftarrow$ root of $Q$
2: queue $L \leftarrow [r]$
3: **while** $L$ is not empty **do**
4:     remove the first node $x$ from $L$
5:     **if** $x$ is a leaf node **then**
6:         invert the value of $x$ (0 to 1 or 1 to 0)
7:     **else**
8:         **for** each descendant $y$ of $x$ **do**
9:             add $y$ to the end of $L$
**Output:** A binary quadtree that represents $Q'$, the complement of $Q$

*Case 1*: If $q$ or $r$ (or both) is a leaf node representing a white area, then
$S$ is constructed with a white leaf node in the position of $q$ and $r$
(lines 5–6 in Algorithm 6.5).

*Case 2*: If $q$ is a leaf node representing a gray area then $S$ is formed using
a copy of the subtree with $r$ as its root (lines 7–8).

*Case 3*: If $r$ is a leaf node representing a gray area then $S$ is formed using
a copy of the subtree with $q$ as its root (lines 9–10).

*Case 4*: If both nodes $q$ and $r$ are non-leaf nodes, the algorithm is applied
recursively to the subnodes of $q$ and $r$ after creating a new non-leaf
node in $S$ (lines 11–14).

Other Boolean operations can be composed from the complement
and intersection operations, using equational relationships, such as De
Morgan's laws. For example, $Q \cup R = (Q' \cap R')'$ and $Q \backslash R = Q \cap R'$.

A link with the previous section on two-dimensional orderings becomes clear from Figure 6.15. If the checker pattern is structured as a

---

**Input:** Binary quadtrees $Q$, $R$

1: $q \leftarrow$ root of $Q$, $r \leftarrow$ root of $R$

2: queue $L \leftarrow [(q, r)]$

3: **while** $L$ is not empty **do**

4:   remove the first node pair $(x, y)$ from $L$

5:   **if** $x$ or $y$ is a white leaf **then**

6:     add white leaf to output quadtree $S$

7:   **if** $x$ is a non-white leaf **then**

8:     add $y$ and all subnodes to output quadtree $S$

9:   **if** $y$ is a non-white leaf **then**

10:     add $x$ and all subnodes to output quadtree $S$

11:   **if** $x$ and $y$ are non-leaf nodes **then**

12:     add a new non-leaf node to output quadtree $S$

13:     **for** pairwise descendants $x'$ of $x$ and $y'$ of $y$ **do**

14:       add $(x', y')$ to the end of $L$

**Output:** A binary quadtree $S$ that represents the intersection $Q \cap R$

---

**Algorithm 6.5:** Quadtree intersection algorithm (breadth-first traversal)

quadtree and the leaf nodes of the quadtree are read from left to right, then the pattern that the trail follows in the plane corresponds to the Morton ordering of the plane.

The quadtree is simply generalized to higher dimensions. Thus, in three dimensions we have the *octree*, in which each non-leaf node has eight ($2^3$) descendants.

octree

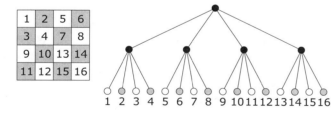

**Figure 6.15:** A region quadtree that is a Morton order of its leaf nodes

Despite its clear strengths, the region quadtree structure has some associated problems. The structure is highly dependent on the embedding of the spatial objects in the raster space: a small translation or rotation of the object will usually result in a quite different quadtree structure. Similarly, dynamic raster data sets that are subject to regular updates are inefficient if stored as a quadtree, because each update may require major reorganization of the quadtree structure. Highly inhomogeneous rasters, such as the alternating checker pattern in Figure 6.15, are not efficient to store in a quadtree because there are no homogeneous regions to be amalgamated. Rasters largely composed of point and linear features also have low storage efficiency.

## 6.4  POINT OBJECT STRUCTURES

This section considers data structures that are designed specifically with point objects in mind. The three structures that are considered are grid files, point quadtrees, and 2D-trees. As usual, assume that the point objects are located in the plane by means of two coordinates, and may have other attributes.

### 6.4.1  Grid structures

As an introduction to grid structures, and to the bucket methods that underlie them, we begin with the fixed grid structure, which allows binary search methods to be applied in more than one dimension. We then introduce the variable grid file structure, which overcomes some of the disadvantages of the fixed grid. We will use the example data set from the Potteries given in Table 6.1 and Figure 6.6.

*Fixed grid structure*

fixed grid

The *fixed grid* or *fixed cell* structure is a partition of the planar region into equal sized cells; squares in our example. Each packet of data with point locations sharing the same cell of the grid is stored in a contiguous area of secondary storage. The usual term to describe a cell of this structure is a *bucket*. Each bucket represents a physical location in storage where the complete records are held. Thus the grid cell structure is a way of partitioning objects that have planar point references so that neighboring objects are more likely to have their attribute data stored in the same or nearby areas of storage. The performance of range queries is thus improved.

bucket

An example of a fixed grid structure for our Potteries point data set is given in Figure 6.16. The region has been partitioned into 16 squares, each containing between zero and two points. Boundary ambiguities are resolved by making the convention that each square owns the points on its south and west boundaries, but not points on the other two boundaries. Thus the point labeled "6" belongs to the square containing the point labeled "7."

The ideal size for the partition is dependent upon at least two considerations. First, the larger the total number of points, the more cells will be required (because the buckets correspond to blocks of secondary storage that can hold only a limited number of records). Second, the size of a cell is dependent on the magnitude of the range in the average range query supported by the system. There is no purpose to be served by making the cell size much smaller than the average range: in such a situation range queries degenerate to near-linear searches.

Looking at Figure 6.16 reveals a major disadvantage of the fixed grid method. The population of cells with points varies, and in some cases there are no points in cells. This problem becomes more acute for less uniformly distributed points. If the distribution is too far from

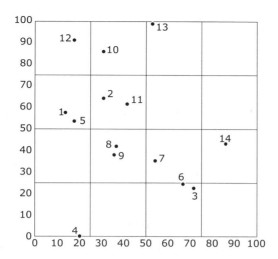

**Figure 6.16:** Fixed grid partition of Potteries point data (labels refer to Table 6.1)

uniform, there will be many empty cells, while other cells will be full to overflowing. Instead, we would like a partition that takes account of and adapts to the density of points in particular local areas. This leads us to the next structure, the grid file (and later point quadtrees).

If the points are uniformly distributed, then the fixed grid structure provides a simple and easy-to-implement solution. The fixed grid is applicable to an arbitrary number of dimensions. It may be appropriate to use a tile planar ordering (e.g., Morton order) for the cells of the grid.

*Grid file*

The *grid file* is an extension of the fixed grid structure that allows the vertical and horizontal subdividing lines to be at arbitrary positions, taking into account the distribution of the points. This structure is designed for dynamic data, entering and leaving the system as time passes. A new level of indirection, called the *grid directory*, is introduced. Grid cells may have their data sharing the same bucket only if their union is a rectangle. Figure 6.17 shows an example, in which a bucket size of two records is assumed (much smaller than a real application). The total area is divided as a three-by-three grid. The two grid cells in the southwest region are amalgamated into a single bucket, because there is spare bucket capacity at that location and the amalgamation results in a rectangle. The grid directory shows the relationships between cells and buckets. The two linear scales, dynamically updated, show the positions of the partitions.

The grid cell is designed to expand and contract as new data is inserted and deleted. A rectangle may be divided if it becomes too full and cells may be amalgamated if the space becomes too empty. Note that in our example, we have reduced from 16 buckets in the fixed grid structure to eight buckets with the grid file structure. Like the fixed grid, the grid file is designed to be applicable to an arbitrary number of dimensions.

grid file

grid directory

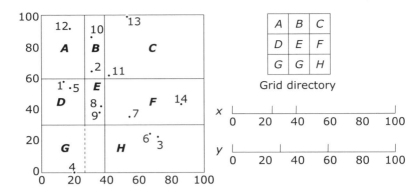

**Figure 6.17:**
Grid file
structure for
Potteries point
data (labels
refer to Table
6.1)

### 6.4.2  Point quadtree

point quadtree    The *point quadtree* combines the grid approach with a multidimensional
generalization of the binary search tree. It has many characteristics in
common with the region quadtree, described earlier in this chapter. Assuming as usual that the data is planar, each non-leaf node is associated
with a data record for a point location and has four descendants (NW,
NE, SW, SE). Thus, each data record comprises two fields to hold the
coordinates, four fields pointing to the four descendants, and further fields
to hold the data associated with the point, such as the name of the city at
that point location (see Figure 6.18).

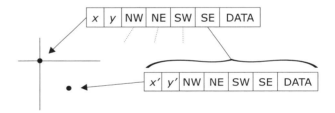

**Figure 6.18:**
Point quadtree
records and the
points to which
they refer

The point quadtree has the property that the position of each partition
into quadrants is centered on a data point, unlike the region quadtree in
which a quadrant is always partitioned into four equal-sized subquadrants.
A data structure in which the positions of subdivisions are independent of
trie    the data points is called a *trie* structure, in opposition to a *tree* structure
in which the positions of subdivisions are dependent. Thus, a more strict
region quadtrie    term for the region quadtree is the *region quadtrie*. The point quadtree is a
true tree structure, because the center of a subdivision into four is always
a data point.

The point quadtree is described with reference to an example. Consider once again the point data from the Potteries given in Table 6.1 and
Figure 6.6. Assume that the points are to be entered into a quadtree index
in order of numerical identifier. The first point (Newcastle Museum) is
placed at the root of the tree, and the plane is divided into four quadrants
as shown in Figure 6.19a. The second point (Waterworld) is compared

with the root. We assume the usual ordering of the cardinal directions, NW, NE, SW, SE. Point 2 is northeast of point 1, and so is placed as the NE (second) descendant of the root.

The northeast quadrant of the plane is divided into four quadrants, as shown in Figure 6.19b. Point 3 (Gladstone Pottery Museum) must now be placed in the tree. It is first compared with the root (SE), and since there are no further nodes along the SE branch, it is placed as the SE descendant of point 1. The southeast quadrant of the plane is divided into four quadrants, as shown in Figure 6.19c. Point 4 is processed next: first it is compared with the root (SE), and then with point 3, the SE descendant of the root. There are no further nodes for comparison, so point 4 is placed as the SW descendant of point 3. The appropriate quadrant of the plane is divided into four quadrants. Figure 6.19d shows the insertion of points 4 and 5 into the tree and the resulting planar subdivisions. Figure 6.20 shows the full quadtree and planar subdivisions for this data set.

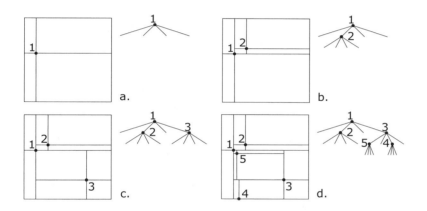

**Figure 6.19:** The first stages of the quadtree construction for the Potteries point data

Algorithm 6.6 presents the procedure for inserting a new point $p$ into the quadtree $Q$. At each iteration, the algorithm selects the appropriate branch of the quadtree based on the relative positions of $p$ and the current node $n$ (lines 5–14). The algorithm iterates, burrowing down through the quadtree, until it reaches the lowest level of the tree, when it inserts the point $p$ and an empty subtree into the quadtree (line 17).

The shape of a quadtree is highly dependent upon the order in which the points are inserted into it. Figure 6.21 shows the quadtree associated with an insertion of the Potteries data in inverted numerical order: 14–1. This dependence has implications for dynamic data. If a point near the top of the tree is deleted, the resulting tree may be substantially changed. Therefore, point quadtrees are not well suited to dynamic geospatial data.

Some worst-case performance measures can be given for the point quadtree. Let $n$ be the number of points structured by the quadtree. Then, the quadtree build time is proportional to the total path length $O(n \log n)$, and the point query search time is $O(\log n)$. We have not presented the algorithms for point and range query for this structure: these operations

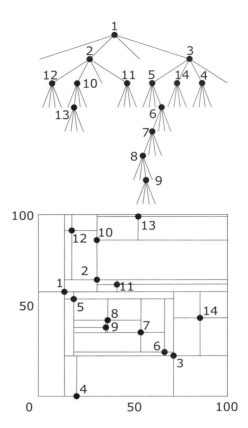

**Figure 6.20:**
Point quadtree
for the
Potteries point
data (labels
refer to Table
6.1)

---

**Input:** Point $p$ and quadtree $Q$

1: **if** $Q$ is an empty quadtree **then**
2:     $Q \leftarrow$ a new quadtree with root $p$ and four null tree descendants
3: node $n \leftarrow$ the root of $Q$
4: **repeat**
5:     **if** $x$-coordinate of $p < x$-coordinate of $n$ **then**
6:         **if** $y$-coordinate of $p < y$-coordinate of $n$ **then**
7:             $n' \leftarrow$ the SW node from $n$
8:         **else**
9:             $n' \leftarrow$ the NW node from $n$
10:     **else**
11:         **if** $y$-coordinate of $p < y$-coordinate of $n$ **then**
12:             $n' \leftarrow$ the SE node from $n$
13:         **else**
14:             $n' \leftarrow$ the NE node from $n$
15:     $n \leftarrow n'$
16: **until** $n =$ null tree
17: insert new node $p$ at position of $n$ with four null tree descendants

**Output:** The quadtree $Q$ updated by the insertion of the point $p$

**Algorithm 6.6:**
Quadtree insert
algorithm

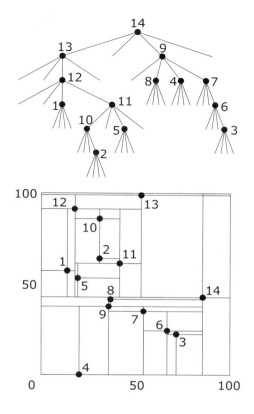

**Figure 6.21:**
Point quadtree
for a reversed
insertion of the
Potteries point
data (labels
refer to Table
6.1)

will be given in detail for the next structure, the 2D-tree, and can be
modified for the point quadtree.

### 6.4.3   2D-tree

The point quadtree takes full advantage of the embedding of the points
in a Euclidean plane. One problem facing the point quadtree is the
exponential increase in the number of descendants of non-leaf nodes as
the dimension of the embedding space increases: for $k$ dimensions, each
node has $2^k$ descendants. The *2D-tree* (in general, $k$D-tree) solves these
problems at the expense of a deeper tree structure. The $k$D-tree is a binary
tree (each non-leaf node has two descendants) regardless of the dimension
$k$ of the embedding space.

2D-tree

$k$D-tree

For planar embeddings, the 2D-tree does not compare points with
respect to both dimensions at all depths, but compares $x$-coordinates at
even depths and $y$-coordinates at odd depths (assume the root is at depth
0). Each record with a point location has two fields storing the coordinates
of the point, two fields pointing to the descendant records, and any
subsequent fields holding additional data. For the binary tree structure,
we make the convention that the left descendant is less than the right
descendant, when compared with respect to the appropriate coordinate
(see Figure 6.22).

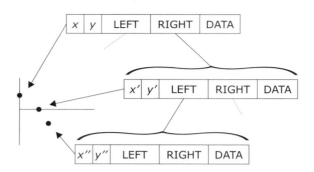

**Figure 6.22:**
2D-tree records
and the points
to which they
refer

The 2D-tree and planar decomposition for our Potteries point data inserted in numerical order is given in Figure 6.23. Point 1 is inserted at the root. Point 2 has a greater $x$-coordinate than point 1 and so is inserted as the right descendant of the root. Point 3 has a greater $x$-coordinate than point 1 and a lesser $y$-coordinate than point 2, so is inserted as the left descendant of point 2. The process continues with each of the 14 points in turn. As for point quadtrees, dynamic data leads to problems restructuring the tree. However, the 2D-tree is a simpler structure, and leaves fewer "dangling" nodes (about half as many).

**Figure 6.23:**
2D-tree for the
Potteries point
data (labels
refer to Table
6.1)

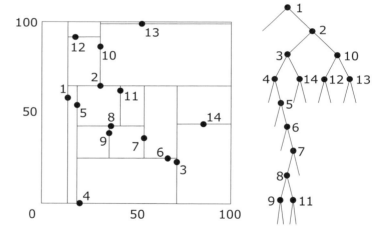

Algorithm 6.7 gives a modified version of Algorithm 6.6 to insert a new point $p$ into a 2D-tree $T$. The algorithm must keep track of the level $l$ of the current node $n$ to ensure the correct test is applied (lines 6–11).

Algorithm 6.8 performs a range search of a 2D-tree $T$ using a rectangular search range specified by diagonal points $p_1$ (southwest extreme) and $p_2$ (northeast extreme). The range search in Algorithm 5.7 is a based on a breadth-first traversal: if the range rectangle contains every point in the 2D-tree, then Algorithm 6.8 simply performs a breadth-first traversal of the entire 2D-tree.

Figure 6.24 shows an example of the algorithm in action with the usual Potteries point data in a 2D-tree. The rectangular range is shown

**Input:** Point $p$ and 2D-tree $T$
1: **if** $T$ is an empty 2D-tree **then**
2: $\quad$ $T \leftarrow$ a new 2D-tree with root $p$ and two null tree descendants
3: node $n \leftarrow$ the root of $T$
4: tree level $l \leftarrow 0$
5: **repeat**
6: $\quad$ **if** $l$ is even **then**
7: $\quad\quad$ $a \leftarrow x$-coordinate of $p$
8: $\quad\quad$ $b \leftarrow x$-coordinate of $n$
9: $\quad$ **else**
10: $\quad\quad$ $a \leftarrow y$-coordinate of $p$
11: $\quad\quad$ $b \leftarrow y$-coordinate of $n$
12: $\quad$ **if** $a < b$ **then** $n' \leftarrow$ the LEFT node from $n$
13: $\quad$ **else** $n' \leftarrow$ the RIGHT node from $n$
14: $\quad$ $n \leftarrow n'$
15: $\quad$ $l \leftarrow l + 1$
16: **until** $n =$ null tree
17: insert new node $p$ at position of $n$ with two null tree descendants
**Output:** The 2D-tree $T$ updated by the insertion of the point $p$

**Algorithm 6.7:**
2D-tree insert
algorithm

**Input:** A non-null 2D-tree $T$ and a rectangular range specified by diagonal points $p_1$ (southwest extreme) and $p_2$ (northeast extreme)
1: node $r \leftarrow$ the root node in $T$
2: queue $Q \leftarrow [r]$
3: initialize set of nodes $R$ as the empty set
4: **while** $Q$ is not empty **do**
5: $\quad$ remove the first node $n$ from $Q$
6: $\quad$ **if** $x$-coord of $p_1 < x$-coord of $n < x$-coord of $p_2$ **and**
$\quad\quad$ $y$-coord of $p_1 < y$-coord of $n < y$-coord of $p_2$ **then**
7: $\quad\quad$ add node $n$ to $R$
8: $\quad$ **if** node $n$ occupies an even level in $T$ **then**
9: $\quad\quad$ $a \leftarrow x$-coord of $p_1$, $b \leftarrow x$-coord of $n$, $c \leftarrow x$-coord of $p_2$
10: $\quad$ **else**
11: $\quad\quad$ $a \leftarrow y$-coord of $p_1$, $b \leftarrow y$-coord of $n$, $c \leftarrow y$-coord of $p_2$
12: $\quad$ **if** $a < b$ and LEFT node of $n$ is non-null **then**
13: $\quad\quad$ add LEFT node of $n$ end of $Q$
14: $\quad$ **if** $b < c$ and RIGHT node of $n$ is non-null **then**
15: $\quad\quad$ add RIGHT node of $n$ end of $Q$
**Output:** The set of points $R$ within the specified range

**Algorithm 6.8:**
2D-tree range
query
algorithm

with a dashed boundary. The search begins with point 1 at the root of the tree. This point is not in the rectangle, so it is not retrieved (Algorithm 6.8, line 6). Point 1 occupies an even level in the tree, so the $x$-coordinates of point 1 and the extremes of the rectangular search range, $p_1$ and $p_2$, are compared (lines 8–9). The $x$-coordinate of $p_1$ is larger than the $x$-coordinate of point 1, so the left subtree of point 1 is not examined (lines 12–13). In any case, this subtree is null. However, the $x$-coordinate of point 1 is smaller than the $x$-coordinate of $p_2$, so the right subtree of point 1 is examined (lines 14–15). The algorithm is applied to the root of this subtree, point 2. Point 2 is within the search rectangle, so its record is retrieved (lines 6–7) and its left and right subtrees are searched according to the algorithm, which continues recursively to the bottom of the tree. The figure shows only the part of tree directly examined by the algorithm. Points 2, 7, 8, 9, and 11, shown circled in the figure, are all retrieved.

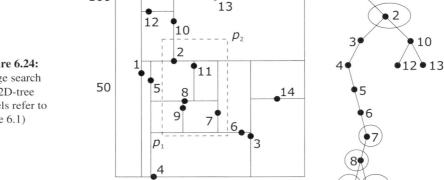

**Figure 6.24:** Range search in a 2D-tree (labels refer to Table 6.1)

As with the point quadtree, the 2D-tree suffers from the problem of its structure being dependent upon the order in which points are inserted. In the worst case, the insertion of each point will result in a further level of the tree and a finished tree with the same number of levels as points.

## 6.5  LINEAR OBJECTS

The spatial data structures described so far have handled points and raster areas. In this section we introduce an index used for linear vector structures, like polylines, linear networks, or the boundary of an areal object. For these linear objects, a first thought might be to use the region quadtree structure, considering linear objects as very thin regions. Unfortunately, the unmodified region quadtree is not usually appropriate, using too much space and resulting in unnecessarily deep tree structures. Instead the *PM quadtree* adapts the region quadtree idea to storing linear objects.

### 6.5.1  PM quadtrees

The PM quadtree is a variant of region quadtree specially designed for the     PM quadtree
structuring of polygonal objects. There are several types of PM quadtree,
including $PM_1$, $PM_2$, and $PM_3$ quadtrees. We will describe the $PM_1$
quadtree, which can be formulated in the general setting of a planar
network of vertices and edges, all of which are straight-line segments.
Assume such a network of vertices and edges enclosed in a square region
of the plane. The region is divided into quadrants as for the region
quadtree (i.e., the trie subdivision). The subdivision is such that vertices
and edges are separated into distinct leaf nodes. To be precise, the region
is divided into the minimum number of quadrants, sub-quadrants, etc.,
such that the quadtree satisfies the following constraints:

1. Each leaf node of the quadtree represents a region that contains at
   most one vertex of the network.

2. If the leaf node of the quadtree represents a region that contains one
   vertex of the network, then it can contain no part of an edge of the
   network unless the edge is incident with that vertex.

3. If the leaf node of the quadtree represents a region that contains no
   vertex of the network, then it can contain only one part of an edge
   of the network.

Figure 6.25 shows the partial construction of a $PM_1$ quadtree for
the boundary of the raster region in Figure 6.9 after the first two subdivisions have been made. In the first stage shown (Figure 6.25a), the
highlighted quadrant is an example violating condition 1, and so must
be further subdivided. In the next stage, in Figure 6.25b, the highlighted
quadrant violates condition 2. To make this quadrant visible, it has been
exaggerated in size by a factor of two. In Figure 6.25c, the highlighted
quadrant (exaggerated in size by a factor of four) violates condition 3,
and so further subdivision is still required. Figure 6.26 shows the final
PM quadtree, in which all conditions are satisfied for all cells.

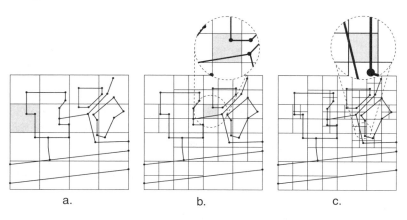

a.                              b.                              c.

**Figure 6.25:**
Stages in the
construction of
the $PM_1$
quadtree

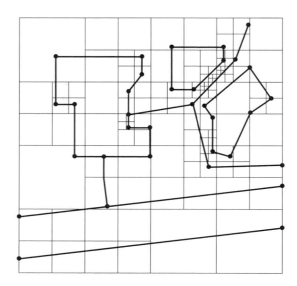

**Figure 6.26:**
Final stage of
construction of
PM$_1$ quadtree

q-edge

The structuring of records in the PM$_1$ quadtree allows different field structures for vertices and edges. In fact, entire edges are not stored in a quadtree node, only parts of edges clipped by the appropriate subquadrants, termed *q-edges*. When an edge is inserted into the PM$_1$ quadtree, the tree is searched for the appropriate places to insert parts of the edge. Moving down the levels of the tree, the edge is successively clipped by the boundaries of the subquadrants. Deletion is a similar but inverse process.

The PM$_2$ and PM$_3$ quadtrees are variants of the PM$_1$ quadtree, where the constraints are altered to provide advantages under some conditions. The PM$_2$ quadtree has a structure that is more stable under translations and rotations of the planar graph. Both PM$_2$ and PM$_3$ quadtrees have a smaller tree structure but correspondingly more complex associated records.

## 6.6    COLLECTIONS OF OBJECTS

A very common requirement in spatial databases is to structure a large collection of objects so as to facilitate good performance on point and range queries (retrieving all those objects located at a given point or within a given range). We have already seen examples of such structures when the constituent objects in the collection are all points (point quadtree and 2D-tree). This section introduces some of the structures designed to index collections containing more general classes of constituent objects, like rectangles, polygons, and complex spatial objects. Only a few such indexes are introduced here; references to related indexes may be found in the bibliographic notes and in the inset "Indexes for complex spatial objects" (shown on the facing page). An important index for interval-based data, the *segment tree*, is discussed in the context of temporal indexes later on in this book (see Chapter 10).

**Indexes for complex spatial objects** *A great many different tree structures have been devised over the years for complex vector-based spatial data. One of the earliest examples, the field tree (Frank and Barrera, 1989), relies on a collection of regular rectilinear grids overlaying the plane to index complex vector-based objects (polylines and polygons). The grids have different resolutions and displacements. Each cell of a grid acts as a container for objects. An object may be inserted into a grid-cell if it does not overlap the grid-cell boundaries and there is no finer-meshed grid that will hold the object. The cell tree (Günther, 1988) is another structure designed to improve the efficiency of retrieval of polygonal objects. As with the $R^+$-tree, the space is hierarchically partitioned into non-overlapping regions. However, while the R- and $R^+$-trees structure objects as contained in their MBBs, the leaves of the cell tree contain convex polygons (polyhedra in the higher-dimensional cases). Arbitrary polygons may be handled by decomposing them as the union of convex polygons. The BANG file (Freeston, 1989) is another index structure similar to the B-tree and $R^+$-tree. Freeston (1993) adapts the BANG file structure to index nested objects with non-intersecting boundaries. Nested objects make it impossible to place a dividing line such that it separates at least one object from the rest of the pack.*

### 6.6.1 Rectangles and minimum bounding boxes

Efficient indexes for rectangles are important because rectangles can be used to approximate bounded planar spatial objects. Each geometric object is enclosed in its MBB (minimum bounding box, the smallest bounding rectangle with sides parallel to the axes of the Cartesian frame, introduced in the previous chapter). Figure 6.27 shows a simple polygon with its MBB. An efficient indexing of MBBs facilitates queries on the objects themselves.

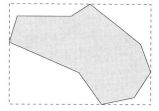

**Figure 6.27:** A minimum bounding box (dashed line) for a simple polygon

Normally, the MBB will be stored separately from the detailed geometry of the bounded object, for example, in the form of a quadtree. The geometry of the bounded object can then be referenced by the record associated with its bounding box. The advantage of using an MBB is that some queries may be answered simply by processing the MBB of an object, rather than retrieving the entire bounded object. For example, consider the range query "Find all the objects which lie in their entirety in a specified disk," applied to the objects shown in Figure 6.28. In order to answer this, the following steps are required:

1. Identify all the MBBs that lie wholly inside the given circular range. The objects inside these bounding boxes must lie in their

entirety in the disk and therefore are to be retrieved. Figure 6.27 shows this case applying to object $A$. Conversely, the MBB of object $E$ lies entirely outside the range and need not be considered further.

2. Identify all the MBBs that intersect with, but do not lie wholly inside the given circular range. The objects inside these MBBs may or may not lie in their entirety in the disk. In these cases, further computation on the detailed geometry of each object is required to determine whether the object lies in its entirety within the given range. In Figure 6.28, $B$ (entirely in the range) should be retrieved, while $C$ (partially in the range) and $D$ (wholly outside the range) should not be retrieved.

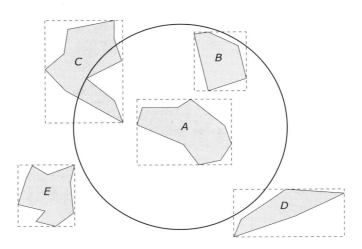

**Figure 6.28:** Using minimum bounding boxes for a range query

## 6.6.2   R-trees and R$^+$-trees

R-tree

Having seen that the MBB can be a useful descriptor of a geometric extent, it is important to find ways of indexing rectangles. The *R-tree* is a rooted tree in which each node represents a rectangle. The leaf nodes represent containers for the actual rectangles to be indexed. Each higher-level node represents the smallest axes-parallel rectangle that contains the rectangles represented by its descendants. It should be noted that the containing rectangles at any given level may overlap one another.

The R-tree is essentially a multidimensional extension of the B-tree (see section 6.1.4), where in the two-dimensional case point pairs define axes-parallel rectangles. The R-tree is a dynamic structure, constructed in a similar way to a B-tree. As rectangles are inserted into the structure, and leaf nodes become full, the effects are back-propagated through the tree nodes, so the tree grows. Conversely, the tree contracts as rectangles are dropped from the collection. The overall tree structure remains balanced and is always at least half-full.

The primary question is how to divide the space of rectangles into groups (represented by higher-level nodes) as the rectangle collection expands, and conversely, how to coalesce groups as the collection contracts. A good subdivision will minimize the total area of the containing rectangles and minimize the area of overlap of the containing rectangles. Figure 6.29 shows an example of a subdivision and simple R-tree for a set of MBBs for the towns in the Potteries (see Figure 1.2). We assume a fan-out ratio of two, such that each node may contain at most two rectangles and thus have at most two descendants.

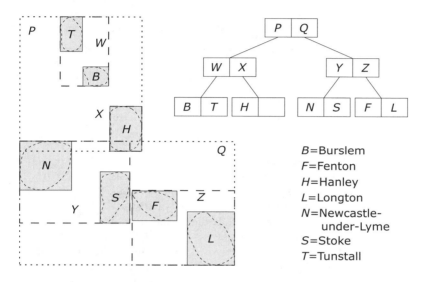

**Figure 6.29:** R-tree structure for minimum bounding boxes of the Potteries towns

$B$=Burslem
$F$=Fenton
$H$=Hanley
$L$=Longton
$N$=Newcastle-
under-Lyme
$S$=Stoke
$T$=Tunstall

A problem with the R-tree structure is caused by overlap of containing rectangles, particularly for rectangles that are large compared with the total space. Point and range searches will be inefficient if the search for an object has to take place in many different subtrees, even though the object is stored in only one of them.

The $R^+$-*tree* is a refinement of the R-tree that does not permit overlapping rectangles associated with non-leaf nodes, although it of course cannot prevent overlapping of leaf rectangles. The $R^+$-tree achieves no overlaps by partitioning rectangles and storing parts in different nodes of the tree. Studies have shown that it admits more efficient searches for large objects. However, while the $R^+$-tree improves the efficiency of point and range queries, more complex insertion and deletion algorithms are needed to achieve efficient use of disk space, and to ensure the tree is always at least half-full.

$R^+$-tree

The Potteries example is structured using an $R^+$-tree in Figure 6.30. When compared with Figure 6.29, $X$ has been extended to contain part of $N$ (the MBB for Newcastle-under-Lyme) that before was contained in $Y$.

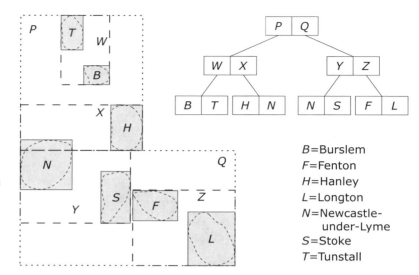

**Figure 6.30:** $R^+$-tree structure for minimum bounding boxes of the Potteries towns

B=Burslem
F=Fenton
H=Hanley
L=Longton
N=Newcastle-
    under-Lyme
S=Stoke
T=Tunstall

### 6.6.3   BSP-tree

BSP-tree

The *binary space partitioning* (BSP) tree is a binary tree that hierarchically decomposes the plane into polygonal regions. The BSP-tree is widely used in first-person perspective computer games, because it is a fast index for searching and sorting the polygons of the spatial environments through which gamers guide their characters. The BSP-tree may be extended to $n$-dimensional space. In two dimensions, the BSP tree hierarchically structures sets of directed line segments. Given a sequence of directed line segments $s_1, s_2, ..., s_n$ in the plane, the construction of the two-dimensional BSP-tree takes place as follows:

1. Place segment $s_1$ at the root of the tree.

2. Extend segment $s_1$ in both directions to form the infinite directed line $l_1$. If any other segment in the sequence is cut by $l_1$, then replace that segment with two segments, taking account of the cut.

3. Examine the next segment, say $s_2'$ in the (possibly revised) sequence. Determine whether $s_2'$ is to the left or right of $l_1$ and place $s_2'$ in the tree as the left or right descendant of $s_1$.

4. Extend segment $s_2'$ in both directions either indefinitely or up to its intersection with $l_1$ to form the directed line $l_2$. If any other segment in the sequence is cut by $l_2$, then replace that segment with two segments taking account of the cut.

5. Continue in this way until every line segment is added to the tree.

An example of this construction is given in Figure 6.31. The segment sequence is initially $a$, $b$, $c$, $d$, $e$, $f$, $g$, as shown in Figure 6.31a. Segment $a$ becomes the root of the BSP-tree, and its associated line (shown as a dashed line) cuts segment $e$ into parts $e_1$ and $e_2$. Segment $b$ is processed next: it falls to the left of $a$ and thus becomes the left descendant of $a$ in the tree. The process continues, eventually giving the planar subdivision, shown as Figure 6.31b, and tree, shown as Figure 6.31c.

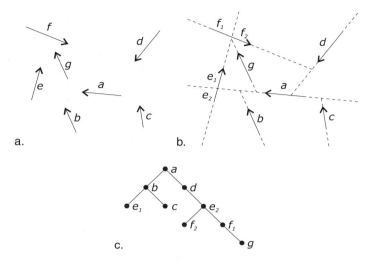

**Figure 6.31:** Construction of BSP-tree from a collection of directed line segments

A balanced tree structure implies minimal depth for the leaves and efficient tree searching. As the figure shows, the BSP-tree may be unbalanced. The structure of the tree is dependent upon the order of insertion of the segments. It is possible to find cases, such as edges of a convex polygon with all segments oriented in the same direction, where there exists no ordering of the nodes that results in a balanced tree. Another factor influencing efficiency of searching and retrieval is the number of edge cuts that are made. In a BSP-tree, the more edge cuts, the more nodes in the tree.

## 6.7  SPHERICAL DATA STRUCTURES

We end this chapter by looking at spherical data structures. Spherical tessellations provide better approximations to the surface of the Earth than planar models over large areas. Chapter 5 introduced some regular spherical tessellations. The octahedral tessellation in section 5.4.6 is the most convenient of these, as it is the only regular spherical tessellation that can be oriented with two vertices at the poles and a set of edges at the equator.

The *quaternary triangular mesh* (QTM) recursively approximates locations on the surface of a sphere by a nested collection of equilateral triangles. A particular case of a QTM arises from the central projection of the edges of an octahedron onto the surface of the globe, termed an *inscribed octahedron* (see Figure 6.32a, similar to Figure 5.22). The initial eight faces can be further subdivided, illustrated in Figures 6.32b and c where the vertices of each triangle are shown joined by the arcs of great circles.

quaternary triangular mesh

inscribed octahedron

**Triangular point quadtree**  *The QTM construction leads to triangular versions of other standard quadtree structures. For example, it is possible to imagine a triangular version of the point quadtree, as shown below left. A triangular frame abc surrounds the data points. These points are inserted into the triangular frame in the order 1, 2, 3, ... as labeled. The first insertion induces a partition of the triangle into three sub-triangles and a consistent labeling of these sub-triangles as shown; further insertions induce further subdivisions. The corresponding tree is shown below right. This structure has the advantage over the standard point quadtree that each node has only three descendants rather than four.*

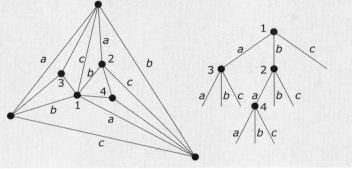

**Figure 6.32:**
Initial levels of the QTM based on the inscribed octahedron

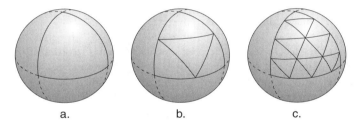

a.                               b.                               c.

For simplicity, each face of the inscribed octahedron may be represented as a planar equilateral triangle, which may be recursively subdivided as in Figure 6.33. At the first level (Figure 6.33a), the nodes of the triangle are labeled 1, 2, and 3. The triangle is partitioned as shown into four equilateral triangles. Each sub-triangle is labeled according to the label of the node nearest to it, except that the inner sub-triangle is labeled 0.

At the second level (Figure 6.33b), the unlabeled nodes of the three sub-triangles of Figure 6.33a are labeled so that each edge contains the labels 1, 2, and 3 in some conventional order. Each sub-triangle is now further divided into four. Every sub-sub-triangle is then labeled with two digits: the left digit is the label of the level one triangle to which it belongs; the right digit is the label of the node nearest to it. Again, the innermost triangle has right digit 0. This process continues for subsequent levels, as far as desired.

The QTM may be used as a triangular version of the region quadtree. An example of this use is shown in Figure 6.34, where a region is decomposed into its maximal triangular pieces. In the traditional rectangular

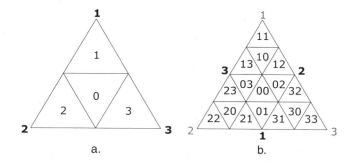

**Figure 6.33:**
Two levels of QTM cell numbering (bold numbers outside triangle indicate labeling scheme)

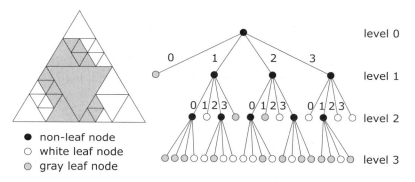

**Figure 6.34:** QTM region quadtree

region quadtree, each node has four descendants, corresponding to the northwest, northeast, southwest, and southeast subregions represented by its descendant nodes. Similarly with this "QTM region quadtree," each node has four subnodes, corresponding to the sub-triangles labeled 0, 1, 2, and 3. As with the traditional region quadtree, the QTM region quadtree is a trie structure.

Triangular versions of other standard quadtree structures are also possible, such as a triangular point quadtree (see "Triangular point quadtree" inset, on the preceding page). However, determining which triangle contains a point is much more computationally expensive than the corresponding point quadtree operation (determining which rectangle contains a point), and so it is unlikely that such an approach would have any more than curiosity value.

## BIBLIOGRAPHIC NOTES

6.1 Most general database texts, including Elmasri and Navathe (2003), Connolly and Begg (1999), and Ramakrishnan and Gehrke (2000), contain more information on physical file organization, hashing, and basic indexes, such as B- and $B^+$-trees. Comer (1979) is a classic reference on B- and $B^+$-trees.

6.2 Samet (1990b, 1995), van Oosterom (1993, 1999), Ramakrishnan and Gehrke (2000), Shekhar and Chawla (2002), and

Rigaux et al. (2001) are all useful sources of further information on spatial indexes. Egenhofer and Herring (1991) contains a concise overview of spatial data structures in GIS.

6.2.1 Further information and references on planar orderings are given in Samet (1990b), van Oosterom (1993, 1999), Ramakrishnan and Gehrke (2000), and Shekhar and Chawla (2002). Abel and Mark (1990) consider the suitability of the orderings with regard to spatial query types, such as point and range queries. Some further analysis of the properties of tiling indexes is given by Goodchild (1989).

6.3.2 Region quadtrees are discussed in Samet (1990b) and van Oosterom (1993, 1999).

6.4.1 The fixed grid structure is introduced in Knuth (1973) and Bentley and Friedman (1979). The grid file can be found in Nievergelt et al. (1984).

6.4.2 The point quadtree is due to Finkel and Bentley (1974).

6.5.1 The class of PM-quadtrees for structuring polygonal maps may be found in Samet and Webber (1985) and Nelson and Samet (1986), and are described in detail in Samet (1990b).

6.6.2 The R-tree structure for rectangles was originally given by Guttman (1984). The non-overlapping $R^+$-tree structure is described in papers by Stonebraker et al. (1986) and Faloutsos et al. (1987). A variant of the R-tree, the R*-tree, is used as the basis for a query processor in Kriegel et al. (1993).

6.6.3 The BSP-tree is discussed in Fuchs et al. (1980, 1983). Extensions to the BSP-tree, including the object BSP-tree are considered in van Oosterom (1993).

6.7 Dutton proposed the Quaternary Triangular Mesh (QTM) to consistently reference any location on a planet, explored in Dutton (1984, 1989, 1990, 1999). Goodchild and Shiren (1992) and Otoo and Zhu (1993) describe work closely related to that of Dutton. Fekete and Davis (1984) introduced a structure similar to that of Dutton but based upon the central projection of the icosahedron.

# Architectures

7

**Summary**

The **architecture** of an information system is the structure and organization of the system's constituent components. Two distinguishing characteristics of a computer-based information system architecture are its levels of **interoperability** (the ability to share data, information, and processing) and **modularity** (the extent to which a system is composed of independent units with clearly defined functions). **Distributed** architectures are often used to achieve high levels of interoperability and modularity in many types of system, including GISs, databases, and **location-based services**.

The data models, structures, and access methods discussed in earlier chapters are fundamental to the efficient storage and analysis of geospatial data. However, these are not the only considerations: the overall structure and organization of the different parts of the system, referred to as the system *architecture*, is also critical. *Modularity* and *interoperability* are two important characteristics that can be used to distinguish different GIS architectures.

architecture

Modularity is the extent to which an information system can be constructed from independent software units with standardized or clearly defined functions. Modularity makes complex GIS software easier to develop, maintain, and adapt to meet the requirements of particular users or specific application areas. Interoperability is the ability of two or more information systems to share data, information, or processing capabilities. Interoperability is especially important to users of GIS, as geospatial analysis often relies on the integration of data from different sources. Planning where to locate a new supermarket, for example, demands the ability to integrate geospatial data from a variety of sources, such as the location of competitor stores, transportation infrastructure, and population density.

modularity

interoperability

**Hybrid GIS** *An example of an early hybrid GIS architecture based on the georelational model is ArcInfo, produced by ESRI (Environmental Systems Research Institute). In the first ArcInfo systems, Arc was the graphics and spatial data engine while Info was the non-spatial database. More recent ESRI GIS architectures are not so simple, and can be used with a variety of different spatial database architectures. Nevertheless, evidence of the importance of the hybrid architecture and the georelational model can still be seen in the ESRI Shapefile format, commonly used for storing and transferring geospatial data. The Shapefile format consists of three separate files—one to store geometry (a .shp file), one to store the non-spatial data tables (a .dbf file), and one to store an index to relate the geometry to a tuple in the data table (a .shx file).*

In this chapter we explore past changes, current trends, and likely future developments in GIS architecture by focusing on the interoperability and modularity of different architectures. The chapter starts by looking at modularity and interoperability independently. First, three basic GIS architectures with different modularity characteristics are explored in section 7.1. Section 7.2 introduces the barriers to interoperability. Section 7.3 explains the role of networks for interoperable GIS architectures, and examines the key system architectures used in GIS to achieve high levels of both interoperability and modularity. Section 7.4 introduces a major application area for interoperable networked computer systems, termed *distributed databases*. Finally, section 7.5 investigates the importance and impact of the increasing use of *location-aware* computing devices on GIS architectures.

## 7.1   HYBRID, INTEGRATED, AND COMPOSABLE ARCHITECTURES

hybrid GIS     A *hybrid* GIS architecture manages geospatial data independently and in a different software module from the non-spatial data. Figure 7.1 shows schematically a hybrid architecture. The left-hand box represents the geometric and topological engines, while the right-hand box holds the non-spatial data. Thus for a land parcel, the geometry of the parcel area along with topological relationships (e.g., adjacencies to other areas) is held in the left unit, while the name, address, owner, and other information about the parcel is held in the right. Typically, a hybrid system is based on georelational     the *georelational model*, in which spatial data is stored in a set of system model     files and non-spatial data stored in a relational database. Spatial and non-spatial data are related to each other using a set of common keys, linking records in the spatial files to tuples in the non-spatial relational database (see "Hybrid GIS" inset, on this page).

The primary motivation for using the hybrid architecture is that it is modular. The special models and structures needed for storing and processing spatial data, covered in earlier chapters, often demand special software and database management strategies. Using a hybrid architecture has the practical benefit of allowing the performance of spatial and

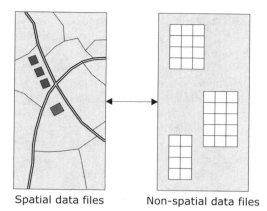

**Figure 7.1:**
Hybrid GIS
architecture

Spatial data files          Non-spatial data files

non-spatial data management modules to be optimized independently. However, the modularity of the hybrid architecture is achieved at a cost. By separating spatial and non-spatial data, the hybrid architecture makes maintaining database integrity, security, and reliability more difficult. The problem lies with separating the storage of spatial and non-spatial data into separate modules, when in fact both the modules are performing rather similar functions, albeit with different data types.

As a result, many GISs utilize an *integrated* architecture, whereby all data, spatial and non-spatial, are stored in a single database (Figure 7.2). Object-oriented databases are ideal for integrated GIS architecture because objects in an object-oriented database may have both spatial and non-spatial references. Integrated object-oriented databases can still achieve high levels of modularity, because individual objects are essentially modules that can perform data integrity, security, and reliability functions.

integrated GIS

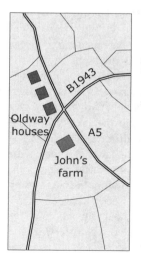

**Figure 7.2:**
Integrated GIS
architecture

Integrated GIS can also be constructed using relational database technology. In the past, integrated relational architectures have been associated with performance problems. The normalization of tables in a relational database can be unsatisfactory for spatial data, as discussed in Chapter 2. A large number of relational accesses and joins are needed to reconstruct spatial objects (points to chains to polygons) and to connect spatial objects to their non-spatial attributes. In practice, many commercial GIS software products rely on object-relational database technology (introduced in Chapter 2), which can directly support geometric data types, such as points, chains, and polygons, as well as offering specialized indexing mechanisms for spatial data, like quadtrees and R-trees.

component

Modularity is a factor not only in GIS database architecture, but also in overall GIS architecture. A *component* is a software module that uses a standardized mechanism for interacting with other software modules. As a result of this standardization, complex software applications can be rapidly assembled from software components. In this respect, components in software engineering are rather like components in mechanical engineering. Mechanical components, such as wheels, pedals, and seats, appear in different configurations and combinations in a range of machines, such as bicycles, tricycles, tandems, and go-carts. Similarly, a software application can be built from multiple different software components.

composable system

Such a software application is referred to as a *composable system*. The process of building software applications using components is sometimes called *mega-programming*.

Figure 7.3 shows a composable GIS architecture comprising four components: a data storage component, a user interface component, a network analysis component, and a digital communications component. In such a composable architecture, additional components might be added as required, for example, further spatial analysis components, or specialized mapping and display components.

## 7.2   SYNTACTIC AND SEMANTIC HETEROGENEITY

The ability to exchange, share, and integrate geospatial data from different sources are fundamental functions of any GIS architecture. There are two main barriers to data sharing, which any interoperable system must overcome: *syntactic heterogeneity* and *semantic heterogeneity*. Syntactic

syntactic heterogeneity

heterogeneity occurs when two or more information systems use incompatible encoding or formats for information. Syntactic heterogeneity may arise as a result of differences in data file formats, software incompatibility, or even incompatible storage media. Whatever the source of syntactic heterogeneity, its consequence is that data must be converted into a compatible format before the two systems can interoperate.

transfer format

Rather than convert directly between different heterogeneous formats, syntactic heterogeneity is usually overcome by adopting a standard intermediate data format, called a *transfer format*. Translating data between the wide variety of GIS software packages and spatial data formats that

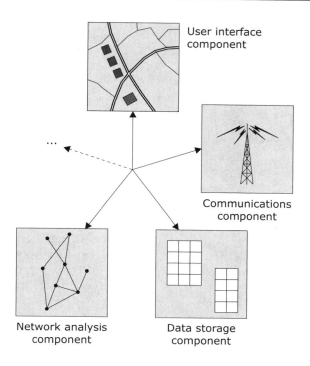

User interface
component

Communications
component

Network analysis
component

Data storage
component

**Figure 7.3:**
Composable
GIS
architecture

exists can present a real problem for GIS projects. Converting directly
between pairs of data formats is inefficient, because it produces many
different *conversion paths*. Each conversion path requires that computer
programs need to be written and maintained to perform the data conversion.

conversion path

In the Figure 7.4a, converting between six data formats demands 30
different conversion paths (the figure has 15 double-headed arrows). In
general, $n$ data formats will require $n(n-1)$ different conversion paths.
Instead of converting directly between pairs of data formats, a transfer
format ensures that in general only $2n$ conversion paths are needed. In
Figure 7.4b, using a transfer format reduces the number of conversion
paths from 30 to 12. The greater the number of data formats, the greater
the efficiency gains from using a transfer format.

Semantic heterogeneity occurs when two or more information systems use different or in some way incompatible meanings. For example,
some different words, like "road" and "street," have the same or very
similar meanings, termed *synonymy*. Other words, like "bank," may have
multiple different meanings (a river bank and a savings bank), termed
*homonymy*. Unfortunately, deciding on exactly what is meant by a piece
of information can be difficult, as meaning is highly dependent on context. The statement "Let's go to the bank" can assume rather different
meanings depending on whether it is followed by the statement "I need
to cash a check" or "I want to feed the ducks." Often, the context for a
statement is implicit.

semantic
heterogeneity

synonymy

homonymy

**Figure 7.4:**
Transfer
formats reduce
the number of
required
conversion
paths

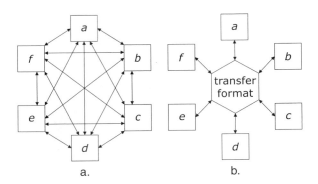

Unlike syntactic heterogeneity, which is largely a technical issue and can usually be solved by technical means, semantic heterogeneity is often difficult to reconcile. The meaning we attach to data is very important, but very difficult to encode. This problem is particularly acute for GIS, because geospatial data is used in such a rich variety of different disciplines, application domains, and communities, each of which has its own idiosyncratic conventions and terminology. Groups that use semantically heterogeneous conventions and terminology are often called *information* *communities*. An information community can arise as a result of a wide range of factors, including different professions, academic disciplines, languages, nationalities, or culture.

information
community

In summary, the distinction between syntactic and semantic heterogeneity mirrors the distinction between data and information, made in Chapter 1. Syntactic heterogeneity is concerned with the properties of data, namely, format and syntax of data. Semantic heterogeneity is concerned with the properties of information, primarily the context and meaning that we associate with data to form information.

### 7.2.1   Standards and transfer formats

All transfer formats address syntactic heterogeneity by providing a standard intermediate format for data conversion. A *de jure* standard is administered by a standards organization with no particular bias or commercial interests in the standard. For example, the British Standards Institute administers a transfer format for geospatial data (BS7567, also known as neutral transfer format, NTF). NTF provides a detailed syntax that ensures any data conforming to this standard will be syntactically homogeneous. A *de facto* standard is a data format that is widely used as a standard, even though it has not been approved by a standards organization. For example, the Shapefile format (see "Hybrid GIS" inset, on page 260) is often used as a *de facto* transfer standard, although the format is published by a commercial GIS software company.

*de jure* standard

*de facto* standard

In addition to syntactic heterogeneity, many transfer formats begin to address semantic heterogeneity issues. For example, Spatial Data Transfer Standard (SDTS) provides a precise syntax for data transfer, but also

**XML** *Like all markup languages, XML uses tags to annotate data with structure that can help in interpreting or analyzing the data. Tags provide a name for the thing the data is describing. XML tags may optionally have attributes that further refine the meaning of data. For example, data about the first edition of this book in XML might look like:*

```
<book>
  <author>Michael F. Worboys</author>
  <title edition="First">GIS: A Computing Perspective</title>
  <publisher location="London">Taylor and Francis</publisher>
  <year>1995</year>
</book>
```

*An important feature of XML is that it is* human readable. *Without knowing the precise syntax of XML, most people would still be able to look at the XML data for the book, above, and understand what information is being represented. XML is not used as a transfer format directly; rather XML is the basis for defining new* vocabularies *for describing and transferring data. Using XML provides a number of powerful additional features. For example, XML provides a mechanism to define what types of structures are allowed within a particular vocabulary (the Document Type Definition or DTD). Software that parses XML can use the DTD for a particular vocabulary to ensure the XML data conforms to the structure described in the DTD, termed* validation. *XML also provides a mechanism for defining* templates, *which can be used to transform XML from one vocabulary into another (eXtensible Stylesheet Language Transformation, XSLT).*

allows users to include information about the definition and meaning of terms used in the data set. Such information comprises a *data dictionary*.     data dictionary
SDTS users can either develop their own data dictionary, or adopt the data dictionary commonly used in a particular information community. A transportation data dictionary, for example, might contain precise definitions of spatial features like "road" and "street," so helping users from other information communities to better understand the meaning to associate with the data. Several other transfer standards have similar capabilities for including definitions or data dictionaries.

The costs of spatial data transfer are high enough so that many countries have developed national strategies for sharing and coordinating geospatial data, including the USA (National Spatial Data Infrastructure, NSDI), Australia (Australian Spatial Data Infrastructure, ASDI), Canada (Canadian Geospatial Data Infrastructure, CGDI), and India (National Geospatial Data Infrastructure, NGDI). These initiatives are often based on the use of particular transfer formats to ease data sharing. In fact, there are so many different standards for geospatial data that converting between different transfer formats can itself become a barrier to interoperability. Standards organizations are increasingly coordinating their efforts in an attempt to minimize such barriers, but the problem is common to standardization generally. Heterogeneity is a natural consequence of the wide variety of different information communities that use geospatial data. Consequently, standard transfer formats cannot eliminate all barriers to data sharing.

XML

XML vocabulary

Partly in response to such issues, a somewhat different approach was used as the basis for defining the eXtensible Markup Language (XML) standard, developed by the World Wide Web Consortium (W3C). XML is not itself a transfer format; rather it is a standard *meta-language* used for defining other languages and transfer formats, termed *vocabularies*. For example, Geography Markup Language (GML) is an XML vocabulary that has been developed as a transfer format for geospatial data (see "XML" inset for more information, on the preceding page).

## 7.3   DISTRIBUTED SYSTEMS

The use of transfer formats represents a flexible, but highly data-oriented and asynchronous approach to interoperability. Transfer formats are data-oriented, in the sense that it is usually only possible to share data using transfer formats, but not directly share processing of data. Transfer formats are asynchronous in the sense that once a data set has been converted into a particular transfer format it may be seconds, hours, days, or years before that data is subsequently converted into another format and used in a target application.

distributed system

By using computer networks, information systems are able to achieve highly processing-oriented, synchronous forms of interoperability. A *distributed system* is a collection of multiple information systems connected via a digital communication network that can synchronously co-operate in order to complete a computing task. A distributed system is therefore a special type of interoperable system, because the different elements in a distributed system must be able interoperate with one another to complete some task and do so using a network.

### 7.3.1   High-level distributed system architecture

mainframe

In addition to the low-level data transmission properties of a network, discussed in Chapter 1, the high-level architecture of a network also affects its communication capabilities. There are three different types of high-level network architecture. The *mainframe* network architecture (sometimes called the *terminal* network architecture) connects multiple terminals to a central computer server. Early information system architectures, including GIS, were dominated by mainframe architectures. These large, multi-user systems represent a very centralized model of computing, with control over computing facilities exercised by professional support staff rather than individual users. This centralized model does have significant advantages, as it tends to lead to increases in data security, data integrity, and data sharing.

The mid-1980s saw a rapid expansion in the use of desktop personal computer (PC) systems for GIS and many other software applications. PCs are able to offer more decentralized personalized applications than mainframe systems. While the move to PC-based systems offered users increased control, it was, initially at least, to the detriment of data sharing.

The *peer-to-peer* network architecture (sometimes called *p2p* networks or *file-sharing* architectures) was popular for early PC networks as it is an inexpensive way of connecting a small number of computer systems. Early peer-to-peer networks were not suitable for connecting large numbers of computers, and could not offer the high levels of connectivity we expect today. More recently, peer-to-peer architectures have enjoyed something of a revival, as their highly decentralized nature is appealing for data-sharing applications.

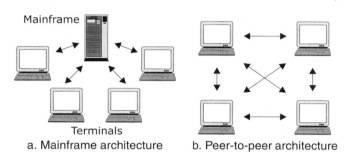

**Figure 7.5:** Mainframe and peer-to-peer architectures

a. Mainframe architecture    b. Peer-to-peer architecture

Figure 7.5 summarizes mainframe and peer-to-peer network architectures. For some years, the most popular network architecture has been the *client-server* architecture, which has proved an extremely flexible and versatile way of organizing communication between information systems. The client-server architecture may be thought of as an intermediate architecture between mainframe and peer-to-peer architectures. As we shall see, many of the features of client-server architectures, discussed next, are common to mainframe and peer-to-peer architectures. Consequently, an understanding of client-server architectures provides a solid basis for understanding both mainframe and peer-to-peer architectures as well.

### 7.3.2 Client-server systems

The client-server architecture provides a clear delineation between the responsibilities of different information systems within a particular application. A *server* is an information system that can offer a particular service to other information systems on the network, while a *client* is an information system that consumes these services. The services offered by a server may include resources such as files, software applications, and hardware devices. Client-server interaction usually follows a *request-response* protocol. Clients request a service from a server, which then responds with the appropriate resource. A familiar example of client-server computing is surfing the WWW. A web browser acts as a client requesting web pages and other files from web servers on the Internet.

The client-server architecture differs from the mainframe architecture in that a client may consume services from multiple different servers. At the same time, the client-server architecture differs from the peer-to-peer architecture in that there is still a clear distinction between the different

peer-to-peer

client-server

server
client

request-response

roles that client and server fulfill within the context of a particular application.

The services provided by a server are defined by a server's *interface*. An interface is a like a contract or agreement, clearly stating what it is a server can do for a client and how a client can access those services. The interface for the WWW is defined by hypertext transfer protocol (HTTP). The term *protocol*, meaning a standard format for communication, is closely related to *interface*. Web browsers use HTTP to communicate with web servers. When you click on a hyperlink in a web page, your web browser sends an HTTP request for the required web page to the web server at the Internet address indicated by the hyperlink. Assuming the link refers to a valid address, the web server will respond with the appropriate web page also using HTTP.

protocol

**Figure 7.6:**
Client-server
architecture

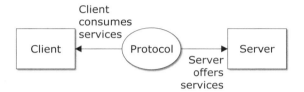

An interface specifies only what services a server offers, not how those services will be performed. The client-server architecture is not only interoperable but also modular, in the sense that clients and servers may be modified, upgraded, or even replaced as long as the interface remains constant. For example, when you upgrade or change your web browser software it is still generally possible to access the same web pages on the WWW using the same addresses, because all web browsers use HTTP. Figure 7.6 summarizes the basic client-server architecture, often called a *two-tier* client-server because every information system in the architecture is either a client or a server. Client-server architectures can be used both for exchanging files, for example, using HTTP or file transfer protocol (FTP), and for sharing processing between clients and servers (see "Parallel processing and Beowulf" inset, on the next page).

two-tier

*Multi-tier client-server architectures*

multi-tier

A *multi-tier* client-server system, also often termed a *three-tier* or *n-tier* client-server system, extends the basic two-tier client-server architecture with additional intermediate tiers. Multi-tier client server systems are essentially composed of chains of two-tier systems. Figure 7.7 illustrates a multi-tier architecture with three tiers. As in Figure 7.6, client and servers communicate with each other. However, in Figure 7.7 an intermediate "middle tier" acts as both a client (for the right-hand server using protocol B) and a server (for the left-hand client using protocol A).

Multi-tier architectures like that illustrated in Figure 7.7 are commonly used for making geospatial data processing available over the WWW. A client web browser requests information from a geospatial data

**Parallel processing and Beowulf** *The traditional von Neumann architecture of a computer, introduced in Chapter 1, assumes sequential processing, in which each machine instruction is executed in turn before control passes to the next instruction. Most computers today follow this architecture, and have a single CPU (central processing unit) responsible for the majority of the computation the computer performs. It is possible to use multiple CPUs working in concert to complete tasks more quickly than is possible with one CPU, termed parallel processing. Parallel processing can be particularly useful for processing geospatial data, because many spatial processing algorithms are computationally intensive. Unfortunately, the complexities of communication between individual processors working on the same task, termed latency, introduces inefficiencies into parallel processing: $n$ parallel processors will always complete a task less than $n$ times faster than a single CPU. Parallel processing systems are often expensive and usually demand specialized programming techniques to enable algorithms to take advantage of the multiple processors. Beowulf (named after the 11th century Old English poem) is an open source parallel processing system based on a client-server architecture, which is simpler and cheaper than many dedicated parallel architectures, at the cost of increased latency. Beowulf (http://www.beowulf.org) uses a cluster of networked computers instead of specialized hardware to achieve parallel processing. A single server uses clients in the network to form a "virtual supercomputer" capable of parallel processing.*

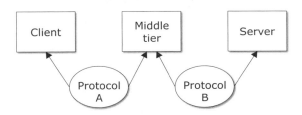

**Figure 7.7:** Multi-tier Client-server architecture

server via a web server middle tier (see "Mapping websites" inset, on the following page). The primary advantage of additional tiers within a client-server architecture is that they allow increased modularity, with specific functions allocated to particular tiers.

### Server- and client-side strategies

A significant factor in determining the characteristics of any client-server system is the allocation of processing responsibilities between the different tiers. In a *server-side* strategy, the server performs the bulk of the computation needed to complete a task. The terms *thin client* and *thick server* are often used to refer to the client and server in a server-side strategy. In a *client-side* strategy, the client performs the bulk of the computation needed to complete a task. Client-side strategies use *thick clients* and *thin servers*. Figures 7.8 and 7.9 illustrate the components within server- and client-side strategies.

server-side

client-side

Because the server performs most of the computation, server-side strategies enable geospatial data and processing to be accessed by clients with minimal computational capabilities (such as a handheld computer or

**Mapping websites** *There exists a variety of websites offering maps for all sorts of purposes. For example, MapQuest* (http://mapquest.com) *offers travel directions, journey planners, and road maps at a variety of different scales from around the world. In 2003 MapQuest received an estimated 2 million visitors per day. The USGS GEODE website* (http://geode.usgs.gov, *example map pictured below) provides access to land cover, geology, and elevation data. Such websites normally utilize a multi-tier client-server architecture, which ensures greater modularization than could be achieved using two-tier architectures. Users accessing the site with a simple web browser may be unaware that the map web server is itself a client for a geospatial data server.*

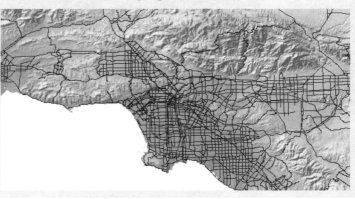

**Figure 7.8:**
Server-side
strategy

Thin client                    Thick server

mobile phone) or using generic software (such as a web browser). Client-side strategies demand clients with higher computational capabilities, but conversely require less powerful server systems. Client-side strategies are also able to offer increased user interface flexibility and interactivity.

For example, map servers on the WWW (see "Mapping websites" inset, on this page) commonly adopt a server-side strategy as it allows users to gain access to online maps using even basic handheld computing devices. However, such sites are usually only able to offer basic display functions for interacting with these maps. In contrast, client-side mapping software relies on the availability of a more powerful computing platform, but might provide a range of options for presenting, querying, and processing geospatial data in addition to basic display functions.

In general, server-side strategies make more efficient use of network bandwidth than do client-side strategies, as the server only needs to

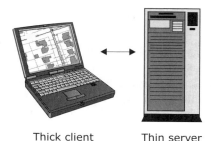

**Figure 7.9:**
Client-side
strategy

Thick client    Thin server

respond with processed geospatial data. Additional data that does not form part of the final response, but might have been used while processing the request, is retained by the server and does not need to be transmitted. For example, when displaying a map on a website it is important not to display too much detail or the map would quickly become illegible. Maps need to be generalized to remove excess detail but retain the essential features of the map. By adopting a server-side strategy, the server can perform the cartographic generalization. The server sends data to the client only after generalization has removed excess detail, so decreasing the bandwidth needed for each map. Client-side systems usually also require an initial download of specialized client software, with additional bandwidth implications.

Finally, server-side strategies ensure higher data security, as control over how data is processed remains with the service provider. In this respect server-side strategies are rather similar to the mainframe network architecture discussed above. A server-side strategy is needed by many mapping websites, because the geospatial data upon which their maps are based is expensive to obtain and maintain. Using a server-side strategy allows websites to provide maps of a particular area at a particular scale to clients, at the same time as ensuring that the underlying geospatial data used to build each map never leaves the server. Client-side strategies mean that service providers may have less control over how their geospatial data is used.

|  | Server-side | Client-side |
|---|---|---|
| Client functionality requirements | Lower | Higher |
| User interface flexibility | Lower | Higher |
| Bandwidth usage | Lower | Higher |
| Data security | Higher | Lower |

**Table 7.1:**
Server- and
client-side
strategies

Table 7.1 summarizes the key features of the server- and client-side strategies. Client-side strategies are ideal for specialized applications, when smaller numbers of advanced or expert users need flexible or personalized data manipulation capabilities. Server-side strategies are generally well suited to applications in which large numbers of non-specialist users only require access to simple geoprocessing capabilities, or if data security is very important.

### 7.3.3    Distributed component systems

There is a clear analogy between client-server architectures and the object-oriented model (Chapter 2). Like clients and servers, objects provide and consume services (using behaviors or methods) via an interface, while ensuring that the actual mechanism for providing a service is hidden by encapsulation. Multi-tier client-server architectures are even more similar to the object-oriented model as they provide increased modularization of clients and servers. A natural evolution of this analogy is to develop distributed systems in which individual components, or even individual objects, act as clients and servers. A *distributed component* architecture uses just this approach, in which individual components or objects can interoperate as part of a highly decentralized client-server architecture.

distributed
component

The dividing line between distributed component and multi-tier client-server architectures is somewhat blurred. The term "distributed component" is generally reserved for systems with many (more than three or four) distinct components. As a result, distributed component architectures are closely related to the peer-to-peer architecture introduced at the beginning of this section. In a distributed component architecture, individual components may be involved in a complex network of interactions, acting as client or server with many other distributed components.

Any distributed component architecture has three main parts. First, each component must have an interface defining what services a server component offers (termed a *server skeleton*) and what services a client component consumes (termed a *client stub*). As components may act as both client and server it is possible for components to possess both a server skeleton and a client stub. Second, servers register their services with a *registry*, which any client may then access in order to find servers offering compatible services. Third, to ensure different components can successfully communicate, a distributed component architecture must use a standard protocol for communication between clients and servers. Figure 7.10 illustrates the key elements of a distributed component architecture.

server skeleton
client stub

registry

There are many examples of distributed component technologies, including CORBA (common object request broker), DCOM (distributed component object model), and Java RMI (remote method invocation). These technologies provide a standard infrastructure that it makes it easier to achieve interoperability between networked software components. In CORBA, for example, components may be written as software for any platform. An *object request broker* (ORB) then acts as a registry, as well as helping to mediate communication between components. A standard protocol is used to transmit requests and responses between components (Internet Inter-Orb Protocol, IIOP). DCOM and RMI follow a similar pattern to CORBA. A group of XML vocabularies has also been developed as an XML-based distributed component architecture standard, collectively known as *web services* (see "Web services" inset, on the facing page).

web services

**Web services** *Web services are distributed component systems that take advantage of XML as a standard for communication between components. The web services infrastructure provides an XML vocabulary, Simple Object Access Protocol (SOAP), for communication between components. Two other XML vocabularies, Web Services Description Language (WSDL) and Universal Description, Discovery, and Integration (UDDI) protocol, are used as the basis of a registry to help to make web services easier to find on the Internet. Web services (and component architectures generally) help make software truly modular, leading to easy-to-build composable systems. For example, a variety of geocoding web services can be found on the WWW. Geocoding is the process of converting a place name into geographically referenced coordinates and is a fundamental operation in many marketing and geodemographics applications. Using web services, it is a relatively simple task to write a software application that reuses existing geocoding services. Geocoding requires processing to match addresses with query locations in addition to valuable data about the locations of addresses. As a result, such services would be extremely difficult for most users to implement unaided.*

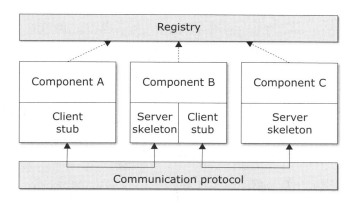

**Figure 7.10:** Generalized distributed component architecture

## 7.4 DISTRIBUTED DATABASES

In the discussion of distributed systems above, each component is expected to be functionally distinct from other interoperating components. For example, Figure 7.11 illustrates a three-tier client-server distributed system architecture for a mapping website, in which the three tiers of the architecture have distinct functions. The spatial database server stores geospatial data; the web browser client provides a user interface to the geospatial data; and the web server middle tier makes the data available on the WWW, providing a bridge between the spatial database server and the user interface client.

**Figure 7.11:** Example three-tier client-server application

The architecture in Figure 7.11 is widely used, and based on a single logically centralized spatial database system, like those introduced in Chapter 2. For some applications, however, it is necessary to use logically related data at multiple sites connected by a computer network, termed a *distributed database*. Many organizations collect and maintain their own data sets, so distributed databases are needed in order that these geographically dispersed data sets can be shared. Geospatial data in particular is often stored at a geographical location close to that to which the data refers. For example, a local government environmental department might collect data about environmental pollution for its administrative region. This data would usually be stored at the department itself, located somewhere within the administrative region. Connecting related data from multiple remote local government departments in different administrative regions requires a distributed database.

distributed
database

For large, geographically dispersed data sets, distributed databases offer several potential advantages over conventional databases:

*Decentralization*: Distributed databases are decentralized. Local database units within a distributed database can be controlled directly by those people or organizations who are primarily responsible for collecting or using the data, termed *local autonomy*.

local autonomy

*Availability and reliability*: Distributed databases offer improved fault tolerance. If one unit within the distributed database becomes unavailable, perhaps because of network problems or power failure, other units should still be accessible.

*Performance*: Local autonomy means that data is stored at a location that is physically close to those users who have the greatest demand for it. As a result, that database may be optimized to offer improved performance for the expected queries of local users. Local users are also more likely to be connected to the local database unit by a high-speed LAN, rather than depending on a possibly slower or more variable WAN.

*Modularity*: Distributed databases are inherently modular. As a result, distributed databases offer improved scalability and maintenance, as individual units can be added or upgraded without disruption to the entire database system.

Figure 7.12 shows a revised version of Figure 7.11, with a distributed database providing access to data stored at multiple different locations. A *distributed DBMS* (DDBMS) is the software system that manages a distributed database. A key goal for a DDBMS is to make access to distributed databases transparent, so that users are shielded from the details of data distribution, and can access a distributed database as if it were a conventional centralized database.

DDBMS

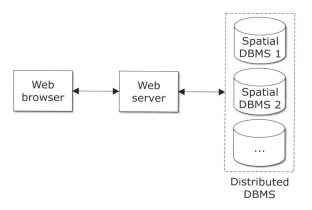

**Figure 7.12:** Example three-tier client-server application based on a distributed database

### 7.4.1 Homogeneous and heterogeneous DDBMS

A major distinction in DDBMSs is between *homogeneous* and *heterogeneous* DDBMSs:

- A homogeneous DDBMS is composed of multiple data storage units each using the same DBMS software and data model.

  *homogeneous DDBMS*

- A heterogeneous DDBMS is composed of multiple data storage units each using different DBMS software or different data models.

  *heterogeneous DDBMS*

Figure 7.13 illustrates the distinction between homogeneous and heterogeneous DDBMSs. In Figure 7.13a the homogeneous DDBMS uses a single data model and DBMS software to provide access to data stores at different sites. In Figure 7.13b the heterogeneous DDBMS maintains multiple different data models and/or DBMSs at different sites. Unified access to the heterogeneous databases is provided through a *gateway* interface, which manages the task of responding to queries using data from the different site data storage and data models. The task of unifying access to heterogeneous databases can become highly complex, especially when the data models exhibit high levels of semantic heterogeneity. The study of systems capable of automatically performing this complex task, called *mediators*, is an important research topic.

*gateway*

*mediator*

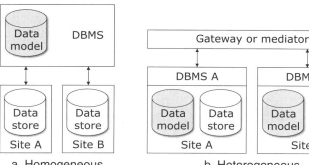

a. Homogeneous    b. Heterogeneous

**Figure 7.13:** Homogeneous and heterogeneous DDBMSs

### 7.4.2   Relational distributed databases

Two key differences between conventional relational database design, covered in Chapter 2, and relational distributed database design are

fragmentation

*fragmentation* and *replication*. Fragmentation occurs when a relation is divided into sub-relations, called *fragments*, which are then distributed among the different database units. There are two types of fragmentation:

*Horizontal fragmentation*:  occurs when fragments are composed of subsets of the *tuples* of a relation

*Vertical fragmentation*:  occurs when fragments are composed of subsets of the *attributes* of a relation

Figure 7.14 shows the horizontal and vertical fragmentation of a relation diagrammatically.

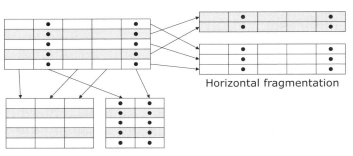

**Figure 7.14:**
Horizontal and vertical fragmentation of a relation

Horizontal fragmentation

Vertical fragmentation

replication

Replication  occurs when data fragments are duplicated across different database units within the distributed database. Replication can improve reliability and performance, because replicated fragments make it more likely that a query can be answered using data from a single site. However, replication makes data updates more complex, as inconsistencies will result if an update to a fragment is not propagated to every copy of that fragment.

To illustrate some of these ideas, recall the relational Potteries CINEMA database introduced in Chapter 2. The database holds information about films showing at various Potteries cinemas. Imagine that we wish to extend this database to a nationally distributed system, with Leeds and London joining Stoke in the enterprise (Figure 7.15). Some specific aspects of this example are not realistic. The reader is asked to be charitable and suspend disbelief temporarily, since the principles are still valid.

The idea is to keep the film details (FILM, CAST, STAR) replicated at each node, but horizontally fragment the CINEMA, SCREEN, and SHOW relations so that nodes only hold tuples relating to their regions. Thus the London site would hold all tuples of the FILM, CAST, and STAR relations. In addition it would hold a subset of the CINEMA, SCREEN, and SHOW tuples, based on that data relevant to the London location. Table 7.2

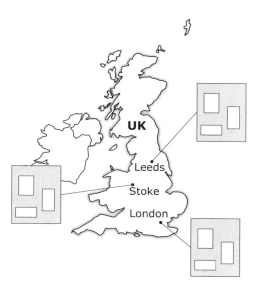

**Figure 7.15:**
CINEMA
distributed
relational
database

### a. London

| CINEMA_ID | SCREEN_ID | TITLE | STD | LUX |
|---|---|---|---|---|
| 7 | 1 | X2 | £9.00 | £11.00 |
| 7 | 2 | Malcolm X | £9.00 | £11.00 |
| 8 | 2 | A Bug's Life | £7.50 | £8.50 |
| 8 | 3 | The Hours | £8.00 | £9.00 |
| 9 | 1 | American Beauty | £9.50 | £10.50 |

### b. Leeds

| CINEMA_ID | SCREEN_ID | TITLE | STD | LUX |
|---|---|---|---|---|
| 4 | 1 | X2 | £6.00 | |
| 5 | 1 | Die Another Day | £5.50 | £7.00 |
| 5 | 2 | X2 | £5.50 | £7.00 |
| 6 | 1 | A Bug's Life | £4.50 | £5.00 |

**Table 7.2:**
Extra data for
the SHOW
relations at the
London and
Leeds nodes of
the CINEMA
distributed
relational
database

provides the SHOW fragments for London and Leeds (similar information for Stoke can be found in Chapter 2, Table 2.4a).

This fragmentation would be efficient, as most user queries could be satisfied with reference to one site's fragment. Local film goers would normally only need find out information about films currently showing at London or Leeds, but not both. However, queries across all sites are still possible in a distributed database, for example, to ascertain which shows are currently popular countrywide, or to perform certain specialized applications, such as cross-site management functions.

### 7.4.3  Summary

Distributed spatial databases have the potential to improve data sharing, modularity, reliability, and performance for geographically dispersed spatial data. As we shall see in the following section, distributed spatial databases are increasingly important to real applications as low-cost connectionless and wireless computer networks become more commonplace. However, distributed databases remain an active research area. Distributed databases may not be practical in some application areas for a variety of reasons:

*Complexity*: Distributed databases and DDBMSs are inherently more complex than conventional centralized databases. Distributed database design, for example, entails all the difficulties of centralized database design, with the additional problems posed by fragmentation, replication, and possibly heterogeneity. Increased complexity also leads to higher costs of development and maintenance.

*Security*: Integrating databases at dispersed sites using a computer network may open sites to unauthorized access and increased security problems.

*Integrity*: Enforcing consistency constraints across multiple databases is more difficult than for conventional databases, because local autonomy may lead to updates in one database unit that are not consistent with other units.

## 7.5   LOCATION-AWARE COMPUTING

location-aware

A *location-aware* system utilizes information about a user's current location to provide more relevant information and services to that user. Location-aware computing is a particular type of *context-aware* computing. In general, *context-aware* computing concerns the use of sensors and other sources of information about a user's context to provide more relevant information and services. The term "context" here implies any information that can be used to characterize a user's physical, social, physiological, or emotional circumstances, for example, whether we are driving a car, in a meeting, tired, or angry. Location is a critical element of a user's physical context, because where individuals are located strongly affects their access to services and so the information they may find useful. For example, a hungry London tourist may be interested in information about the best places to eat in London, but the same information is unhelpful to someone looking for a good restaurant in New York.

context-aware

In addition to context-aware computing, two other closely related areas of research in computer science are pertinent to location-aware systems: pervasive and mobile computing. *Pervasive computing* (also termed *ubiquitous computing*) describes the idea that networked computers embedded throughout everyday objects can become unseen personal assistants, helping us with many of our daily tasks. In relation to location-aware systems, *mobile computing* is primarily concerned with information systems that can move around with us as we go about our daily business (but see "Mobile computing" inset on the next page).

pervasive computing

mobile computing

**Mobile computing** *At least three different types of mobility are relevant to information systems: software, device, and user mobility. Software mobility concerns the migration of software applications or agents between different computing devices or platforms, usually with the aim of completing some specific task on behalf of a user (see Chess et al., 1995; Rothermel and Schwehm, 1999). Device mobility concerns the ability of computing devices to move around, usually while retaining a network connection (see Forman and Zahorjan, 1994; Want and Pering, 2003). Portable devices, such as handheld computers, are examples of mobile devices. Some mobile devices, such as in-car navigation systems, may be mobile but not portable. The term "mobile" may also refer to the mobility of the user through an environment. Indeed, user mobility is an implicit feature of location-aware systems: if a user never moves, then that user's locational context is static and trivial.*

Most location-aware systems are either pervasive, mobile, or both (Figure 7.16).

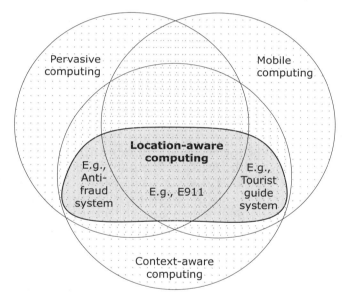

**Figure 7.16:** Relationship between location-aware computing and context-aware, pervasive, and mobile computing

As Figure 7.16 indicates, all four topics, location-aware, context-aware, pervasive, and mobile computing, have a large overlap. Location-aware systems are by definition also context-aware (because location-awareness is a type of context-awareness). Examples of different types of location-aware system from Figure 7.16 include:

*E911 system*: E911 (enhanced 911) is a program in the USA aimed at ensuring that cell phone users who call the emergency services can be rapidly located. The E911 system is clearly *location-aware*. The system is also *mobile*, because cell phones may be carried at all times by users. Finally E911 is *pervasive*, because E911 technology is embedded within the cell phone system and aims to automatically locate users in an emergency.

*Tourist information guide*: Mobile location-aware tourist information guides designed to provide information to tourists about nearby attractions have already been developed. These systems need to be *mobile*, so that they can be carried around by a tourist, and *location-aware*, so that users can select attractions close to their current location. However, these systems are not necessarily *pervasive*, as they commonly rely on specialized devices made available to visitors by a local tourist information authority.

*Credit-card anti-fraud system*: Credit card companies commonly detect fraud by analyzing the patterns of credit card usage. In particular, the location of each purchase is recorded by the credit card company. If a particular card is used in unexpected or widely different locations in rapid succession, this unusual locational pattern may trigger the credit card company to suspect fraud and block that card. Such a system is certainly *location aware*, because it tracks shoppers' locations via their credit card usage. It is also *pervasive*, because this protection is embedded within the credit card system, but not mobile, since the computing devices involved in the system are static (i.e., the network of retail "swipe" machines and the credit card company information system are at fixed locations).

wearable
computing

One further topic relevant to location-aware computing is *wearable computing*. Wearable computing is concerned with computer systems that are worn about a user's body and are always accessible to the user even when walking around or engaged in other activities. As a result, wearable computing lies within the intersection of pervasive and mobile computing.

Location-aware systems are able to fundamentally alter the way we interact with GIS. Instead of working with GIS at a desktop or workstation, location-aware systems allow us to actively interact with the geographic environments about which we are receiving information. As with many important developments in computing, these new possibilities have arisen as a result of technical developments:

- The number and variety of forms of computing devices commonly encountered have dramatically increased in recent years. Computing hardware, such as notebook, tablet, and handheld computers, continues to become smaller, cheaper, more powerful, and more energy efficient. Computing capabilities are also embedded within many everyday objects, such as cars, televisions, white goods, and telephones.

- The use of wireless communication networks, which allow mobile users to access remote information and services related to their location, is increasingly widespread.

- The use of sensors capable of determining a mobile user's location has become more commonplace.

Before examining location-aware computing in more detail, the following sections give an overview of wireless computer networks and location sensors. Hardware issues have already been covered in Chapter 1.

### 7.5.1 Wireless computer networks

Wireless networks are vital to mobile computing, as they allow computing devices to communicate with one another without the need to be physically connected. Wireless technology can be classified into three categories based on the spatial extents over which they operate.

*Wireless WAN*: Wireless WAN (wide area network) operates over large-scale geographical regions, such as states and countries.

*Wireless LAN*: Wireless LAN (local area network) operates over medium to small-scale geographical regions, for example, from city centers down to individual buildings.

*Wireless PAN*: Wireless PAN (personal area network) operates over very small geographic or sub-geographic areas, typically only a few meters in size.

PAN

A wireless WAN operates over the widest geographical extents of any wireless network, allowing users to roam nationally or even internationally. Cell phones are typical of wireless WAN technology. Cell phones use high-frequency radio and microwave signals to communicate with terrestrial or satellite relay stations. In common with cable-based telephone networks, newer cell phone technologies offering connectionless services are replacing older connection-oriented services.

Wireless LAN technology operates over smaller distances than wireless WANs, typically 100 m or so from the LAN access point. Isolated wireless LAN access points, such as those found in some airports, cafes, and hotels, are termed *hotspots*. Small clusters of wireless networks are referred to as Neighborhood Area Networks (NANs). Despite their short range, larger clusters of wireless LAN networks can achieve good coverage, especially in densely populated metropolitan areas, termed Metropolitan Area Networks (MANs). The first MANs were based on experimental projects in major cities like San Francisco, London, and Seattle. The most common example of current wireless LAN technology is usually referred to as Wi-Fi (wireless fidelity, based on the IEEE 802 family of standards).

hotspot

NAN

MAN

Wireless PAN technology operates over the shortest distances of any wireless network. Wireless PAN is ideal for connecting multiple small computing devices together, such as telephones, keyboards, mice, speakers, printers, and personal computers, without the need for wires. An example of a low-power radio-wave wireless PAN technology is *Bluetooth*, named after the 10th century Danish king. Infrared signals can also be used for line-of-sight wireless PANs (see Chapter 1). Figure 7.17 summarizes the different types of wireless network. Wireless networks

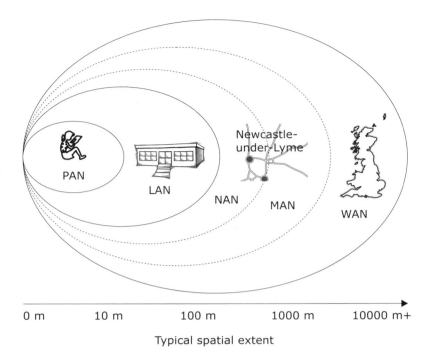

**Figure 7.17:**
Typical spatial
extents of
wireless
networks

Typical spatial extent

rely on a variety of techniques to ensure that wireless signals from different devices do not interfere with each other (see "Combating wireless
interference" inset, on the facing page).

### 7.5.2   Location sensors

There exists a wide range of sensors that can be integrated with computing
devices to automatically obtain information about a user's context; these
include sound, light, temperature, motion, and location sensors. Figure
7.18 shows a taxonomy of location-sensing techniques.

active location sensor

*Active* location-sensing techniques determine location based on signals from transmitters or *beacons*. Location can be determined using the
*proximity* to beacons or by *triangulation* of beacon signals. For example:

*Proximity*:  Proximity to beacon signals is used by all cell phone networks
    to locate cell phones. A cell phone network uses an array of
    wireless transmitters at known locations to relay phone calls. At
    the most basic level, each transmitter serves a small geographical
    area where its signal is stronger than any other transmitter, termed a
    *cell*. Signal strength decays with distance from the transmitter, so in
    the simplest case the geometry of a cell forms a Thiessen polygon
    (see Chapter 5), each cell containing the proximal locations to the
    transmitter. A cell phone can determine in which cell it is located
    by identifying the transmitter with the strongest signal. The size
    of cells varies across a cell phone network, depending on terrain,

**Combating wireless interference** *The problem facing all wireless network technologies is how to ensure that signals from different devices do not interfere with one another, causing data loss. There are essentially three increasingly sophisticated mechanisms for achieving this. The simplest option is to ensure that each device only uses a narrow frequency range for communication. This is rather like the way radio stations broadcast only on particular frequencies, into which you need to tune to hear the broadcast. The second option is to ensure that each device only transmits for a certain amount of time on its frequency range. This allows the same frequency to be shared by several devices, rather like the way that different radio programs occur at different times on the same radio station. The third option is to ensure that each device transmits at a range of frequencies, hopping between frequencies at different times in some sequence agreed by transmitter and receiver, termed spread spectrum technology. This would be rather like starting to listen to your favorite radio program on one station, and part way through retuning your radio to another station for the remainder of the program. This "frequency hopping" makes the network more tolerant to interference from other devices and environmental background noise: if some of the signal is lost to interference, the chances are that the signal will be regained next time it hops to a new frequency. Spread spectrum technology commonly switches hundreds of times every second between dozens of randomly chosen frequencies.*

transmission frequency, and expected volume of calls. Typically, terrestrial transmitters are spaced from about 200 m apart in urban areas to more than 5 km apart in rural areas.

*Triangulation*:  There are two types of triangulation: *lateration* and *angulation*. Lateration is the process of computing the position of an object based on its *distance* from other known locations. Angulation, is the process of computing the position of an object based on its *angle* from other known locations. *Global positioning systems* (GPSs) rely on lateration to determine location (discussed in the following section). Angulation is used, for example, by some cell phone networks to determine location.

lateration

angulation

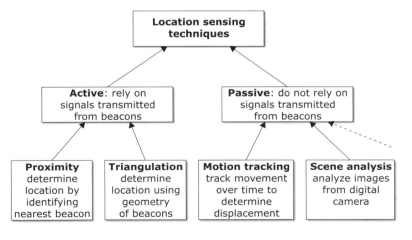

**Figure 7.18:** Taxonomy of location-sensing techniques

Some active location-sensing techniques, such as GPSs, use a mobile sensor to receive signals from external beacons at known locations. Other techniques operate using external sensors at known locations to receive signals from a mobile beacon. For example, the *Active Badge* system is an early example of location-sensing technology, which uses an array of infrared proximity sensors distributed at known locations throughout a building to locate individuals. Each individual wears a small badge that identifies the wearer by transmitting a coded infrared signal every 10 seconds.

passive location
sensor

*Passive* location-sensing techniques do not rely on signals transmitted from locater beacons. Instead passive location-sensors determine location indirectly, by relating sensor-based information to other spatial information about the environment. The geographic environment is highly heterogeneous, so almost any type of sensor measurement can potentially be used in some form of passive location-sensing (hence the dotted arrow in Figure 7.18). Imagine being transported from your home to an unknown location on the Earth's surface. As long as your watch accurately records the time at your home and you are able to "sense" the time at your location (perhaps by observing the position of the sun) you can obtain a crude estimate of your location in terms of your longitude. In the 18th century, sailors used a similar method to help them navigate. Two important classes of passive location-sensing are:

*Motion tracking*: Speed and direction sensors are often used in robotics to track the velocity of a mobile robot over time. This information can be used to calculate a robot's location at a particular point in time relative to its starting point. Motion tracking techniques are sometimes referred to as "dead reckoning."

*Scene analysis*: A digital camera is a sensor array that can be used to determine location in certain circumstances. Scene analysis can be used either to determine the location of objects or people in a digital image, or to determine what the location of the camera must have been in order to have taken the picture.

*Global positioning system*

A GPS receiver uses radio wave signals, transmitted from an array of GPS satellites orbiting the Earth, to calculate location of the receiver on the surface of the Earth. GPSs rely on lateration (defined above as the process of computing the position of an object based on its distance from other known locations). For example, assume for simplicity that the world is planar, and that we wish to discover the identity of an unknown city in the USA. If we are told that this city is 3000 km away from Los Angeles, then the locus of the city's possible locations is a circle with radius 3000 km centered on Los Angeles. If the city is also known to be 1000 km from New York, only two possible locations fit this information: Chicago and Atlanta. Finally, if the city is known to be 2000 km from Miami, then only one location fits this information: Chicago (see Figure 7.19).

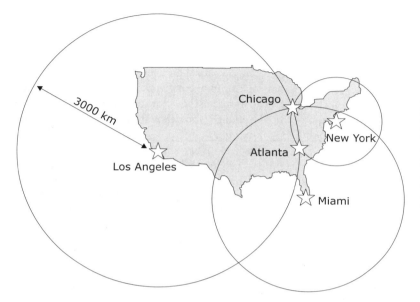

**Figure 7.19:** Lateration of location, based on distance from Los Angeles, New York, and Miami

GPS uses lateration in three dimensions, whereby the locus of locations a fixed distance from a point forms a sphere rather than a circle. As a result, four spheres are needed to pinpoint a location in three-dimensional space. Since the Earth itself is roughly spherical, most GPS receivers can obtain a fix using signals from only three satellites by assuming that the receiver is located on the surface of the Earth. GPS receivers calculate the distance to each satellite by measuring how long it takes a radio wave signal to travel from each satellite. Each satellite in the GPS constellation continually transmits a radio signal with the exact time and that satellite's position. The lag between the time stamp on the satellite signal and the current time at the GPS receiver is used to calculate the distance to each satellite, assuming the signal travels at or near the speed of light. GPS satellites measure time using very accurate atomic clocks. While the clocks in GPS receivers are not so accurate, any inaccuracies will affect all measurements equally and so can be accounted for in the GPS lateration algorithm.

GPS is a complex technology, and using conventional GPS carries some disadvantages. The precise frequency of transmission varies slightly as a result of the high speed of orbiting satellites. Each satellite uses a different encoding for its signal, which lasts about 30 seconds before starting again. As a result, the process of searching for signals and achieving an initial fix is complex, increasing the cost, size, and power consumption of GPS receivers. Further, the *time to first fix* (TTFF) usually lasts at least 1 minute, often several minutes. In some applications, such as emergency services, this delay can be critical.

time to first fix

*Sensor accuracy and precision*

Two important features of any location-sensing technique are its accuracy and precision. Sensor accuracy is the closeness of data from a sensor to the correct value(s). Sensor precision is the level of detail of the data generated by a sensor (see Chapter 9 for further information about accuracy and precision). For example, active proximity location-sensing techniques are generally accurate, but their precision depends upon the density of beacons. In a basic cell phone network, where beacons may be located several kilometers apart, precision is low because it is not possible to determine exactly where a phone is located within a cell (Figure 7.20a).

error propagation

Motion tracking techniques tend to be precise, but inaccurate if used for extended periods, because relatively small measurement errors can be compounded over time, a process known as *error propagation*. For example, if you travel at 3.0 kmh$^{-1}$ on a bearing of 068°, but your sensors inaccurately measure your velocity as 3.3 kmh$^{-1}$ on a bearing of 074°, then the location calculated from the sensor data diverges from your actual location over time (Figure 7.20b). After 5 seconds the inaccuracy in location will be about 60 cm. After 5 hours the inaccuracy in location will be more than 2 km.

**Figure 7.20:**
Sensor
accuracy and
precision

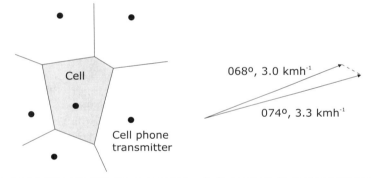

a. Imprecision in cell phone location    b. Inaccuracy in motion tracking

GPS can achieve high levels of both accuracy and precision. However, obtaining an initial fix (TTFF) can be slow, and the high-frequency low-power GPS microwave signals cannot be received inside or in the shadow of obstacles such as buildings. Improvements in accuracy and precision, in addition to other benefits, can often be achieved by using combinations of different location-sensing techniques. For example, most in-car navigation systems use a combination of GPS and motion tracking to accurately and precisely determine location. When a GPS fix is available, the car's location can be determined with high accuracy and precision. When local conditions, like tunnels or proximity to tall buildings, block GPS signals for short periods, tracking the speed and orientation of the car can fill in the "gaps" left by the GPS sensor.

Similarly, *assisted GPS* combines proximity-based location sensing with GPS. This combination results in greater precision than proximity-based location sensing and at greater speed than GPS-based location sensing can offer unaided. Assisted GPS dramatically simplifies the process of obtaining satellite signals. By using the proximity-based location information generated by the cell phone network, assisted GPS can predict the location of GPS satellites and their signal parameters, reducing the TTFF to a few seconds. For emergency applications such as E911, where conventional GPS is too slow and proximity-based location sensing too imprecise, assisted GPS may offer a viable alternative.

*assisted GPS*

### 7.5.3 Location-based services

*Location-based services* (LBS) are specific applications that require location-aware computing to operate. Some examples of LBS have already been given in Figure 7.16. LBS may also be classified according to their functional characteristics as positioning, tracking, and mobile resource allocation services.

*Location-based services*

*Positioning*: Positioning services are concerned with providing access to information and resources based on where users are currently located. A simple example of a positioning service is a location-aware map, such as a street map or weather map, which automatically centers itself on a user's current location.

*Tracking*: Tracking services are an extension of positioning services, concerned with providing access to information and resources based on a user's current and past locations. Examples of tracking services include automated road tolls and congestion charging, in which users pay for their road usage based on a history of where they have driven in their car.

*Mobile resource allocation*: Mobile resource allocation services are an extension of tracking services, concerned with providing access to information and resources based on a user's current and past locations, as well as planning where users need to be located in the future. Navigation, such as location-sensitive tourist directions, and fleet management, such as coordinating the real-time movements of vehicles in a delivery business, are examples of mobile resource allocation services.

Two additional features are required by many LBS. First, some LBS need to be *collaborative*, involving groups of interacting users rather than isolated individual users. Examples of collaborative LBS include enabling a group of friends to meet up; enabling players to engage in a location-aware collaborative game; or enabling visitors to leave "virtual" messages on tourist attractions. Second, some LBS require the integration of other non-locational contextual data with data derived from location-sensing technology. For example, many health and safety applications can benefit from integrating biosensors, such as heart monitors, with LBS. At-risk patients can be monitored for medical emergencies, such

**Websigns LOBS**  *The Websigns system pioneered several foundational ideas in LOBS, developed as part of the CoolTown project (http://cooltown.com). The Websigns system enables users to access a wide range of WWW-based information and services connected with physical objects in their vicinity. Websigns use GPS and digital compass sensors connected to an Internet enabled wireless device, such as a mobile phone or personal digital assistant (PDA). Users point their PDA at a feature of interest in their environment, such as a building, statue, or other landmark. The PDA connects to a Websigns server, which uses the location and orientation of the device to calculate what object is being pointed at. The Websigns server can then respond with the information or services available for that object. For example, pointing the Websigns PDA at a restaurant might cause the PDA to display information about tonight's menu or allow you to book a table for the evening. The system is based on a modular client-server architecture and uses XML to allow different devices to interoperate. Pradhan et al. (2001) contains more information on the Websigns system, while Egenhofer and Kuhn (1998) suggest some potential applications of LOBS.*

as a heart attack, and emergency services dispatched automatically to the patient's location at the first sign of a problem. Of particular importance are applications based on the combination of orientation and location sensors (location/orientation-based services, LOBS). LOBS enable users to access information about their environment simply by pointing at features of interest (see "Websigns LOBS" inset, on this page).

It should already be clear that LBS are inherently distributed, requiring architectures with high levels of modularity and interoperability. LBS rely on multiple independent computing devices, including databases, sensors, and mobile computers. Individual computing devices need to be able to integrate and process information from a variety of sources, interoperating with other networked sensors and computing devices. The constraints of size and power usage mean that mobile computing devices have lower computational capacity than static systems. As a result, client-server systems and server-side strategies are used by many emerging LBS (see "Websigns LOBS" inset, on this page). However, the highly decentralized nature of LBS means that distributed component and peer-to-peer network architectures are even more suited to LBS and location-aware computing generally.

### 7.5.4  Privacy and location-aware systems

privacy

While mobile location-aware systems have the potential to revolutionize many aspects of everyday life, a significant challenge is the question of how to protect an individual's *privacy* when using location-aware services. In an emergency most of us would be grateful for technology that could automatically inform the emergency services of our location, such as the E911 system in the USA. At the same time, we might feel our privacy and safety were being compromised if this information were to be broadcast to anyone who wanted to know.

We might have mixed feelings about other organizations knowing our location. For example, a landmark court case in the USA concerned a car rental company that used an enhanced GPS system to track their vehicles and charge any drivers who exceeded the speed limit. The company argued that their system was a safety feature that helped discourage dangerous driving. One customer argued that such tracking was an invasion of his privacy. Following a lengthy legal battle the customer won the case and substantial damages. There is clearly a balance to be struck between *pervasive* and *invasive* uses of LBS.

Privacy is internationally recognized as a basic human right, for example, in Article 12 of the United Nations Universal Declaration of Human Rights. Protecting digital information about individuals is an important component of the right to privacy, usually termed *data protection* or *fair information practices*. The goals of data protection are to ensure that organizations:

data protection

- only collect and use personal data for specific purposes;

- only collect personal data with the consent of the individuals involved; and

- take adequate steps to ensure that personal data is secure, accurate, and available to the individuals it concerns.

However, data protection must reach a compromise between protecting the individual's right to privacy and enabling new and potentially beneficial technologies to be developed and used.

Data protection is an issue that concerns any personal digital data, but there are some specific problems posed by LBS. For example:

- Knowledge of an individual's location can be used to infer other personal information about that individual, such as what an individual is doing or what interests the individual. For example, locational information about frequent visits to a hospital might lead an organization to infer an individual was seriously ill, perhaps adversely affecting insurance or employment prospects for that person. Such inferences may be incorrect. Whether or not they are correct, most people would regard such a practice as unfair and an invasion of privacy.

- Mobile location-aware systems do not always give a good indication of an individual's location. Location-aware sensors vary in accuracy and precision, sometimes depending on local environmental conditions. Further, location-aware sensors record the location of personal devices directly, but only record the location of people "by proxy." If I accidentally leave my location-aware device on a train, it rapidly provides a very inaccurate indication of my location, even though it may continue to accurately record its own (and the train's) location.

- It may not be immediately evident to a user when a location-aware sensor is collecting information about that user's location, just as the driver was apparently unaware that his movements were being tracked by the car rental company. When a user is aware of location sensors, he or she will usually be able to control if and when those sensors are collecting locational information. Location-sensing devices can usually be turned off, or simply left at home. However, when locational information is generated as a by-product of some other service a user is accessing, such as the proximity-based location sensing generated by cell phone networks, protection for individual privacy is especially important.

The question of how to safeguard personal privacy at the same time as allowing new location-aware technology and innovation to develop remains one of the most important unresolved issues facing LBS.

## BIBLIOGRAPHIC NOTES

7.1 Waugh and Healey (1987) provide a description of the GEOVIEW system based on the georelational model. Samet and Aref (1995) survey many of the important issues in spatial database architecture, including extensible, integrated, and object-oriented databases. Several papers survey the applications of the object-oriented approach to GIS, e.g., Worboys et al. (1990), Egenhofer and Frank (1992), Worboys (1994a). Component-based software is discussed in Clements (1996), with more detailed history and treatment in Garlan and Shaw (1993) and Szyperski (2002).

7.2 Wegner (1996) provides a concise introduction to the important issues surrounding interoperability. Vckovski (1998) provides more detail on interoperability in theory and practice from a geoinformatics perspective.

7.2.1 Moellering (1991, 1997) provides a snapshot of spatial data transfer standards from around the world.

7.3 Kurose and Ross (2002) is a general textbook on networking and distributed system architectures. Coleman (1999) and Sondheim et al. (1999) provide a GIS-based overview of distributed system architectures.

7.4 Most recent database textbooks will contain material on distributed databases. Elmasri and Navathe (2003) and Connolly and Begg (1999) are particularly recommended. A chapter on distributed databases can be found in Ramakrishnan and Gehrke (2000). Oszu and Valduriez (1999) is a recommended specialized text on distributed databases. Shekhar and Chawla (2002) contains a section on distributed spatial databases. Goodchild (1997) discusses the relationship between the location where geospatial data is stored and the region to which that data refers.

7.4.1 Wiederhold (1992) introduced the idea of information system mediators. A taxonomy of different types of distributed database is contained in Sheth and Larson (1990).

7.5 A clear and readable introduction to location-aware systems and location-based services is given by Werbach (2000). Theodoridis (2003) discusses spatial data structures and queries for location-aware systems. Davies et al. (2001) provides an overview of a location-aware tourist guide system. Weiser (1991) introduced the idea of ubiquitous computing.

7.5.2 An excellent overview of different location-sensing techniques is provided by Hightower and Boriello (2001). The Active Badge system is described in Want et al. (1992).

7.5.4 The location privacy case brought by James Turner against the Acme car rental is well documented, for example, in the *Chicago Tribune* (2001). One approach to safeguarding privacy in location-aware systems based on statistical analysis is described in Beresford and Stajano (2003). Onsrud et al. (1994) provides an overview of privacy issues in GIS, enumerating several key principles for handling personal data in a GIS. Monmonier (2002) provides a readable introduction to location privacy and surveillance technology.

# Interfaces

8

**Summary**

The field of **human-computer interaction** (HCI) is concerned with the design, evaluation, and implementation of effective **user interfaces** between humans and computing devices. A good interface should be both **intuitive** (easy to learn and use) and **expressive** (able to specify and perform the desired tasks efficiently). GIS interfaces are often based on the **map metaphor**: they exhibit similar characteristics to conventional paper maps. However, GIS interfaces can extend the map metaphor in several ways, for example, using animated, three-dimensional, non-visual, and multimodal displays.

The ability of computer-based systems to interoperate is a core feature of any GIS architecture, covered in the previous chapter. However, the success of any computer-based system ultimately rests on whether it can be used effectively by *people*. Just as GIS architectures are designed to promote interoperability between different computers, so GIS interfaces are designed to ensure that GISs and people can "interoperate." In this chapter, the essential characteristics of GIS interfaces are explored. Section 8.1 introduces the basic principles of computer interfaces. The main interface styles that may be used in a GIS are then explored in more detail, starting with conventional map-based interfaces in section 8.2 and moving on to the roles of animation, three-dimensional displays, non-visual interfaces, and feedback in section 8.3. Section 8.4 addresses the development in GIS interfaces for different types of user tasks.

## 8.1 HUMAN-COMPUTER INTERACTION

The term "human-computer interaction" (HCI) was first used during the 1980s to describe the interaction between computer systems and people. The term is also used to refer to the field of HCI: the study of the

human-computer interaction

293

design, evaluation, and implementation of the interfaces between computing devices and people. Figure 8.1 summarizes the key components of HCI: the human, the computer, and the interaction between them. This section explores in more detail how information is exchanged between the computer and the human, and what types of interface are used to facilitate interaction.

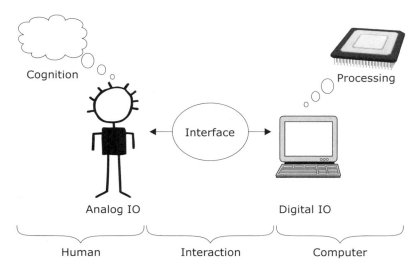

**Figure 8.1:**
Components of human-computer interaction

### 8.1.1   Input-output channels

IO channel

haptic

display

input

multimodal

Humans and computers are able to send and receive information in several different modes, termed *input-output* (IO) channels. Computers send information on IO channels that are detectable by human senses (see "Human senses" inset, on the next page). Computers are equipped with devices to detect information sent by human users on similar IO channels. For example, a PC receives information from a human user primarily on the *haptic* IO channel ("haptic" means related to touch), via the keyboard and the mouse. In return, information is sent from the PC to its human user on the visual IO channel, via a VDU (visual display unit) or monitor.

The same IO channel can be used for sending information in both directions, from computer to human and from human to computer. For example, auditory information can be sent via a computer speaker and received by human hearing, or sent by the human voice and received by a computer microphone. To avoid confusion, the term *display* is often used to refer to output from a computer to a human for any IO channel, while *input* is normally reserved for input to a computer from a human. Information displayed on more that one IO channel at a time may be useful in reinforcing important messages. For example, to alert a PC user to a warning, a visible message will often be accompanied by an audible "bleep." Systems that enable display or input of information on more than one IO channel (mode) at a time are termed *multimodal*.

**Human senses** *Humans have five senses that receive information from outside the body, termed exteroceptors: sight (visual sense), hearing (auditory sense), touch (haptic sense), smell (olfactory sense), and taste (gustatory sense). A fire alarm, for example, provides us with auditory information about a dangerous fire, ideally long before we receive similar information with our visual or olfactory senses. Most humans rely primarily on sight for receiving information from the environment. As a result visual information is the most commonly used sense in HCI, for example, in PC monitors and handheld-computer screens. Auditory and haptic information normally plays a secondary role in HCI, reinforcing visual information. However, such information may become the primary mode of HCI in certain circumstances. For example, some computing devices are too small for effective visual display. Olfactory and gustatory senses play a less significant role in HCI. Humans have two other senses relating to the movements of the body, termed proprioceptors: balance (vestibular sense) and kinesthesia, the ability to sense our own bodily movements and tensions. These senses are important in HCI, although normally used in support of visual information. For example, flight simulators can be mounted on a mobile base that stimulates the vestibular sense, mimicking the movements of a real aircraft. In the future, new IO channels for HCI may be developed, like neural interfaces, which enable computers directly to send and receive information from the human nervous system. In 2002, UK cybernetics academic, Professor Kevin Warwick, had a tiny electrode array surgically implanted into his arm as part of an experiment into neural interfaces (Warwick, 2000). Such HCI research may prove highly beneficial for people with severe physical disabilities who are unable to operate a computer using a keyboard or similar devices.*

There are many different devices used by humans to interact with computers, like keyboards, mice, VDUs, and speakers, and more will undoubtedly be developed in the future (see "Human senses" inset, on this page). Irrespective of which IO channel a device uses, it must convert information between digital and analog formats (Figure 8.1). Humans are organisms that send and receive information in a continuously varying analog format, whereas computers are digital machines that require discrete digital information to operate. For example, a computer mouse converts the continuous analog movements of our hands and arms into a discrete digital format that can be used by the computer. A speaker converts discrete digital audio information stored or generated by a PC into audible analog sound waves.

Finally, an increasingly important consideration in user input is the distinction between *implicit* and *explicit* input. In conventional information systems, most user input is explicit, entered consciously by the user via devices like a keyboard or mouse. Context-aware systems, like LBS (introduced in the previous chapter), make use of implicit input, whereby the system automatically gathers and interprets sensed information about a user's context. For example, a traveler wishing to find out about bus arrival and departure times might explicitly input a query for this information into a mobile computing device, perhaps using a keyboard or a handwriting recognition system. Instead, a location-aware system

explicit input

implicit input

might interpret the user's location at a bus stop as an implicit query for information about the time of the next bus at that stop.

### 8.1.2  Cognition and processing

We use information systems because a computer's information processing capabilities can support and enhance our own information processing capabilities. A GIS, for example, can reliably store data about millions of geographic features, calculating the areas or lengths of those features in a fraction of a second; this is something that any human would find impossible. At the same time, many information processing tasks that humans find simple are difficult to accomplish using a computer. The information processing capability of a human is rather different from that

*cognition*  found in a computer system, and is often termed *cognition* to distinguish between the two. It is possible to identify at least two general areas in which human cognition capabilities exceed computer-based information processing capabilities: *reasoning* and *problem solving*.

*Reasoning*

*reasoning*  Reasoning is the process by which information is used to infer new information about a problem domain. There are three different types of in-

*deductive*  ference that are used in reasoning. *Deductive inference* involves applying
*inference*  rules to specific examples. *Inductive inference* is a process of generalizing
*inductive*  from specific examples to general rules. Note that inductive inference
*inference*  should not be confused with mathematical induction, which is essentially
a deductive process. *Abductive inference* involves generating explanations
*abductive*  
*inference*  for some state of affairs. Table 8.1 summarizes with examples each type of inference.

| Inference | Form | Example |
|-----------|------|---------|
| Deductive | rule + case ⇒ result | All British people like tea, Mike and Matt are British, So, Mike and Matt like tea. |
| Inductive | case + result ⇒ rule | Mike and Matt are British, Mike and Matt like tea, So, all British people like tea. |
| Abductive | rule + result ⇒ case | All British people like tea, Mike and Matt like tea, So, Mike and Matt are British. |

**Table 8.1:** Deductive, inductive, and abductive inference processes

Assuming that the rules of deduction preserve truth, then deductive inference guarantees that if the premises are true, the conclusion must also be true. Inductive and abductive inference cannot guarantee true conclusions even if the premises are true. In the example of inductive inference in Table 8.1, even if Mike and Matt are British and like tea (we are and we do), it does not necessarily follow that all British people like tea.

Despite this unreliability, inductive and abductive inference are central to human reasoning: they allow us to reach conclusions and generate hypotheses that exceed the information contained within the premises. Humans possess several mechanisms for regulating potentially unreliable reasoning, such as incorporating implicit contextual knowledge into the reasoning process, or by revising or retracting knowledge that is found to be unreliable. For example, if having used the inductive inference in Table 8.1 to infer that "all British people like tea," I subsequently meet Mary, who is British but intensely dislikes tea, I would revise my original inference to accommodate this new information.

In general, computers rely solely on deductive inference processes, although inductive and abductive reasoning are used in some artificial intelligence-based systems. As a consequence, processing in a computer system is deductively valid, but this mode of reasoning prevents computers from generating new conclusions and hypotheses. We shall return to this topic in later sections, and to logic more generally in the next chapter.

### Problem solving

While reasoning involves using information about a familiar problem domain, problem solving refers to the ability to design solutions to problems in unfamiliar problem domains. Humans commonly use *heuristics*, in which informed trial and error based on rules of thumb are used to solve problems. Some examples of heuristics used in algorithms were encountered in Chapter 5, like the A* heuristic in section 5.7.3.

heuristic

Another human problem-solving technique is the use of *analogy*: adapting solutions from one domain to problems in a different domain. Object-orientation, discussed in Chapter 2, is an example of a successful analogy used in software development. Treating software modules as if they were physical objects, with identity, state, and behavior, makes sophisticated software easier to design and program. In fact, analogy is synonymous with *metaphor*, a basic technique in user interface design for making the operation of a computer more intuitive. We shall meet several different metaphors in the course of this chapter.

analogy

metaphor

Humans are also adept at *learning*, improving their performance over time by acquiring and refining skills with repeated exposure to an initially unfamiliar problem. Many artificial intelligence techniques have been developed to enable computers to imitate human reasoning and problem-solving capabilities, some of which are covered in more detail in Chapter 9. However, these techniques are generally only effective within narrow problem domains. No computer has yet been developed that can match the flexibility and adaptability of human cognition.

learning

### 8.1.3   Interfaces and interaction

A *dialog* is a process of interaction between two or more agents, whereby agents cooperate to resolve conflicts and complete some task. For example, humans commonly use dialog to communicate with each other. In a

dialog

**Intuitive interfaces** *In the classic book* The Design of Everyday Things, *Don Norman (1988) sets out in detail what it means for an interface to be intuitive. Norman's analysis is quite general and may be applied to computerized user interfaces, like a GIS interface, as well as physical interfaces, like a door handle. Four key features are central to Norman's analysis: visibility, mappings, affordances, and feedback. Visibility is the extent to which the features of an interface are prominent and easy to interpret. An interface in which key functions are hidden will not be intuitive. Affordances are those properties of an object that facilitate some action. A button is for pressing (i.e., a button affords pressing); a door handle is for turning or pulling. Mappings are closely related to metaphors and are concerned with whether the properties and affordances of an object conform to "natural" patterns. For example, in a GIS interface, the "up" cursor key might be expected to zoom in to a map, or perhaps pan up. Other mappings seem strange and counterintuitive: for example, using the "left" cursor key to pan up. Feedback involves sending information back to the user about what has been done and whether it was successful. Feedback is addressed in more detail in later sections of this chapter.*

human conversation each interlocutor will listen to the other, responding to what has previously been said. The process of interaction between a computer and its human user also normally takes the form of a dialog, although HCI dialogs usually have a much more rigid structure than human conversations. The dialog between humans and computers is mediated

*user interface*

by a *user interface*. The user interface in HCI is conceptually similar to an interface in a distributed system, introduced in Chapter 7 (and to an interface in object-oriented modeling, introduced in Chapter 2). A user interface in HCI is the mechanism by which a human user accesses the services offered by a computer.

Two characteristics of any user interface are how *intuitive* and how *ex-*

*intuitive interface*

*pressive* the interface style is. An intuitive interface is easy to use, requiring minimal effort to learn to operate (see "Intuitive interfaces" inset, on

*expressive interface*

this page). An expressive interface enables users to achieve specific tasks efficiently. Expressive interfaces allow users to specify commands more precisely, using a range of options that modify a command's behavior. The ideal interface style is both expressive and intuitive. Unfortunately, it is not usually possible to achieve both these goals. Different interface styles must strike a balance between being expressive and intuitive. Several common interface styles are discussed below and summarized in Table 8.2.

**Table 8.2:**
Basic interface styles

|  | Interface style | Example |
|---|---|---|
| *More expressive* | Command entry | UNIX "date" command |
| ⇕ | Menu | Microsoft Word menu |
| | Forms | Website journey planner |
| | WIMP | Mac OS and applications |
| *More intuitive* | Natural language | |

## Command entry interface

In a command entry interface, a human user issues commands to the computer directly. Command entry offers many different options that may be used to modify the behavior of basic commands. Command entry interfaces are therefore expressive but not intuitive, because the human user needs to learn and remember large numbers of commands and command options. The command prompt in Windows is an example of a command entry interface. Typing the command "dir" will result in the contents of the current directory being listed, while the command "dir /a:d /p" will list only the subdirectories of the current directory one page at a time. Figure 8.2 provides an example command entry interface, showing the command at the top and the results of executing the command below.

command entry

```
S:\work\book\gisacp>dir /a:d /p
 Volume in drive S is Shared
 Volume Serial Number is 1620-5919
08/19/2003  09:24 PM    <DIR>          chapter1
08/13/2003  09:29 PM    <DIR>          chapter10
08/20/2003  09:39 PM    <DIR>          chapter2
08/19/2003  08:28 PM    <DIR>          chapter3
08/19/2003  09:20 PM    <DIR>          chapter4
08/19/2003  09:35 PM    <DIR>          chapter5
08/19/2003  10:01 PM    <DIR>          chapter6
08/20/2003  03:10 PM    <DIR>          chapter7
08/27/2003  03:21 PM    <DIR>          chapter8
08/13/2003  09:29 PM    <DIR>          chapter9
08/22/2003  10:56 AM    <DIR>          misc
               0 File(s)              0 bytes
              13 Dir(s)   7,009,910,272 bytes free
```

**Figure 8.2:** Example command entry interface

## Menu interface

In a menu interface, commands are organized into logical groups that make it easier for users to access and remember commands. Menu interfaces are often found on automated telephone systems, usually referred to as *interactive voice response* (IVR) systems. A user of an IVR must select the desired function from a list, presented aurally over the telephone. Individual functions may lead to a further list of related functions being presented to the user, termed a *submenu* (Figure 8.3). Menu interfaces are less expressive than command entry interfaces, because the menu constrains the possible commands. On the other hand, menu interfaces are more intuitive as menus provide a logical arrangement for commands.

menu

submenu

## Form interface

In a form interface, the computer presents specific questions to which a user must respond in order to perform some task. For example, when booking travel tickets on the WWW, a form interface is often used to collect information about destination, date, and time of travel. Forms are intuitive, as users are led step by step through the interaction. Forms are not as expressive as menu and command entry interfaces, because the form interface can only offer access to a few specialized commands.

form

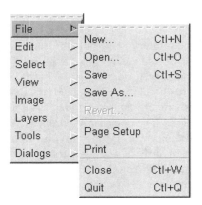

**Figure 8.3:**
Example menu
interface with
submenu

*WIMP interface*

WIMP    WIMP interfaces are familiar to PC users as they are the basis of most modern desktop-computer operating systems, including Microsoft Windows and Mac OS. WIMP interfaces have been successful because they are intuitive at the same time as being relatively expressive.

window    WIMP stands for *windows*, *icons*, *menus*, and *pointers*. Windows are used as independent containers for particular processes or applications

icon    (see Figure 8.4). Icons are small pictures that provide a metaphor for a command or function. For example, a pair of binoculars may be used as the icon for a "search" command (Figure 8.4). Icons make accessing individual commands more intuitive for users. Menus, introduced previously, are used to provide a logical structure for commands and are important

pointer    components of the WIMP interface. Finally, pointers enable users to point to a feature of interest and activate some function related to that feature by clicking on it. Pointers are highly intuitive, because pointing is a simple gesture that humans often use to draw attention to something of interest.

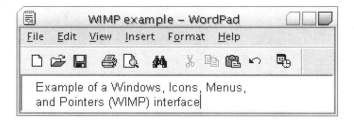

**Figure 8.4:**
Example
WIMP
interface

desktop metaphor    Groups of windows in the WIMP interface are often arranged and organized using the *desktop metaphor*. The desktop metaphor suggests a likeness between the user interface and an office desktop, with windows taking the place of sheets of paper, calculators, clocks, and so forth on an office desktop (see "Metaphors" inset, on page 302). Using the desktop metaphor helps make coordinating and switching between different software applications more intuitive.

## Natural language interface

Most humans use spoken or written languages, termed *natural language*, to communicate with one another. Natural language is potentially ideal for HCI interfaces, because it is both highly intuitive (once it has been learned) and highly expressive. However, natural languages are often ambiguous; as a result natural language interfaces are not yet advanced enough to be generally used in HCI (hence no example in Table 8.2). For example, the sentences "Time flies like an arrow" and "Fruit flies like a banana" are at first glance structurally similar. Only once the meaning of each sentence is grasped does it become clear that the two sentences are ambiguous and can assume quite different structures depending on the interpretation. In the most commonsense interpretation, "flies" acts as a verb in the first sentence and a noun in the second sentence, while "like" acts as an adverb in the first sentence and a verb in the second sentence.

*natural language*

It is important to note that, with the exception of the WIMP interface, the interface style is independent of the IO channels used for HCI. Although the WIMP interface relies primarily on vision, other basic interface styles (command entry, menus, forms, and natural language interfaces) can be based on visual, auditory, or to a lesser extent haptic IO channels.

## 8.2 CARTOGRAPHIC INTERFACES

With the exception of natural language interfaces, examples of all the interface styles described above can be found in almost any common computer system. The dialog between a human user and a GIS usually involves some additional, more specialized interface styles. One topic that has contributed strongly to the development of current GIS interfaces is *cartography*. Cartography is the art, science, technology, and history of maps and map making. Cartographic design principles have been developed over centuries of map use, and are still highly relevant to effective HCI in GIS.

*cartography*

There is a distinction, not always made, between the *spatial* and *graphical* representation of objects. Spatial objects exist to directly model the application domain, while graphical objects are the presentational form of spatial objects. Chapter 1 introduced the idea that maps conflate the data storage and data presentation functions of a GIS. A fundamental difference between maps and GISs, then, is that in a map the spatial and graphical representations of objects may be the same; in a GIS spatial and graphical representations are always separate. In the remainder of this section we are concerned solely with the graphical representations of objects in a GIS or on a map, but not the spatial representation, which was a major topic of Chapters 3–6.

Maps are effective tools for presenting geospatial data because they provide abstract graphical representations of the geographic world, high-lighting important information and suppressing unimportant information.

**Metaphors** *In human language, a metaphor is the use of a word or phrase denoting one kind of idea or object in place of another, in order to make a figurative comparison between the two. For example, the statement "an interface is a contract" is a metaphor, intended to suggest an imaginative likeness between a computer interface and a contract (i.e., an interface defines and restricts the types of interaction that may take place between two agents, the computer and the human). More generally, metaphor has been defined by Johnson (1987) as "a pervasive mode of understanding by which we project patterns from one domain of experience in order to structure another domain of a different kind." In this respect, metaphors are very similar to models: a source domain is projected by the metaphor into the target domain. (Indeed the analogy between metaphor and model itself constitutes a metaphor.) For example, in the desktop metaphor the source desktop domain is projected into a target domain of computer file-handling. Members of the source and object domains in a metaphor are usually taken from different linguistic categories. Madsen (1994) sets out some of the major linguistic categories as animate versus inanimate, and human versus animal versus physical. It is the ingrained nature of metaphoric understanding in humans that makes metaphors a powerful vehicle for human-computer interaction. A challenge for GIS research is to find appropriate metaphors for spatial data handling. Kuhn and Frank (1991) discuss a formal approach to metaphoric functions within the context of spatial applications. Johnson (1987) uses the term image schema to describe entire metaphoric structures. Mark (1989) analyzes geographic concepts in the context of image schema and discusses the relation between users' views of space and GIS operations.*

Map users are therefore able to focus on the salient relationships represented in the map without becoming distracted by irrelevant details.

Maps achieve this abstraction using three primary mechanisms. First, maps are *simplified*, providing only a limited the amount of detail about the geographic world. For example, we would not usually expect a road map to represent the most minor details about each road, such as a slight narrowing of a road or the smallest twists and turns. Second, maps are *classified*, only providing information on certain types of relevant features. For example, information on soil types or underground pipelines would not normally appear on a driver's road map. Third, maps use *symbolization* to represent different geographic features. Map symbolization is discussed in more detail in the following section.

The term *cartographic generalization* is used to describe the process of generating maps at appropriate levels of abstraction using operations like cartographic simplification, classification, and symbolization. To illustrate, Figure 8.5 shows two topographic maps of the Orono region of Maine, USA, at different scales. The extents of the map in Figure 8.5b are shown by the dashed line on Figure 8.5a. However, the two maps are not simply enlargements and reductions of each other; they are at different levels of abstraction. Generalization operations can be used to describe the relationships between the two maps. For example, individual houses in Figure 8.5b (shown as black dots) have been removed in Figure 8.5a (a type of simplification process known as *elimination*). Examples of line simplification, introduced in section 5.2.3, can also be seen in the edges

*simplification*

*classification*

*cartographic generalization*

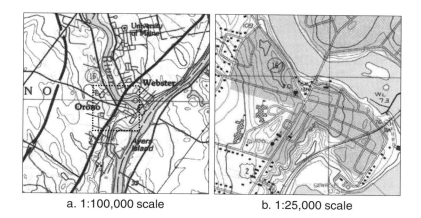

a. 1:100,000 scale          b. 1:25,000 scale

**Figure 8.5:** Cartographic generalization in topographic maps (Source: USGS)

and boundaries of Figures 8.5a and b. Further references to literature on cartographic generalization can be found in the bibliographic notes.

### 8.2.1   Map symbolization

Maps use an array of visual symbols to represent different geographic features, a process known as *map symbolization*. Several graphical aspects of these visual symbols, termed *visual variables* or *graphic variables*, may be manipulated to represent and distinguish different map features. Six visual variables, *position*, *size*, *orientation*, *shape*, *color*, and *pattern* are summarized in Table 8.3.

In addition to being used to distinguish different symbols, variations in each visual variable are identified with common associations or messages. The most obvious visual variable on a map is position. The relative positions of different symbols on a map communicate information to a user about the relationship between the different geographic features represented by those symbols. The size of map symbols is often used to signify relative quantity or importance: *larger* symbols usually indicate *more* of something. Orientation is normally used to communicate direction or flow. Shape is often used iconically: a square with a cross on top is conventionally used to indicate a church.

Color is normally modeled as the combination of three components, *hue*, *saturation*, and *value*, known as the HSV model. Hue refers to the shade of a color, for example, the amount of "redness" or "greenness." Certain hues have particular associations in map symbolization. Blue, for example, is often reserved for water bodies, while red might be used on a map to indicate hazardous areas. Value refers to the intensity of a color; its lightness or darkness. Like size, value is often used to signify relative quantity or importance: *darker* colors normally indicate *more* of something. Saturation (also termed *chroma*) refers to the purity of a color; how intense a color appears. Saturation is often used in combination with value, but may be used independently to control the prominence or impression of certainty of symbols.

*(margin notes: map symbolization, visual variables, hue, value, saturation)*

| Visual variable | Example | Example associations |
|---|---|---|
| Position | 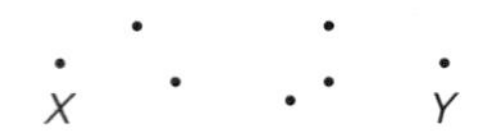 | Indicates position of features, e.g., $X$ is west of $Y$. |
| Size | 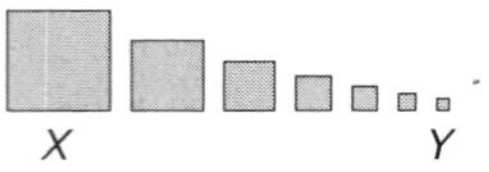 | Indicates quantity or importance, e.g., $X$ is more important than $Y$. |
| Orientation |  | Indicates direction or flow, e.g., $X$ runs perpendicular to $Y$. |
| Shape |  | Often used iconically, e.g., $X$ represents a church, $Y$ represents a triangulation point. |
| Color (hue, saturation, and value) | 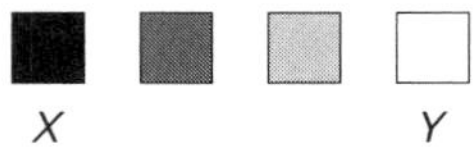 | Value indicates quantity or importance, e.g., there is more of something at $X$ than $Y$. |
| Pattern (texture, focus, and arrangement) |  | Texture (density) indicates quantity or concentration, e.g., something is more concentrated at $X$ than $Y$. |

**Table 8.3:**
Visual
variables

Finally, pattern concerns the *texture* (density), *arrangement* (e.g., ordered or random), and *focus* of symbols (e.g., crisp or fuzzy). Pattern may be used to communicate a variety of information, including relative quantity (e.g., denser symbols indicate higher concentrations of something) or certainty (e.g., fuzzier symbols indicate less certainty).

map metaphor

Just as WIMP interfaces often use the desktop metaphor to make interaction more intuitive, so many GIS interfaces use the *map metaphor*: they are designed to utilize map symbolization and exhibit map-like characteristics. Maps and map symbolization are immediately familiar to many people. Consequently, adopting the map metaphor usually leads to intuitive GIS interfaces, at least for those people who are already comfortable with conventional maps.

The map metaphor is a powerful visual presentation technique that needs to be used and interpreted with care. Mark Monmonier's book *How to Lie with Maps* provides many examples of the manipulation and misrepresentation of information using maps. For example, Figure 8.6 shows a map of reported UFO sightings in Ontario, Canada. The

relative sizes of the major cities and the UFO sighting symbols have been
deliberately chosen to exaggerate the importance of the UFO sightings.
Further, the lengthy time period over which the sightings were made
has been suppressed, giving the impression that Toronto is besieged with
space invaders.

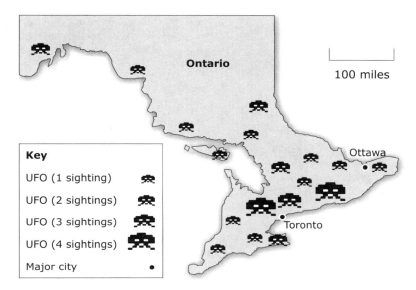

**Figure 8.6:**
UFO sightings
in southern
Ontario
Province,
Canada

Despite its advantages, the map metaphor does has four main disadvantages that limit how expressive and intuitive a map-based GIS
interface can be. Specifically:

- Maps are *static* and poor at representing change and evolution in
  geographic features.

- Maps are two dimensional, so using the map metaphor makes it difficult to represent complex three-dimensional geographic features.

- Maps and map symbols are based exclusively on visual IO channels, and do not take advantage of auditory, haptic, or other IO
  channels.

- Maps offer only limited opportunities for *feedback*. Maps allow
  visual display of information, but map users cannot engage in more
  sophisticated interaction with maps.

The topic of *geovisualization*, introduced in the following section, aims
to address these limitations of the map metaphor.

## 8.3 GEOVISUALIZATION

Geovisualization (also termed *geographic visualization*) is the process of
using computer systems to gain insight into and understanding of geospatial information. Experimental evidence suggests that humans use two

geovisualization

**Cartography cube**  *The cartography cube (below) has been suggested as a way of concep-*
*tualizing the relationship between conventional map use and geovisualization (MacEachren,*
*1994b). The cartography cube has three axes: interactivity, goals, and audience. Interactivity*
*is the degree to which users can manipulate and redefine a map. Goals refer to the degree*
*to which a map is designed to help users discover new information. Audience refers to the*
*degree to which a map is targeted at a specialized audience. In general, conventional maps*
*are located in the lower corner left-hand corner of the cube, presenting known information to*
*a public audience using low levels of interactivity. Conversely, geovisualization techniques*
*are located in the upper right-hand corner of the cube, using high levels of interactivity to*
*help private audiences discover new information.*

*For example, a conventional topographic*
*map might be useful in presenting known*
*information, such as the route to the top*
*of a mountain, but on its own is less likely*
*to help users discover new information,*
*such as understanding the land-forming*
*processes within that region. A conven-*
*tional topographic map is also public in*
*the sense that anyone, from hill walkers*
*to town planners, might use it. Many of*
*the views of information we encounter*
*using a GIS are private in the sense that*
*they might only exist on our computer*
*screen for a few seconds (or less) before*
*being altered or replaced.*

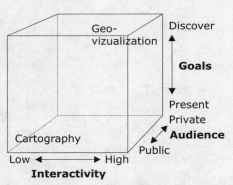

Cartography cube, after MacEachren (1994b)

fundamentally different types of information in thought processes: verbal
and visual. Verbal thinking is important for reading and writing, engaging
in conversation, and logical thought. Visual thinking is important for
reasoning about groupings, parts, and spatial configurations of objects.

scientific
visualization

*Scientific visualization* is the process of using information systems
to represent and interact with information in a way that enhances visual
thinking. It is important to emphasize that the "visual" in visualization
refers to "visual thinking," and not necessarily to the visual IO channel
(visualization may be based on any IO channel, e.g., visual, auditory,
haptic).

Geovisualization has emerged in recent years as a branch of scientific
visualization that deals specifically with geospatial information (see "Cartography cube" inset, on this page). Geovisualization design principles
emphasize the importance of using the dynamic, interactive, and multimedia capabilities of computers to help users gain insight to geographic

geographic
thinking

problems, sometimes referred to as *geographic thinking*. Geovisualization
techniques can extend the map metaphor in several ways, resulting in
interfaces that are both more intuitive and more expressive. The following
sections examine in more detail four visualization techniques with par-

ticular relevance to geospatial information: animation, three-dimensional displays, non-visual displays, and feedback.

### 8.3.1 Animation

Conventional paper maps are static and, aside from wear and tear, their appearance is fixed at the time of manufacture. In contrast, by using the computer to display a series of static images of data it is possible to convey the impression of motion or change over time, a technique known as *animation*. Each static image within the sequence that makes up an animation is called a *scene*.

animation
scene

Animation may be used simply to highlight or emphasize features on a static map, for example, by using a flashing arrow to indicate "You are here!" on an animated tourist information map. However, the most important function of animation is for visualizing *change* in geographic phenomena. Static maps can sometimes be used to depict change, for example, using static symbols like arrows to represent flow or movement. Animation is a more intuitive mechanism for visualizing complex dynamic phenomena than static maps. Weather reports, for example, often use a chronological series of maps of atmospheric conditions to show the movement of a storm front.

Animations are not restricted to depicting change over time. Animations can also be constructed from a sequence of spatial or attribute changes. One example of an animated spatial change is *fly-by* animations, such as those found in flight simulators, in which the user's view of a static data set is gradually moved. Attribute change, also called *attribute re-expression*, involves an animated logical sequence of scenes constructed from ordered attributes. Figure 8.7 illustrates the difference between animations of chronological change and attribute change. In Figure 8.7a, the animation can be interpreted as showing a gradual northerly migration of the population over a 30-year period. In Figure 8.7b, the animation can be interpreted as highlighting a spatial variation in population age, with younger people comprising a greater proportion of the population in the southern regions.

fly-by

attribute
re-expression

Animations are built from sequences of static scenes, so it follows that animations of geospatial data may use all of the static symbols in conventional maps, encapsulated by the six visual variables discussed previously. In addition, there are six *dynamic visual variables* that may be used to distinguish different features in an animated map: moment, frequency, duration, magnitude of change, order, and synchronization.

dynamic visual
variable

The point in time at which a change occurs in an animation is referred to as *moment*. Moment may be thought of as the temporal counterpart of the static visual variable, position. *Frequency* is the rate at which a change occurs in an animation, and is akin to the static visual variable *pattern*. *Duration* refers to the length of time that each static scene is visible. Duration gives an animation its pace. Scenes that are visible for longer move at a slower pace and appear to have greater importance or relative

moment
frequency

duration

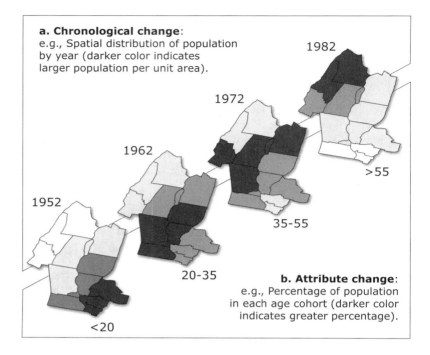

**Figure 8.7:**
Chronological
and attribute
changes in
animation

value when compared with those faster-paced scenes that are visible for a
shorter length of time.

magnitude of
change

    The *magnitude of change* is the amount of change that occurs in
moving from one scene to the next. Sequences of scenes in which the
magnitude of change from one scene to the next is small yield smooth
animations. Sequences in which the magnitude of change is large yield
abrupt or jumpy animations. The ratio of magnitude of change to duration
in an animation is termed the *rate of change*. The fifth dynamic visual

order

variable is the *order* in which scenes appear. The order of scenes in an
animation is normally dictated by the chronology of the data (for example,
in animating the spread of an epidemic over time) or the natural order
of the attribute used to construct the animation (for example, the age

synchronization

classes in Figure 8.7b). Finally, *synchronization* is closely related to order,
and refers to the relative timing of changes in two or more phenomena
represented in an animation.

    Order and synchronization are particularly significant in animation as
they can be used to suggest a causal relationship between phenomena. As
long as the magnitude of change is sufficiently small, users may interpret
features in one scene as being the cause of features in a subsequent scene.
In an animated map of the spread of an infectious disease epidemic, for
example, a region of disease that enlarges through a sequence of scenes
would normally be interpreted as contagion, in which individuals with
the disease in one scene are the cause of further cases of infection in
subsequent scenes.

Table 8.4 summarizes the dynamic visual variables by showing changes as shapes located on a animation timeline, with time moving forward from left to right.

| Dynamic visual variable | Example | Explanation of example |
|---|---|---|
| Moment | | Moment of change is indicated with a circle |
| Frequency | | Changes increase in frequency, from low to high |
| Duration | | Result of square change lasts longer than circle changes |
| Magnitude | | Change is gradual (circle to square) then abrupt |
| Order | | Order of changes is apparently random |
| Synchronization | | Changes in circle and square phenomena are closely related |

**Table 8.4:** Dynamic visual variables, after MacEachren (1994a)

### 8.3.2   Three-dimensional displays

Conventional maps and most visual computer displays are limited to two spatial dimensions. Yet geographic phenomena, like topography (the shape and features of the land surface), have three spatial dimensions. *Contours*, which connect points of equal height, are the most common mechanism for representing a topographic surface on a conventional map. An example of contour lines is shown in Figure 8.8a, along with an example of a *hypsometric map* in Figure 8.8b. A hypsometric map uses a logical sequence of colors to indicate elevation. Such maps are only useful for displaying surfaces, not true three-dimensional shapes like buildings, cliffs, or caves, because they can only show one (elevation) value at each location (see "Digital elevation models" inset, on page 143, Chapter 4).

contour

hypsometric map

Contour and hypsometric maps provide an abstract view of a three-dimensional surface, with which a skilled map-reader is able to gain an excellent understanding of the shape of a surface. These types of map are expressive but not particularly intuitive. Contour maps can be difficult for the novice map-reader to interpret. A more intuitive representation of three-dimensional surfaces involves using psychological cues to give

depth cue

the impression of depth. There are several such *depth cues* that may be exploited in a two-dimensional map or computer display. *Shading* may be used to give the impression that an object with height is casting a shadow. Figure 8.8c gives an example of hill shading, common in both conventional maps and computer displays. For topographic surfaces, the correct effect is most likely to be achieved when the source of illumination appears to be positioned in the northwest. Other sources of illumination are more likely to result in misinterpretation of the shape of surface, with valleys being mistaken for ridges, and vice versa.

**Figure 8.8:** Three representations of a terrain surface in two dimensions

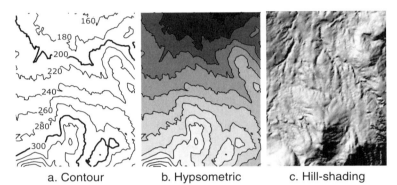

a. Contour       b. Hypsometric       c. Hill-shading

Other depth cues include the *relative size*, *linear perspective*, and *interposition* of the objects depicted, illustrated in Figure 8.9. Familiar objects with a smaller relative size than expected appear to be farther away. For example, we would normally interpret Figure 8.9a as a moose that is farther away than the mouse, rather than as a mouse that is much larger than the moose. Linear perspective concerns the effect that parallel lines appear to converge into the distance, such as the edges of the road in Figure 8.9b. Interposition concerns the effect whereby distant objects are occluded by nearer ones, such as the white building partially obscuring the tall building in Figure 8.9c.

linear perspective

interposition

**Figure 8.9:** Relative size, linear perspective, and interposition depth cues

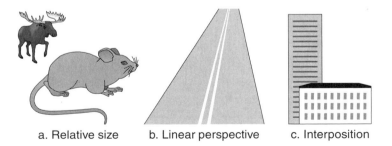

a. Relative size       b. Linear perspective       c. Interposition

The impression of depth can also be achieved using animation, in which closer objects appear to move faster than more distant ones. This effect, which you can observe out of the window of any moving vehicle, is called *motion parallax*. A wide range of techniques and software tools

motion parallax

now exist to build complex three-dimensional scenes using a combination of all these techniques, a process known as *rendering*.

rendering

The depth cues described above are termed *monocular*, because they are effective with only one eye working. Other depth cues rely on *both* our eyes functioning correctly (*binocular* depth cues). As a result of the slightly separated positioning of our eyes, each eye receives a slightly different view of the world around us, termed *retinal disparity*. Like other depth cues, these slight discrepancies are used by the visual system to help determine how far away an object is. One other binocular depth cue, termed *convergence*, occurs as our eyes move to fixate on near objects, such as your finger in front of your face. Convergence only affects depth perception for relatively close objects (less than a few meters away); eye convergence is negligible for objects in the middle to far distance.

monocular

binocular

retinal disparity

convergence

A *stereoscopic* display takes advantage of retinal disparity, by showing precisely controlled different images to each eye using special hardware, like a headset or stereo glasses. The need for specialized hardware means that stereoscopic displays are used less frequently than conventional visual displays. Further, there are perceptual difficulties associated with maintaining the illusion of stereovision, making stereoscopic displays less practical than monocular depth cues.

stereoscopic

Displaying physically three-dimensional phenomena, such as topographic surfaces, is one obvious application of three-dimensional interfaces in GIS. However, not all (or even any) of the dimensions need to be spatial to take advantage of three-dimensional interfaces. In relation to geospatial information, the third dimension is often used to represent some other non-spatial information, such as temperature. For example, Figure 8.10 shows a three-dimensional view of key buildings on a university campus, rendered using linear perspective, shading, relative size, and interposition as discussed above. The buildings are superimposed on a surface, but not a topographic surface. Instead of topography, the surface shows people's perception of how easy it is to move from one place to another, based on a study using human subjects. High peaks reflect the subjects' perception of regions that offer high resistance to movement; low troughs show relative ease of movement.

**Figure 8.10:** Non-topographic surface in three dimensions

### 8.3.3 Non-visual displays

Most people rely to a high degree on their visual sense to gain information about the world around them. However, many people have some form of visual impairment ranging from mild myopia (nearsightedness) to profound blindness. Even for sighted people, there are many situations for which the visual IO channel is inappropriate in an interface. When driving a car, for example, it may be dangerous to distract the driver's visual attention from the surrounding hazards. Conventional maps are only accessible to the human visual channel. GIS interfaces can be accessed via the other human input channels.

The most important non-visual human input channel is sound. The *sonification* process of using sound to represent data is termed *sonification*. Sounds in *sound symbol* HCI can be *symbolic* or *realistic*. Sound symbols are abstract sounds used to represent and distinguish information. For example, most people will be familiar with the Geiger counter, a device for measuring the levels of ionizing radiation. The Geiger counters often seen in movies use a sound interface, with more frequent, louder, and higher pitch clicks indicating increasing levels of radiation. There are a variety of *sound variables* that may be used in sound symbolization, including loudness and pitch. Sound is inherently dynamic, so many of the dynamic visual variables have counterpart sound variables, including duration, order, frequency (in this context meaning the periodic recurrence of sounds), and rate of change.

Realistic sounds may be separated into two categories: sound icons and spoken language. Like visual icons, sound icons, often termed *earcons* *earcons*, are sounds that provide a metaphor for a command or function. For example, in desktop computing the action of deleting a file is often accompanied by the sound of paper being scrunched up as if it were being discarded in a waste paper bin. Spoken language is increasingly a common feature of many interfaces, such as in IVR telephone menus (commonly used in telephone banking, for example). Speech synthesis systems capable of converting text to spoken language (text-to-speech systems, or TTS) are now easily available for many different languages. While computers are able to synthesize spoken language, computerized natural language recognition and generation remain long-term research questions (see section 8.1.3 above). The different types of sonification are summarized in Figure 8.11.

**Figure 8.11:** Taxonomy of sonification techniques

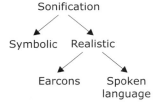

Our sense of hearing also allows us to locate the source of sounds in three-dimensional space. Slight discrepancies between the sounds heard in each ear enable us to determine the direction of a sound source,

**Multimodal GIS** *TACIS (tactile acoustic computer interaction system) is an example of an early multimodal information system specifically designed for the visually impaired. TACIS is able to emboss graphics onto A3-sized paper or plastic. The embossed graphics are then placed on top of an A3-sized touch pad. Users can access additional information about the graphics by applying pressure to different parts of the embossed graphics. This additional information can be displayed using sound symbols or speech synthesis. The TACIS interface has been used as the basis of a tourist information GIS. Sound symbols are produced as a visually impaired user explores an embossed map of a city with his or her hands. The different sound symbols enable users to distinguish different geographic features, such as roads and bus stops. Other geospatial information, such as speech synthesized bus timetables, can be accessed by pressing harder on the relevant parts of the embossed map.*

analogous to retinal disparity in vision. Other audio cues, like loudness, help to determine distance. Many digital sound systems, like those found in modern cinemas, take advantage of this facility and produce sounds that appear to be located in space. However, the precision with which humans are able to determine position in space, termed *spatial acuity*, is lower for hearing than for vision.

*spatial acuity*

Haptic displays are one other important type of non-visual interface. Human haptic interfaces are already used in certain specialized domains, like Braille displays and embossers for the visually impaired, vibrating mobile phones, and computer game controllers. While sound-based displays may be used as a primary IO channel, for example, in IVR systems, haptic displays are almost always used as a secondary IO channel, providing supporting or supplementary information to users as part of a multimodal interface (but see "Multimodal GIS" inset, on this page). Using visual, sound-based, and haptic displays together within a multimodal interface carries three main advantages:

- Multimodal interfaces may be used by a wider variety of users (including visually impaired or hearing-impaired people) than unimodal interfaces.

- Multimodal interfaces are able to operate in a wider variety of conditions than is possible with a unimodal interface, such as when a user's visual attention cannot be directed at a computer display.

- Multimodal interfaces enable users to access complementary information on different IO channels concurrently. Experiments have shown that this can lead to increased efficiency and decreased errors in user interaction.

### 8.3.4   Feedback

Feedback is the process of accepting and responding to a user's actions with information about what a user has done, and what has been achieved. Feedback is a distinguishing feature of GIS interfaces when compared

*feedback*

with conventional maps. While maps are able to present geospatial information, GIS interfaces can respond to user input. Users can formulate, refine, and test different hypotheses during the dialog between user and GIS. As a result, feedback is a key process in promoting reasoning and problem solving with geospatial information.

response time

To ensure the reasoning and problem-solving process is uninterrupted, responses to user input must be rapid. *Response time* is the time taken for a computer to respond to a user's input. As a rule of thumb, a response time of about 0.1 seconds or less is perceived by users as instantaneous, and so is ideal for reasoning and problem solving. A response time of about 1.0 second is still fast enough for a user's flow of thought to be uninterrupted, but is a noticeable delay lessening the impression of interactivity. A response time of 10 seconds is about the limit for feedback, as a user's attention is likely to be transferred to another task if delays exceed this time.

dynamic query

A basic feedback technique is to enable a user to continuously vary the selection criteria for a query and simultaneously view the results of that query, termed a *dynamic query*. Examples of dynamic querying include:

zooming

- changing the level of detail at which part or all of a data set is displayed, termed *zooming*;

panning

- changing a user's viewpoint of a data set, but not the level of detail, termed *panning*;

focusing

- changing a threshold value used as a criterion in a dynamic query, termed *focusing*; and

brushing

- pointing to a location to dynamically query information related to that location, termed *brushing*.

As an example, imagine a real estate website offering information and maps about houses for sale to potential buyers. The website might first allow users to *zoom* and *pan* onto a particular region of interest. Then the site might allow buyers to *focus* in on subsets of properties within a region by interactively setting the threshold distance to a local school or other amenity. Buyers might then further narrow down their search by viewing the price of candidate properties simply by *brushing* their pointing device over the location of each house in turn. Figure 8.12 provides a static illustration of zooming, focusing, and panning techniques for a map of the Newcastle-under-Lyme region in the UK. Note that cartographic generalization operators, like simplification and elimination, are used within the zooming process in Figure 8.12.

linked views

Dynamic query techniques are often used in combination with multiple *linked views* of a data set. Using linked views, changes to highlighted information in one view are mirrored by changes in other views. For example, as users of the real estate website brush their pointer over particular locations, changing information about the price, property taxes, and photographs of each property might appear in a separate window. Similarly, when zooming in on a region of particular interest, it is helpful

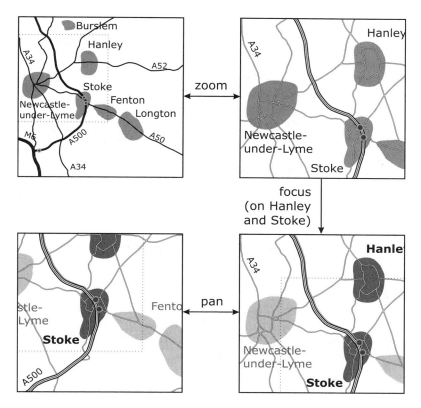

**Figure 8.12:**
Zooming,
focusing, and
panning

to provide a separate view of the entire data set showing the current zoom extents. This linked view ensures users are able to keep track of how the more detailed view relates to the data set as a whole.

Feedback can be achieved using input and display on any combination of IO channels. Input normally relies on the haptic input channel, using devices like a keyboard or a mouse. Audio-based speech recognition input is increasingly common in many interfaces. Reliable computerized speech recognition is more difficult to achieve than speech synthesis, as a result of the high levels of variability in voices, accents, idioms, and vocabulary used by different people. Various techniques exist for interpreting human bodily posture and movement as input to a computer, including *eye tracking* and *gesture tracking*. Eye tracking systems use infrared sensors to detect the direction in which a user's eye is looking. Gesture tracking systems detect and interpret simple gestures, like pointing. Detecting human movements is normally achieved using sensors capable of tracking infrared or magnetic tags attached to the user, or by processing digital images of a user.

eye tracking

gesture tracking

Combining feedback with the different display styles can lead to interfaces that are highly intuitive. For example, dynamic three-dimensional displays are especially intuitive and powerful because they enable users to use the metaphor of navigation to explore and gain a deeper understanding

**Immersive virtual reality** *Immersive virtual reality (VR) is an advanced interface style that simulates as closely as possible the features of a real physical environment. VR systems typically use multimodal display and input. Headsets are commonly used in VR to display three-dimensional graphics and sound. The same headset may track head and eye movements. Visual or sound scenes are rapidly redisplayed in response to head and eye movements to provide high levels of feedback. For example, when a user turns his or her head to one side, the movement is detected, and the visual display is updated to reflect the user's change of perspective. The haptic IO channel is also harnessed in VR systems. Users may don special gloves that track hand movement (input) and provide tactile stimulation to fingertips (display). VR systems may contain representations of real geographic spaces. Such systems can provide a safe environment in which to train emergency staff, like firefighters, to deal with hazardous situations. VR systems may also be used for imagined geographic spaces. Architects sometimes use VR to gain insight into the effects of a planned development. VR is in some ways the antithesis of pervasive computing, discussed in the previous chapter. VR systems are designed to simulate and replace our physical environment, while pervasive computing systems are intended to blend invisibly with our physical environment. Related to both pervasive computing and VR, augmented reality (AR) is the technique of combining real physical environments with computer-generated information. Users of an augmented reality system are able to access relevant digital information overlaid on top of information gained through normal interaction with the world around them. For example, while walking around a city, a tourist wearing special AR glasses might be shown information about opening times, menus, and reviews for restaurants within that user's field of view.*

of a data set. Humans are adept at navigating through physical geographic spaces. We are able to structure geographic space by remembering routes through a space and significant landmark locations. The same techniques can be used for navigation through a virtual environment. Such a virtual

immersive virtual
reality

environment need not be geographic at all. For example, *immersive virtual reality* (VR) is an advanced interface style that provides a user with the impression of a real physical environment by using multimodal input, like gesture tracking and eye tracking; multimodal display, including visual and haptic displays; and rapid feedback (see "Immersive virtual reality" inset, on this page). VR systems are commonly used for exploring real or imagined geographic spaces, but may also be applied to more abstract spaces, like navigating around the Internet using the metaphor of navigation in a city.

## 8.4   DEVELOPING GIS INTERFACES

GISs are used in a variety of different types of task, including presentation, querying, browsing, editing, integrating, analyzing, decision making, and problem solving, defined below:

*Presentation*: Displaying information in an appropriate manner

*Querying*: Retrieving stored information in response to a specific question posed by a human user

*Browsing*: Retrieving stored information when users are unable to formulate a specific question

*Editing*: Correcting, updating, or adding to stored information

*Integration*: Combining or conflating information from different data sets

*Analysis*: Processing information to highlight or reveal relationships and patterns within a data set

*Decision making*: Evaluating and choosing between different options or courses of action

*Problem solving*: Designing solutions for problems in new and unfamiliar domains (see section 8.1.2)

An overriding objective of any user interface, termed *usability*, is to enable users to successfully complete particular tasks. From an HCI perspective, each task for which geospatial information is used places different demands upon the interaction between the computer and its human user. Table 8.5 summarizes, with examples, the different tasks for which geospatial information may be used. Some tasks demand relatively simple forms of interaction, involving only the exchange of information between human and computer. Querying, for example, only requires the computer to interpret a user's query and to communicate the answer to that query back to the user.

Other tasks demand more sophisticated forms of interaction, utilizing a computer's processing capabilities to stimulate and augment a human user's own cognitive processes and promote geographic thinking. Decision making, for example, is primarily a human activity, but interfaces that allow human users to explore and gain insight into the problem domain can support the decision-making process. In general, tasks that require more complex forms of interaction, such as decision making and problem solving, will benefit most from interfaces that draw on geovisualization techniques. Feedback, for example, allows users to formulate and test hypotheses, a key component of human reasoning processes. Similarly, the visualization process of transforming data into multiple different forms, such as three-dimensional displays and animations, supports human analogical and metaphorical problem solving.

To ensure a GIS is usable, the interface must be appropriate to the type of task for which it is designed (see "Collaborative spatial decision making" inset, on page 319). For example, an in-car navigation system should present authoritative information to the user in a form that does not distract attention from the hazardous task of driving. A navigation system that allows the user to explore many different options is likely to be confusing (for example, a navigation system that offers options such as "Take the next turn left, or the second turning on the right").

usability

|                      | Task             | Example                                                                 |
| -------------------- | ---------------- | ----------------------------------------------------------------------- |
| Simple interaction   | Present          | "Show me a map of my home town."                                        |
|                      | Querying         | "Find the location of my house."                                        |
|                      | Browse           | "Let me see what things are located in my neighborhood."                |
|                      | Edit             | "Change the database to reflect the fact I have built an extension to my house." |
|                      | Integrate        | "Combine the street map and local government zoning plans for my home town." |
|                      | Analyze          | "Calculate the travel time from my house to my work."                   |
|                      | Decision making  | "Help me choose a new house to move to."                                |
| Complex interaction  | Problem solving  | "Help me assess whether I need to move to a new house."                 |

**Table 8.5:** Interaction between human user and GIS for different tasks

Similarly, a navigation system that relies on complicated visual output is more likely to distract a user's visual attention from driving and contribute to an accident.

Conversely, a planning system designed to assist urban planners considering whether to give permission for a new building project should be focused on promoting problem solving and geographic thinking. A system that simply provides a "give permission" or "deny permission" answer is unlikely to be usable (even if it were ultimately the correct decision). Sophisticated visual displays, such as three-dimensional animated or virtual reality simulations of the project, may be helpful in gaining a better understanding of the impacts of the development.

### 8.4.1   Usability engineering

usability
engineering

Developing usable interfaces is not an exact science: experience and judgment play a vital role in successful interface development. The process of developing interfaces that maximize usability is termed *usability engineering*. In general, all usability engineering techniques emphasize the importance of considering usability at every stage of the system lifecycle (see section 1.3.4). The usability of an existing interface can be assessed empirically by evaluating its performance either in the laboratory or in actual use in the field. The usability of interfaces is often measured against five basic criteria, listed below (after Shneiderman, 1997). These criteria are, in essence, an attempt to quantify the degree of intuitiveness and expressiveness of an interface.

*Time to learn*: How much time does it take for users to learn how to use a system?

**Collaborative spatial decision making** *An emerging application area for GIS in which innovative interfaces are needed is collaborative spatial decision making. Systems that are designed to aid collaborative or cooperative work are termed groupware. The study of groupware and groupware interfaces is a discipline within HCI, called computer-supported cooperative work (CSCW). Spatial decision making, such as the decision to build a new hospital, often affects many stakeholders within a community. Modern planning methods are usually designed to involve communities within the decision-making process, with planners, developers, and the community working together as a group. Conventional desktop GISs are not well suited to this task, because they are too expressive to be usable by people who are not experienced map or GIS users. Related to CSCW, participatory GIS (PGIS) is intended to help develop GISs and GIS interfaces that promote cooperative decision making. As a result of the wide range of non-specialist interest groups involved in cooperative decision making, PGIS interfaces should be highly intuitive. Feedback allows users to explore for themselves the possible effects of different types of development. Three-dimensional displays may be used in preference to less intuitive map-based displays. A variety of input modes, like sketch-based input, can make it easier for all the stakeholders in the community to express their ideas. For further reading, Dix et al. (1998) contains two chapters on groupware and CSCW and Jankowski and Nyerges (2001) provides material on participatory GIS.*

*Speed of performance*: How quickly can users carry out benchmark tasks?

*Rate of errors*: How many and what type of errors do users make in carrying out benchmark tasks?

*Retention over time*: How well do users retain their skills and knowledge over time?

*Subjective satisfaction*: How much did users enjoy or dislike performing benchmark tasks?

In addition to empirical user evaluation, interface designers often use more analytical techniques to ensure usable interfaces. To end this chapter we provide a brief overview of the three main techniques: prototyping, design rationale, and design analysis. *Prototyping* is one of the sim- prototyping plest usability engineering techniques. Prototyping involves developing a "mock-up" (prototype) of the target interface in order to reveal problems that may be hard to detect with pen-and-paper designs. Prototypes may be used simply to test design ideas, termed a *throw-away* prototype, or developed as a preliminary version of the actual interface, termed an *evolutionary* prototype.

A *design rationale* is a document that provides some explanation of design rationale why the system has been constructed in a particular way. At each stage of development, designers document the decisions they make by attempting to justify the decision and consider alternatives. Design rationales may be written or diagrammatic. Several diagrammatic notations, such as IBIS (issue-based information system), exist for design rationales.

design analysis    *Design analysis* provides a step-by-step framework for interface de-
GOMS    sign. One of the most common design analysis techniques is GOMS
(goals, operations, methods, selection). GOMS recursively decomposes
high-level interface functionality into smaller component functionality.
In a GOMS model, the user *goals* are specified in terms of the basic
*operations* a user can perform. GOMS also models the different *methods*
for achieving a particular goal, and attempts to predict which particular
method will be *selected* and in which cases. GOMS is closely related
to *hierarchical task analysis* (HTA), another common design analysis
technique.

Overall, prototyping is the most informal and low-cost usability engineering technique, and can be valuable for even the smallest projects.
Design analysis techniques are quite specialized and require time to
learn and apply, but are well suited to large projects, which require
collaboration between many designers to produce an effective interface.
Design rationales provide a mid-ground between the relative formality
of design analysis and the relative informality of prototyping. All three
techniques will normally be used in combination with empirical user
testing. References to more detailed material on this topic can be found
in the bibliographic notes that follow.

## BIBLIOGRAPHIC NOTES

8.1 There are many HCI textbooks available. Dix et al. (1998) is a
highly readable introduction to HCI which covers in more detail
many of the topics touched on in this chapter. Shneiderman
(1997) and Preece et al. (2002) are also recommended HCI texts.
An introduction to HCI issues in pervasive computing can be
found in Abowd et al. (2002).

8.2 Robinson et al. (1995) is a standard textbook on cartography.
Jones (1997) tackles the provision of cartographic processes
and display techniques within a GIS, including material on
cartographic generalization. Examples of other sources of information on cartographic generalization are Weibel and Jones
(1998), Buttenfield and McMaster (1991), and João (1998).

8.2.1 MacEachren (1995) provides more information on map symbolization and visual variables. The idea of visual variables
in graphical design originates with Bertin (1967). Monmonier
(1996) aims to teach critical evaluation of maps through an
understanding of their use and misuse.

8.3.1 Di Biase et al. (1992) and MacEachren (1994a) introduce animation in GIS and dynamic visual variables, as well as provide
a detailed account of the different types of animation in geographic information. Peterson (1994) gives an overview of the
uses of animation in GIS.

8.3.2 In addition to being a good all-round scientific visualization textbook, Ware (2000) contains a concise explanation of depth perception and its use in three-dimensional interfaces. Camara and Raper (1999) provide a collection of papers with examples of the use of multimedia and virtual reality in GIS.

8.3.3 Krygier (1994) provides an overview of sound symbolization for geographic information, and enumerates the key sound variables. An example of the use of symbolic sounds in a GIS interface to represent the reliability of geospatial information can be found in Fisher (1994). Gaver (1986) introduced earcons.

8.3.4 Dykes (1997) discusses the use of simple feedback techniques, including brushing and linked views, with geographic information. An example of the use of focusing to explore uncertainty in geographic information is given in MacEachren et al. (1993).

8.4 Egenhofer and Kuhn (1999) put forward a different typology of GIS tasks from that used in this chapter. Timpf (2003) has discussed GIS interfaces from the perspective of task analysis.

8.4.1 More details on usability engineering techniques and interface evaluation can be found in the HCI textbooks mentioned above: Dix et al. (1998), Shneiderman (1997), and Preece et al. (2002).

# Spatial reasoning and uncertainty

9

**Summary**

*Spatial reasoning is concerned with the cognitive, computational, and formal aspects of making logical inferences about a spatial environment. Knowledge of our geographic environment is almost always **imperfect**, so effective spatial reasoning must be able to operate under **uncertainty**. This chapter reviews the basic concepts involved in uncertainty, and goes on to describe some of the most commonly used techniques for managing uncertainty in information systems.*

The theme of this chapter is reasoning with geospatial information under uncertainty. Uncertainty is prevalent in any information system, but this topic is even more important for GIS because geospatial information is closely linked with observations of the world, and these observations are often inherently uncertain. Before engaging directly with the topic of reasoning under uncertainty, the chapter begins with a general discussion of formal aspects of spatial reasoning. Section 9.2 then considers uncertainty in many of its guises. Then the discussion turns to some specific qualitative (section 9.3) and quantitative (section 9.4) techniques for representing and reasoning under uncertainty. Section 9.5 presents some example applications in which uncertain regions are important.

## 9.1 FORMAL ASPECTS OF SPATIAL REASONING

*Spatial reasoning* provides ways for humans and computers to make inferences about spatial aspects of the environment. Various components of spatial modeling were introduced in Chapters 3 and 4. This section introduces reasoning using these models, and lays the foundation for the more practical case where uncertainty is introduced.

spatial reasoning

Spatial reasoning has *cognitive*, *computational*, and *formal* aspects, and draws from disciplines as diverse as psychology, cognitive science, spatial cognition artificial intelligence, and robotics. *Spatial cognition* is an important research topic concerned primarily with the cognitive aspects of spatial reasoning. Formal aspects of spatial reasoning derive from logic. Computational spatial reasoning is concerned with implementing these formal approaches in computational devices.

Logic is an enormous subject, whose beginnings go back to when people first thought about their own reasoning processes, the first major written sources being work in the philosophical schools of ancient Greece. This section indicates some of the most important concepts in logic, showing how they apply to spatial reasoning.

### 9.1.1  Syntax and semantics

A key logical distinction is between *syntax* and *semantics*. Syntax and semantics have already been encountered in the context of interoperability in Chapter 7, but here we make the distinction more precise. Consider the sentence, "Paris is in France." A syntactic analysis would focus on the terms of the sentence; the four-word sentence contains two proper nouns (starting with uppercase letters), a verb, and a spatial preposition. The semantics, or meaning, of the sentence can only be determined by its context. If the term "Paris" refers to that city famous for the Eiffel Tower, and "France" is the name of a European country well known for its wines, then the sentence has a clear meaning, and is in fact true. However, if "Paris" refers to the diamond merchant Paris F. Rocksdoller, currently in Cape Town, South Africa, again the semantics are clear denotational although the sentence is false. In this *denotational* view of semantics, semantics the meaning of a term is determined by reference to the domain entity or relationship it denotes. The meaning and possibly the truth value of a complex proposition are determined by the meaning and structuring of its terms.

A small but important item missing from the discussion so far is a consideration of the preposition "in." Here, "in" denotes a spatial relation, in the same way as "outside," "next to," "near to," and "connected to" are spatial relations. The study of such spatial relations, and their place in formal and natural languages, is at the heart of the topic of spatial reasoning. The temporal context of a denotation is also often important. In our example, if Mr. Rocksdoller were later to go to France, then at the moment in time of his entry to France, the proposition would become true. Temporal terms can be added to the syntax, leading to *temporal logics*.

### 9.1.2  Logic and deduction

Consider the three sentences:

> Paris is a city in France.
> All cities in France are European cities.
> ───────────────────────────────────
> Paris is a European city.

The proposition "Paris is a European city" below the line (the *conclusion*) follows from the sentences above the line (the *premises*) by *deduction*. We can note that the process is quite general, and independent of the meanings of the individual terms. The general form of the deduction is:

> $x$ is a $y$.
> All $y$'s are $z$'s.
> ---
> $x$ is a $z$.

*conclusion*
*premise*

This particular deduction is an example of a classical *syllogism*, whose principles were codified by Aristotle in around 350 BC. Ideally, a deduction will be *valid*; that is to say, it preserves truth. If the premises of a valid deduction are true, then the conclusion is also expected to be true.

*syllogism*

*deductively valid*

In general, deduction is a purely syntactic process, and is therefore amenable to computational methods. However, proving propositions computationally is usually a highly problematic process, as the number of possible deductions from a set of premises may be very large. In general the complexity of the problem is intractable unless guided by human insight.

Premises can be of two basic types:

*Facts*: Simple particular statements, such as "Paris is the capital of France."

*Rules*: General principles, usually in conditional form, such as "All oak trees are broadleaved" or "No US state name begins with the letter B."

In a particular domain, the collection of all premises (facts and rules) forms the *knowledge base* for that domain. The *theory* associated with this knowledge base is the collection of all conclusions that can be drawn from its premises, using the rules of deduction. If a knowledge base already contains within it all its deductive consequences, we refer to that knowledge base as *deductively closed*.

*knowledge base*

*theory*

*deductively closed*

In classical logic, the more premises we add, the more conclusions can be drawn. The technical term for this property is to say that the logic is *monotonic*. We shall see later in the chapter that there are many cases in practical reasoning where we do not want such a property for our logic.

*monotonic logic*

Parallel to and closely associated with the syntax versus semantics dichotomy is that of *proof* versus *truth*. In any knowledge base, it is to be hoped that there is a relationship between those propositions that are provable (the theory) and those that are true. There are two important concepts that capture this relationship:

- A knowledge base is *sound* if all its deductive consequences are true. So, in a sound knowledge base it is never possible to deduce a falsehood, but there may be things that are true and cannot be deduced.

*soundness*

completeness

- A knowledge base is *complete* if all true propositions that may be constructed using the language of terms and relationships are provable by deduction from its premises.

An ideal for which we strive in constructing a deductive system is a system that is both sound and complete. However, many quite basic deductive systems have been shown to be either unsound or incomplete (see, for example, the "Geometric algorithms and computation" inset, on page 170, Chapter 5).

### 9.1.3   Spatial reasoning example

Let us illustrate these ideas using the example depicted by the map in Figure 9.1. Suppose that our knowledge base consists of the following facts:

1.   Aland, Bland, Cland, and Dland are countries.
2.   Eye, Jay, Cay, and Ell are cities.
3.   Exe and Wye are rivers.
4.   City Eye belongs to Aland.
5.   City Jay belongs to Bland.
6.   City Cay belongs to Cland.
7.   City Ell belongs to Dland.
8.   Cities Eye, Ell, and Cay lie on the River Exe.
9.   City Jay lies on the River Wye.

and the following rule:

10.   Each river passes through all countries to which the cities that lie on it belong.

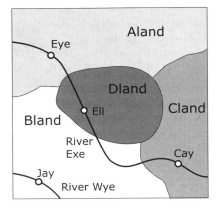

**Figure 9.1:**
Spatial
relations
example

The facts in our knowledge base contain two kinds of propositions. Propositions 1–3 give us information about the *types* of our entities, classified into countries, cities, and rivers. The remaining facts give information about spatial relationships between entities. The rule enables us to work with the further spatial relation "passes through."

If we assume that the map in Figure 9.1 is accurate, then there are truths expressed by the map that are not deducible from the premises in the knowledge base. The truth that Aland and Bland share a common border is an example of such a proposition. Thus, in a general sense, our knowledge base is not complete. However, let us restrict attention to facts about countries, cities, and rivers; which cities are in which countries; which cities lie on which rivers; and which rivers pass through which countries. Using the facts and rules in our knowledge base, we can deduce the following consequences:

11. River Exe passes through countries Aland, Dland, and Cland.
12. River Wye passes through country Bland.

Note that the knowledge base is sound, because every proposition in it is true to the map. It is not complete, even in the specific sense described above, because the map shows the River Exe passing through the country Bland, but we cannot deduce this from the knowledge base.

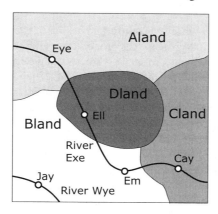

**Figure 9.2:**
Updated spatial relations example

However, if a further city Em were to be placed on the map, as shown in Figure 9.2, then we could add facts:

13. Em is a city.
14. Em belongs to the country Bland.
15. The River Exe passes through city Em.

Now, a further consequence is:

16. River Exe passes through the country Bland.

The knowledge base is complete with respect to the limited vocabulary of cities, rivers, countries, cities belonging to countries, cities lying on rivers, and rivers passing through countries. We note in passing that our example also shows the monotonic nature of the logic, because by adding more premises we add more consequences, and can never remove consequences in this way.

### 9.1.4  Formal notation

Formal logic provides a framework within which the kinds of reasoning processes above can be applied. It also provides a condensed language that can be used to shorten and provide structure to the forms of propositions. We illustrate the use of some specialized logical symbols by rewriting the facts and rules of the map knowledge base example.

1.  $country$(Aland) $\land$ $country$(Bland) $\land$ $country$(Cland) $\land$ $country$(Dland)
2.  $city$(Eye) $\land$ $city$(Jay) $\land$ $city$(Cay) $\land$ $city$(Ell)
3.  $river$(Exe) $\land$ $river$(Wye)
4.  $belongs\_to$(Eye, Aland)
5.  $belongs\_to$(Jay, Bland)
6.  $belongs\_to$(Cay, Cland)
7.  $belongs\_to$(Ell, Dland)
8.  $lies\_on$(Eye, Exe) $\land$ $lies\_on$(Ell, Exe) $\land$ $lies\_on$(Cay, Exe)
9.  $lies\_on$(Jay, Wye)
10. $\forall r.\forall x.\forall c.\ (river(r) \land lies\_on(c, r) \land belongs\_to(c, x))$ $\Rightarrow passes\_through(r, x)$

The system contains *constants* Aland, Bland, Cland, Dland, Eye, Jay, Cay, Ell, Exe, and Wye, acting as proper names for entities in the domain. The one-place predicates $country(\cdot)$, $river(\cdot)$, and $city(\cdot)$ act as typing constraints, and determine which entities are of which type. The two-place predicates $belongs\_to(\cdot, \cdot)$, $lies\_on(\cdot, \cdot)$, and $passes\_through(\cdot, \cdot)$ indicate relationships, in this case spatial relationships between domain entities. So, $belongs\_to(c, x)$ is to be interpreted as "$c$ belongs to $x$"; $lies\_on(c, r)$ means "$c$ lies on $r$"; and $passes\_through(r, x)$ means "$r$ passes through $x$."

The logical language consists of parentheses to ensure no ambiguity, ***logical quantifier*** $\land$ (and), $\lor$ (or), $\neg$ (not), and $\Rightarrow$ (implies), and the *logical quantifiers* $\forall$ (for all) and $\exists$ (there exists). The symbols $r$, $x$, $c$ act as variables on which the quantifiers can act.

The relationship between formal and natural language should be clear from comparing the formal and natural language versions of premises 1–10. As a further example:

$$\exists c.passes\_through(\text{Exe}, c) \land passes\_through(\text{Wye}, c)$$

may be interpreted as "there is a country that both rivers Exe and Wye pass through."

## 9.2  INFORMATION AND UNCERTAINTY

Uncertainty is one of the most difficult concepts to deal with in an information system. Even the most basic ideas are fraught with problems. Consider, for example, the null value, introduced in Chapter 2 as the fundamental database approach to uncertainty. Suppose the capital city of a particular region is entered as NULL. This could be interpreted as meaning that actual data was not entered because the person who populated the database with information did not know the capital city of that region. Alternatively, the null value might be interpreted as meaning that the region (e.g., Antarctica) does not have a capital city. So, even in this simple case, there is ambiguity in the meaning of a fundamental uncertainty value. Uncertainty is closely related to data and information, so before proceeding further, we briefly review some fundamental ideas about data and information.

### 9.2.1   Data and information revisited

It is not difficult to see that data and information are distinct concepts. The string 01101101 contains 8 bits of data but no information unless we have some means of interpreting it. A red light at a railroad crossing can convey a lot of information to a motorist with only a small amount of data (assuming that the light is either on or off, then only a single bit of data). Data is easy to measure: it may be measured in bits and bytes. A more difficult question is "How do we measure information?" Indeed, "What is information?"

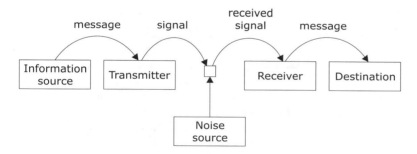

**Figure 9.3:** Shannon-Weaver model of information flow

The dominant theory of information flow has been the channel theory of Shannon and Weaver, and its basic structure is shown in Figure 9.3. Information is seen as a *message* carried along a *channel* by a *signal* from a *source* to a *destination*. The message is sent from the source by means of a *transmitter* and delivered to a *receiver* at the destination. In the course of transmission, the signal may undergo degradation, in the form of *noise*.

The Shannon-Weaver model was originally developed as an aid to analysis of information carried on telephone cables. However, the model has wider applicability. For example, information about a route to a restaurant might be conveyed by a digital sketch map transmitted from a spatial information server to a tourist's personal computer via a wireless network. In this case the source is the server, and the message about the route to the restaurant, is sent to a receiver in the personal computer using wireless technology. If, for example, the receiver is inside a building, then the wireless signal might be degraded and noise will result (but see "Network errors" inset, on the next page).

**Network errors**  *Despite the noise inherent in any transmission system, data transmission in modern computer networks is extremely reliable. This reliability is achieved using error detection. A simple error check is to transmit a parity bit at the end of every block of binary data. The parity bit ensures that each block has an even number of 1's, termed even parity (or an odd number of 1's, termed odd parity). For example, the ASCII code for the letter 'A' is the 7-bit number 1000001. Using even parity the data would be transmitted as the 8-bit number 10000010. Noise in the transmission might lead to one bit being inverted, for example, 10100010, in which case the receiver would detect this odd number of 1's as a transmission error. Parity checking is limited, because if two bits (or any even number of bits) are inverted, the parity check will fail to ensure reliability (e.g., a parity check cannot detect as an error the noisy letter 'A' 10100000). A more powerful error checking technique is the cyclic redundancy check (CRC). Using CRC, the sender appends a binary number to the data block, called a frame check sequence (FCS), such that when the data and FCS are added together, the two are exactly divisible by some prearranged number. The receiver may then check that the received data is indeed divisible by the received FCS with no remainder. Unlike the parity check, CRC makes it very unlikely that random bit inversions due to noise will result in undetectable errors. High-level network protocols ensure that when an error is detected by a receiver, the data is sent again. As a result, noisy data transmission in computer networks leads to slower network performance, because erroneous data must be re-sent, but errors themselves are almost always detected.*

The principal issue that the channel model was constructed to address was a metric for the *quantity* of information communicated. Shannon and Weaver came up with a measure of information quantity (*information entropy*) in terms of its capacity to "surprise" the recipient. Surprise in this context takes on the specialized meaning as the smallness of the chance that a particular signal will be received, and this may often be measured using probability theory. The smaller the chance, the greater the surprise, and the higher the information is valued. In Maine in the winter, there is almost always snow on the ground, so a weather forecast of snow is not unexpected. However, a forecast of a tropical rainstorm would be unusual, to say the least, and therefore might be said to contain more information (for example, it might lead to decision making of greater impact).

The concepts of transmitter and receiver of information are of course metaphors. The image of information leaving the source, being transmitted through a medium (channel) to a receiver, and being the subject of degradation through noise, is compelling. In fact, it is so compelling that sometimes it is easy to forget that it is only an image, and an image with limitations. In particular, in Shannon and Weaver's elaboration of the metaphor, "value" becomes synonymous with "quantity" of information, which is measured by information entropy. However, it is easy to find intuitive examples where value is not measured this way. Suppose a burglar has determined by observation that almost every night the owners of a house forget to set their house alarm system, and only remember on those rare occasions when they take their dog for a walk. On the night planned for the burglary, the lack of a dog being walked provides unsurprising but highly valuable information to the burglar. Context is the key here, and an understanding of context needs to contribute to any metric of quantity of information related to value. A stronger position would be that context is a necessary facilitator of information flow, and without context there is no flow.

### 9.2.2 Uncertainty

We may talk about uncertainty in two ways. First, it may refer to a state of mind. Second, it may be applied directly to the world, or to data or information about the world. So, we might say that the time of a meeting is uncertain, meaning that we are unsure when the meeting is to take place. On the other hand, if we said that the depth of the sea at a particular location was uncertain, it might be a comment on the imprecision of our measuring devices. Uncertainty is an important and unavoidable property of the world, information about the world, and our cognition about the world. There is even a physical principal of uncertainty, formulated by Werner Heisenberg, which says that the more precisely the position of a subatomic particle (such as an electron) is determined, the less precisely its momentum is known. However, the quantum level is not the concern of this book, and so we can neglect such effects. In our terms, uncertainty can lead to doubt and inability to make effective decisions based on the information provided.

*uncertainty*

Uncertainty arises at several different stages in the process of developing a GIS: from populating it with data to using it for analysis in an application domain. Consider the example of capturing the data associated with the boundary of a lake. The following uncertainty issues are among those that might arise:

*Uncertain specifications*: The lake may not have a completely specified boundary. Unless the boundary has been prescribed by human *fiat*, it is unlikely that we can completely determine the lake's edge. There will be temporal variation in the water's edge, and lack of clarity about what we mean by "lake" (*vagueness*). For example, should the sand beach be included, and if so what parts of it, or even, what is a sand beach?

*Uncertain measurements*: Even if the lake's boundary is completely specified in general, we may have problems measuring its location. Our measurements may not be *accurate* because the measuring instrument is incorrectly calibrated, or we misuse the instrument (for example, misread the dial or use it at the wrong temperature). Our measurements may lack appropriate *precision*. Complete and absolute precision for geospatial information is usually not attainable, due to the complex nature of geographical phenomena, such as lakes, or due to limitations of measuring devices. We should aim to achieve a precision of measurement that matches the level of detail required for the application.

*Uncertain transformations*: Apart from the measuring device itself, uncertainty may arise in the *transformations* that are required to convert raw measurements into the specification of the properties of interest. In the case of the lake boundary, the measuring instrument may take readings of the boundary at particular points, and a process of interpolation will be required to specify the boundary as an arc.

This example brings to the foreground several different notions that are key in any discussion about uncertainty. We have seen that uncertainty
imperfection     arises because of *imperfection* in our tools for representing, observing, measuring, and making inferences about the world. The word "imperfect" is not used here in any value-laden sense. Imperfection is contained within the very fabric of our representations of the geographic world, and needs to be faced head-on.

### 9.2.3   A typology of imperfection

Knowledge about reality is gained through observation and representation. Observations are imperfect in the sense that they can never fully or correctly reflect all aspects of reality. Representations, which here include the way in which observations are tied in with the cognitive, formal, or computational model, are similarly bounded by the nature of the world and its relationship to our models of it. Imperfection is therefore the root of our typology, as the concept refers generally to the inevitable deviations from perfection when observing and representing reality. Figure 9.4 shows a basic typology of imperfection, comprising two major distinct orthogonal components: *inaccuracy* and *imprecision*.

inaccuracy     • Inaccuracy refers to a lack of correlation between observations or
error         representations and reality. Inaccuracy is synonymous with *error* (see "Errors of omission and commission" inset, on the facing page).

imprecision     • Imprecision concerns a lack of specificity or a lack of *detail* in an observation or representation.

To see that inaccuracy and imprecision are independent, consider the following. "This chapter is being written in the USA" is accurate but rather imprecise as it does not provide much detail about where in the USA this chapter is being written. In contrast, "This chapter is being written in the attic room of a house called Oldway, in Madeley, Staffordshire, England" is more precise than the previous statement (it provides more detail about where the chapter is being written). But it is hopelessly inaccurate (in fact, that was where the first edition of the book was written, but this material is new to the second edition). So accuracy and precision are *orthogonal* concepts.

**Errors of omission and commission** *There are two types of errors (inaccuracies) that we can distinguish, similar to the distinction between Type I and Type II errors in statistical significance testing. An error of commission is made when we conclude that a proposition is definitely the case, when in fact it is either undetermined or definitely not the case. An error of omission occurs when we do not declare a proposition to be the case, when it is the case. For example, a remotely sensed image may be used to determine land cover. For a particular land cover class, such as "forest," those regions that are actually forest but have been misclassified as some other land use class, such as "agricultural crops," constitute the errors of omission for the forest class. Conversely, those regions that are not in actuality forest, but have been misclassified as forest constitute the errors of commission for the forest class. In the case of land cover maps, errors of omission and commission are inversely related to each other, because every location must be a member of exactly one land cover class. Thus, those "forest" regions that are misclassified as "agricultural crops" constitute errors of omission for "forest," but the same errors are errors of commission for "agricultural crops."*

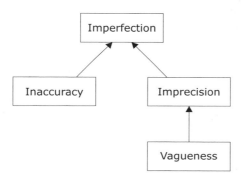

**Figure 9.4:**
A typology of imperfection

## *Granularity and indiscernibility*

Granularity is closely related, but not identical to imprecision. Granularity refers to the existence of clumps or grains in observations or representations, in the sense that individual elements in the grain cannot be distinguished or discerned apart from each other.

    For example, imagine a remotely sensed image of a region of the Earth. In this image, any two locations on the ground that fall within the same pixel cannot be distinguished in characteristics by the observation. So, granularity in an observation makes things *indiscernible* that in principle are distinguishable from each other (Figure 9.5).

    Indiscernibility is a key concept in modeling granularity and imprecision. Intuitively, a finite collection of elements is indiscernible with respect to a particular observation if any pair of elements in the collection cannot be distinguished from each other by the observation. The indiscernibility relation is often assumed to be an equivalence relation (see section 3.2.2). The implications of this assumption are that indiscernibility is reflexive ($a$ is indiscernible from itself), symmetric (if $a$ is indiscernible from $b$, then $b$ is indiscernible from $a$), and transitive (if $a$ is

*granularity*

*indiscernibility*

indiscernible from $b$ and $b$ is indiscernible from $c$, then $a$ is indiscernible from $c$).

**Figure 9.5:**
Granularity and
indiscernibility

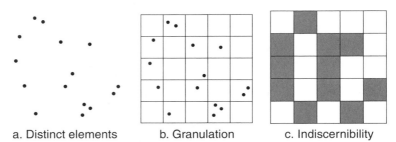

a. Distinct elements      b. Granulation      c. Indiscernibility

More formally, let $\rho$ be a binary indiscernibility relation on set $X$. Any equivalence relation on a set partitions the set into equivalence classes. In our case, the classes will consist of blocks of members of $X$ that are indiscernible from each other. The blocks are pairwise disjoint and completely cover $X$. The set of blocks is termed the *frame of discernment* of $X$ with respect to $\rho$, and is written $X/\rho$. With reference to Figure 9.5, Figure 9.5a represents the set $X$. The frame of discernment in Figure 9.5b results in the blocks (equivalence classes) shaded in gray in Figure 9.5c.

frame of
discernment

Indiscernibility may occur not only by granulation of the physical space but also the thematic space. The granularity of a schema specifies the levels of detail obtainable for an observation using that schema. For thematic classes in an object-oriented GIS, the granularity may be expressed as the level of detail provided by the object classes available in the object inheritance hierarchy. For example, an object-based topographical data model whose only object classes for roads are "major road" and "minor road" could not be used to distinguish a two-lane from a three-lane highway.

So, given a phenomenon under observation, the granularity of the observation's schema induces a mapping (in the mathematical sense) of subsets of a source set into a finite target set. The source set is formed by the collection of constituents of the phenomenon. Elements that are indiscernible by observation are identified with one another in the target set.

In the case of pixelation of an image (as in Figure 9.5), indiscernibility is indeed an equivalence relation. However, in general an equivalence relation is not always a good model of indiscernibility. For example, the sensor arrays used to generate remotely sensed satellite imagery often exhibit more complex forms of indiscernibility, where the transitivity property in particular may not be satisfied. A relation that is symmetric and reflexive but not transitive is termed a *similarity relation*.

similarity relation

*Vagueness*

vagueness

Vagueness concerns the existence of *borderline cases* for a concept. "Mount Everest" is an example of vagueness. There exist many locations that are definitely part of "Mount Everest" (such as the summit and the north face of Mount Everest) and many locations that are definitely not part of "Mount Everest" (such as Paris and Maine). Crucially, there are also many borderline locations in the Himalayas where it is not clear whether they form part of "Mount Everest" or not. The term *crisp* is often used as an antonym for vagueness, to refer to concepts or regions where there are no borderline cases. Similarly, a *crisp set* is simply a normal classical set, in which an element is either a member of the set or not (no borderline cases).

*crisp*

*crisp set*

The question of whether vagueness is a feature of our representations of the world, termed *epistemic vagueness*, or whether the world itself is vague, termed *ontic vagueness*, is a topic of debate within philosophy. For example, a "mountain" may be regarded either as a vague linguistic representation of part of a crisp world, or as a vague object in the world. In general, most theories of vagueness are founded on the former view: that vagueness is epistemic and is a feature of language and other representations of the world.

epistemic vagueness

ontic vagueness

Not all imprecision is vagueness. For example, we could say that a person is located in the USA, which is not a vague statement (assuming for the moment a crisp boundary to the USA), but is quite imprecise. However, every vague statement must also be imprecise, because of the lack of specificity at the boundary. Hence, vagueness is shown as a special type of imprecision in Figure 9.4.

## Reasoning with vagueness

Reasoning using vague assertions is surrounded by difficulties, exemplified by the *sorites paradox*. The sorites paradox concerns the logical problem that arises when attempting to reason about the vague concept of a "heap" (etymologically, "sorites" derives from *soros*, Greek for "heap").

sorites paradox

This paradox can be illustrated in another way, using a geographically vague concept such as "southern Maine" (Figure 9.6). We may be certain that Portland, Maine, is in southern Maine; and that Presque Isle is not. Because the geographical region called "southern Maine" has no clear boundary, it seems reasonable to assert that a single step taken by a person walking north along the road from Portland to Presque Isle cannot make the difference between that person being in southern Maine or not. Given that no single step can make such a difference, it follows that a traveler starting at Portland and walking to Presque Isle could reasonably infer at each step that they were still in southern Maine. Eventually, the traveler would arrive both at Presque Isle and at a paradox: the original assertion that Presque Isle is definitely not in southern Maine is contradicted by the step-by-step reasoning process during the walk from Portland to Presque Isle. The sorites paradox shows that our normal modes of reasoning, in this case the principle of mathematical induction, cannot be applied in the face of vagueness.

**Figure 9.6:**
Maine, USA

### 9.2.4   Dimensions of data quality

data quality   *Data quality* refers to the characteristics of a data set that may influence the uncertainty associated with decisions based upon that data set. The accuracy and precision of a data set are clearly elements of data quality, because both may influence the confidence a decision maker would have lineage   in decisions based on that data set. *Lineage* is another element of data quality, which describes the *provenance* of a data set (its source, age, metadata   intended use). The term *metadata* is often used to indicate data that provides information about the data set. Data quality elements, such as accuracy, precision, and lineage, are examples of metadata. Table 9.1 summarizes some important data quality elements.

Accuracy, granularity, lineage, and precision have been discussed bias   above. *Bias* concerns the existence of systematic distortions within a data set. Such distortion may be introduced deliberately, or as an unforeseen consequence of observational, data collection, or analysis techniques. It is worth noting that precision, accuracy, and bias all have statistical counterparts (see "Statistical precision and accuracy" inset, on page 338). completeness   *Completeness* describes the exhaustiveness of a data set, in terms of what types of features are included and excluded from the data set.

currency   The *currency* of a data set indicates a temporal relationship between data and its source, that is, how "up-to-date" it is. Depending on how time-critical the application is, currency is more or less of an important factor. timeliness   *Timeliness* is more concerned with temporal relevance, and the issue is whether the data arrives at the right moment. Current data may not be timely, and vice versa.

| Element | Concise definition | |
|---------|-------------------|---|
| accuracy | Closeness of the match between data and the things to which data refers | |
| bias | Existence of systematic distortions within data | |
| completeness | Exhaustiveness of data, in terms of the types of features that are represented in data | **Table 9.1:** |
| consistency | Level of logical contradictions within data | Some data |
| currency | How "up-to-date" data is | quality |
| format | Structure and syntax used to encode data | elements |
| granularity | Existence of clumps or grains within data | |
| lineage | Provenance of data, including source, age, and intended use | |
| precision | Level of detail or specificity of data | |
| reliability | Trustworthiness of degree of confidence a user may have in data | |
| timeliness | How relevant data is to the current needs of a user | |

*Reliability* refers to the trustworthiness of data, or the degree of confidence a user may have in a data set. To an extent, reliability will be dependent on the other elements of data quality: low accuracy data sets are likely to be unreliable. However, the relationship is not entirely straightforward. For example, a biased data set may still be "reliable" in the sense that the systematic distortions may be uniformly applied and predictable. Reliability is an important consideration when determining relative preference of a data user for one data set over another, a topic discussed in more detail in later sections.

reliability

### Consistency

Consistency concerns the existence of logical contradictions within the data set. Consistency is violated when information is self-contradictory, either explicitly or by implication. For example, an implicit contradiction can be deduced from the following set of premises:

consistency

> Bangor, Maine has a population of 31,000 inhabitants.
> Only cities with more than 50,000 inhabitants are large.
> Bangor is a large city.

The conclusion "Bangor is not a large city" is a valid deduction from the premises above. However, this conclusion contradicts the premise "Bangor is a large city."

Inconsistency can arise in several ways. Inaccuracy, imprecision, and vagueness can all result in inconsistency. For example, assume the actual population of Bangor is 31,473. Two inaccurate estimates of the population are likely to be inconsistent with each other (for example, inaccurate population estimates of 34,371 and 29,934 inhabitants). Similarly, two imprecise observations of the population, if wrongly interpreted, may

**Statistical precision and accuracy**  *Many of the data quality elements described in this chapter have closely related specific statistical counterparts. Precision is measured statistically using standard deviation:*

$$\sigma = \sqrt{\frac{1}{n} \sum (x_i - \overline{x})^2}$$

*where $n$ is size and $\overline{x}$ is the mean value of some set of observations $\{x_i | i = 1...n\}$. Accuracy is measured statistically using root mean square error (RMSE):*

$$rmse = \sqrt{\frac{1}{n} \sum (x_i - t)^2}$$

*where $t$ is the "true" value for the set of observations, or some value taken to be true. Standard deviation measures the spread of a set of observations of a value, while RMSE measures the discrepancy between a set of observations of a value and the true value. Although standard deviation and RMSE are conceptually distinct, numerically they are often equivalent, as the mean $\overline{x}$ of a set of observations is normally used as an estimate of the true value $t$ of a set of an observable phenomenon.*

be inconsistent with each other (for example, a population of 31,000, accurate to the nearest 1000 inhabitants, and a population of 30,000, accurate to the nearest 10,000 inhabitants). Finally, if we assume for a moment that "Bangor" is a vague concept referring to the Bangor region (Bangor city itself has a precisely defined administrative boundary), then disagreements about where exactly the boundary lies will lead to different measurements of population. This results in inconsistency: some inhabitants excluded from the Bangor region in one observation may be included in another.

Classical logic and indeed most standard logics are *explosive*: they have the important but sometimes frustrating property that, when using them, anything may be inferred from a contradiction. As a consequence, inconsistency must be avoided or resolved at all costs in these standard logics. However, inconsistency should not be regarded as necessarily undesirable. For example, inconsistency may cause us to review, revise, or refine our beliefs. In general, three types of action may be prompted by the discovery of new information that is inconsistent with existing information or beliefs:

*Resolve inconsistency*:  Most approaches to inconsistency aim to resolve inconsistency in some way when it occurs. For example, in traditional databases integrity constraints attempt to prevent inconsistent information ever being inserted into the database (see Chapter 2). If information violates a database's integrity constraints it is discarded, a very simple form of inconsistency resolution. A variety of more sophisticated techniques for resolving inconsistency, such as belief revision and many-valued logics, form the basis of the next section.

*Retain inconsistency*: The disadvantage of resolving inconsistency is that information is usually lost during the resolution process. For example, *least-squares adjustment* is a common statistical technique for resolving inconsistencies in survey measurements. However, once least squares adjustment has been applied, it is not possible to retrieve the original measurements, making it difficult to resolve any subsequent inconsistencies that may arise. One area for future research is the development of knowledge bases that are able to retain inconsistent information, resolving inconsistency only when necessary for a particular application. A few logical systems, such as *paraconsistent logics*, have been proposed for retaining inconsistencies while avoiding the problem of explosive logics (see bibliographic notes).

least-squares
adjustment

paraconsistent
logic

*Initiate dialog*: Finally, inconsistency may lead to a *dialog* between the agents responsible for the different information sources. An important example of dialog occurs when inconsistencies prompt the acquisition of new information, such as further data collection, in an attempt to resolve the original inconsistencies.

### Relevance and fitness for use

Relevance is an important concept, both in information science and studies of human cognition. Relevance describes the relationship between information and its context. In terms of data sets, relevance concerns the connection of a data set to a particular application. Information is relevant if appropriate for one's needs, useful, and close and pertinent to one's requirements. Relevance is not independent of some of the other indicators. For example, if the information is at the wrong level of detail, or if it is not timely, then it may not be relevant. The bibliographic notes contain further references to theoretical work on relevance.

relevance

Relevance is one aspect of data quality that provides a basis upon which to assess the *fitness-for-use* of a data set for a particular application. For example, a study of habitat change in a national park might require a temporal series of data with high precision and accuracy. In addition, detailed lineage and completeness metadata might be needed to allow a comparison of the (possibly different) habitat classifications used at each time. If instead the application is a tourist map to help inform and educate visitors to the national park about habitat types and history, data sets with much lower accuracy and precision may be entirely appropriate. Even with data quality metadata, assessing whether a particular data set is fit for some specified use may be complex, requiring experience and judgment; without such metadata the task is impossible.

fitness-for-use

## 9.3   QUALITATIVE APPROACHES TO UNCERTAINTY

This section begins to outline some key approaches to representing and reasoning under uncertainty. As we have seen, uncertainty can take many guises, and the variety of its representations mirrors this.

A distinction is often made in spatial reasoning between *quantitative* and *qualitative* approaches. An approach is generally referred to as *quantitative* if it is based on analysis of numerical (interval or ratio) data. By contrast, an approach is usually referred to as *qualitative* if it is based on analysis of classifications and ordering (nominal and ordinal data). The terms "qualitative" and "quantitative" are often used as antonyms. However, qualitative and quantitative approaches should not be seen as mutually exclusive, nor should one be seen as inherently preferable to the other. The most appropriate approach is dependent on the phenomena under investigation, and the form in which the observations are made and data collected.

*quantitative*
*qualitative*

### 9.3.1   Possible worlds

In classical propositional and first-order logic, statements are either true or false. Under conditions of uncertainty, we may not be able to determine whether a statement is true or false. A simple way to extend propositional and first-order logic to model uncertainty is by considering all the *possible worlds* that an agent might know about. As a simple example, consider an area of land divided into two regions $A$ and $B$ (Figure 9.7). Suppose that we are interested in knowing whether each region is forested or not. For the moment, we assume that the property of being forested is not vague, and therefore must be either true or false (but not both) for any given region. The states of possible knowledge can be expressed using the following statements:

possible worlds

> $p$: "Region $A$ is forested"
> $q$: "Region $B$ is forested"

For this domain there are four possible worlds, enumerated below:

> World $W_1$: Statement $p$ is true, statement $q$ is true
> World $W_2$: Statement $p$ is true, statement $q$ is false
> World $W_3$: Statement $p$ is false, statement $q$ is true
> World $W_4$: Statement $p$ is false, statement $q$ is false

**Figure 9.7:**
Two regions $A$
and $B$

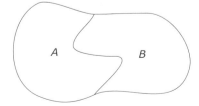

The statements $p$ and $q$ are either true or false (but not both), so exactly one of the possible worlds may hold at any particular time. Our state of knowledge is represented by a set of worlds: those worlds that are feasible, given what we know. Therefore, if we are in a state of complete ignorance, all worlds are possible. Complete ignorance is represented by the set $\{W_1, W_2, W_3, W_4\}$. In contrast, complete knowledge may be represented by a singleton set. For example, if we know that that both regions are forested, then our state of knowledge is represented by $\{W_1\}$. In general, the larger the set of possible worlds, the more imprecision exists in our state of knowledge. The empty set represents a state of *inconsistent* information. Indirect information might lead to increased precision in our state of knowledge. For example, if from a state of complete ignorance, we learn that both regions are of the same land type, our knowledge may be represented by set $\{W_1, W_4\}$.

In the case above, we assumed that the land types in the two regions were independent of each other, and therefore that the truth values of the two statements were *a priori* independent. This need not be the case as the following example shows. Figure 9.8 shows an additional region $C$, which is a part of region $A$. To represent this we add the statement:

$r$: "Region $C$ is forested"

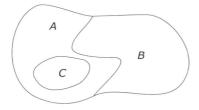

**Figure 9.8:**
Three regions
$A$, $B$, and $C$

Suppose we have the constraint that if region $A$ is forested then the region $C$, contained within $A$, must be also be forested (although the converse need not be true). Then, for the enlarged domain, there are six possible worlds, shown below. Note that as a result of the constraint above, two situations are not possible (those where $p$ is true, $r$ is false, and $q$ is true or false) and so are omitted from the possible worlds below.

World $W_1$: $p$ is true, $q$ is true, $r$ is true
World $W_2$: $p$ is true, $q$ is false, $r$ is true
World $W_3$: $p$ is false, $q$ is true, $r$ is true
World $W_4$: $p$ is false, $q$ is false, $r$ is true
World $W_5$: $p$ is false, $q$ is true, $r$ is false
World $W_6$: $p$ is false, $q$ is false, $r$ is false

### 9.3.2   Belief and knowledge

The possible worlds semantics lead to a useful distinction between *knowledge* and *belief*. Belief is an agent's conviction in the truth of a statement.            belief

knowledge

Knowledge is justified true belief. For example, I might *believe* that pigs can fly, but I cannot *know* this, as it is in fact false. Further, while all knowledge is true belief, not all true beliefs are knowledge. For example, I might believe that there is life on distant stars, and this might indeed coincidentally be the case, but that is not sufficient for knowledge—there must also be some justification in the form of evidence or inference.

modal operator

It is clear, then, that there is a difference between something being the case, and our belief or knowledge of it. One common device for constructing logics of belief is to precede a proposition by a *modal operator* to indicate our knowledge or belief of it. So, as before, if we have statement:

$p$: "Region $A$ is forested"

then:

| | |
|---|---|
| $Kp$ is the statement | "I know that region $A$ is forested" |
| $Bp$ is the statement | "I believe that region $A$ is forested" |

Belief and knowledge might be related by formulas like the following:

| | |
|---|---|
| $\neg K \neg p \rightarrow Bp$ | If I don't know that $p$ is not the case, then I can believe $p$. |

Logics of knowledge and belief incorporate premises expressing properties of the knowledge and belief operators. For example, from the above discussion, we might have:

| | |
|---|---|
| $Kp \rightarrow p$ | If I know $p$, then $p$ must be true. |

and

| | |
|---|---|
| $\neg Kp \rightarrow \neg p$ | If I don't know $p$, then $p$ cannot be true. |

closed world
assumption

The axiom $\neg Kp \rightarrow \neg p$ models the *closed world assumption* of database theory—if the database does not hold a fact, then the fact is not the case. In some cases, depending on the domain, there may be other more complex axioms that are obeyed, such as:

| | |
|---|---|
| $Kp \rightarrow KKp$ | If I know $p$, then I know that I know $p$. |
| $\neg Kp \rightarrow K \neg Kp$ | If I don't know $p$, then I know that I don't know $p$. |

introspection

These last two axioms encapsulate the concept of *introspection*. The former axiom, "I know what I know," is known as *positive introspection*; the latter axiom, "I know what I don't know," is *negative introspection*.

This modal formalism can be extended to model knowledge held by more than one agent. In this case, the modal operators are labeled by names of agents. For example, if $a$ is the name of an agent, then $K_a p$ might express the fact that agent $a$ knows proposition $p$.

### 9.3.3   Belief revision

If new information arises that contradicts our current beliefs, we may
want to review, revise, or retract our old beliefs so as to make way for the
new information. This process is called *belief revision*. Typically, belief          belief revision
revision systems are *non-monotonic*, because the belief set may grow or             non-monotonic
retract as new information is added to the knowledge base. The ability of             logic
non-monotonic logics to retract previously held beliefs in the light of new
information more closely models commonsense reasoning (see section
8.1.2).

A key question facing any non-monotonic belief revision system is
which beliefs to retract and which to retain. Unfortunately, because beliefs
are often founded on other beliefs, the effects of removing one belief may
cascade through the knowledge base in a way that is difficult to predict. A
well-known example, due to Gärdenfors, concerns a knowledge base that
contains the following facts and rules:

> The bird caught in the trap is a swan.
> The bird caught in the trap comes from Sweden.
> Sweden is part of Europe.
> All European swans are white.

Suppose we learn new information that "The bird caught in the trap
is black." This information is implicitly inconsistent with the knowledge
base, as the knowledge base may be used to infer the statement "The bird
caught in the trap is white." The question facing any belief revision system
is which beliefs to retract in order to regain consistency. Discarding any
one of the five pieces of information we now have (the four listed above,
plus the new information) will result in a consistent system.

To decide what to keep and what to remove, we must have some
way of choosing one piece of information over another. Techniques for
deciding what to retract usually involve evaluating the strength of beliefs
held, using a *preference relation*, often taken to be a partial order. For          preference
example, we might prefer a well-established piece of information (e.g.,               relation
"Sweden is part of Europe") than a fact about the current situation (e.g.,
"The bird caught in the trap is a swan"). Sweeping statements such as
"All European swans are white" might also be called into question. A
general principle often used in belief revision to revise beliefs so that
the amount of change is minimized—the *principle of minimal change*.                principle of
Another possible principle is to favor beliefs that arise from evidence               minimal change
obtained spatially or temporally close to the phenomenon, rather than
farther away (the *nearness* principle). For example, when faced with
two inconsistent pieces of information we might prefer the more current
information source, as it is by definition temporally closer to the present.

Belief revision is related to the topic of *default reasoning*. In the exam-          default reasoning
ple above, the first three facts relating to the bird caught in the trap being a
swan from Sweden, which is a part of Europe, are potentially possible to
verify. The fourth universal statement, the rule that "All European swan
are white" would be difficult or impossible to verify beyond doubt (we

would need to check every swan in Europe). Universal statements make powerful rules, but may be subject to occasional counterexamples, such as our Swedish black swan. Instead of an unconditional universal what we really want to say is:

> All European swans are white (except if we have definite evidence to the contrary in the case of a particular swan).

While the default is that European swans are white, default reasoning allows the possibility that some counterexamples may exist. A default predicate is assumed to hold, unless any evidence turns up to the contrary, in which case it is retracted. Default reasoning is therefore a further example of non-monotonic reasoning.

### Revision and update

At this point, it is important to make a clear distinction between knowledge base *revision* and *update*. An information system undergoes an *update* when new information arrives that indicates a change in the application domain. An information system undergoes a *revision* when new information arrives that indicates a change in our beliefs about the application domain, but the application domain itself may not have changed.

For example, if new information from satellite imagery indicates that the land cover in region $A$ is different from that recorded in the database, then two possibilities exist. The land cover for region $A$ may have actually changed from that recorded in the database to that detected by the remote sensing satellite. This might occur when a farmer plants a new crop in a field. In this case, the database must be updated to reflect this change. For a temporal database (discussed in more detail in the next chapter) it may also be necessary to record the timing and history of any changes.

However, if there is no reason to suspect any actual change, then it will be necessary to revise our beliefs about the land cover in region $A$. If we are able to invoke a preference between the information from two sources, this will inform the decision as to which belief to retract. For example, if as a result of new satellite sensor technology the new land cover information is expected to be more reliable than the old information stored in the database, we may decide to prefer the new information.

Figure 9.9 illustrates the distinction between update and revision. Figure 9.9a shows the initial state of the application domain on the left and the database on the right. In Figure 9.9b new information from satellite imagery indicates that the region with stored land cover type "Urban area" is in fact a region of land cover type "Pastoral land." Since the application domain has not changed, a revision operation is needed. A subsequent change in the application domain, shown in Figure 9.9c where part of the forested region has become agricultural land, requires a update of the database. At both stages in Figure 9.9 the database is changed, but at only one stage does the application domain change.

update

revision

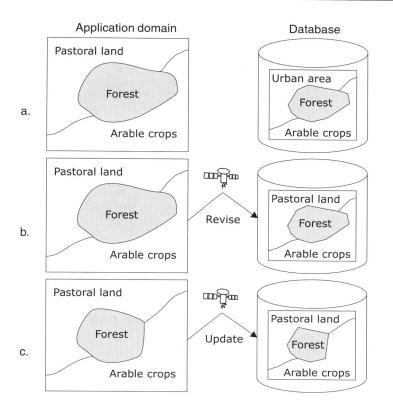

**Figure 9.9:** Update and revision

### 9.3.4   Three-valued and many-valued logics

In the preceding discussion of possible worlds, knowledge, and belief, statements are either true or false, with uncertainty modeled as the state of an agent's knowledge or belief about the world. We might also need to allow for the possibility that a statement is neither true nor false. A statement could be neither true nor false for a variety of reasons, such as when there is insufficient evidence to decide, when it makes no sense to give a value, or when the statement is inherently vague. For example, the statement "Region $A$ is forested" may not be decidable as true or false because no data exists to confirm or deny the truth of the statement (perhaps cloudy weather prevented satellite imagery of the location). Alternatively, the statement "Region $A$ is forested" might be neither true nor false because the concept "forested" is vague: while there may be some trees in the region it might be unclear whether the region can be interpreted as forested. In the former case, more evidence might allow a decision, true or false, to be made; in the latter case no amount of evidence will suffice.

A seemingly simple solution is to introduce a third truth value, resulting in *three-valued logic*. We could call the third truth value "unknown" or "undetermined," and indicate with a symbol such as $U$ or '?'. So we have three truth values, $\top$ (true), $\bot$ (false), and ? (undetermined). Just as

three-valued logic

we can make truth tables for two-valued, classical logic, so we can for three-valued logic. Depending on the interpretation of '?', we can arrive at different truth tables. An example of a three-valued logic is the Kleene system. Truth tables for the logical operators ¬ (not), ∧ (and), ∨ (or), → (implies), and ↔ (equivalent) are shown in Figure 9.10.

**Figure 9.10:**
Truth tables for Kleene's three-valued logic

| ∧ | T | ? | ⊥ |
|---|---|---|---|
| T | T | ? | ⊥ |
| ? | ? | ? | ⊥ |
| ⊥ | ⊥ | ⊥ | ⊥ |

| ∨ | T | ? | ⊥ |
|---|---|---|---|
| T | T | T | T |
| ? | T | ? | ? |
| ⊥ | T | ? | ⊥ |

| ¬ | T |
|---|---|
| T | ⊥ |
| ? | ? |
| ⊥ | T |

| → | T | ? | ⊥ |
|---|---|---|---|
| T | T | ? | ⊥ |
| ? | T | ? | ? |
| ⊥ | T | T | T |

| ↔ | T | ? | ⊥ |
|---|---|---|---|
| T | T | ? | ⊥ |
| ? | ? | ? | ? |
| ⊥ | ⊥ | ? | T |

In Kleene's system, uncertainty is interpreted as a limitation on reasoning or computing resources. So, a statement takes truth value '?' whenever its truth or falsity cannot be determined by the resources at hand, even though we know it has to be either true or false (that is, inherent vagueness is not considered in this system).

For example, suppose we have statements $p$ and $q$ as before, and that "forested" is assumed to be a crisp property of a region. Suppose further that because of cloudy conditions we do not know whether or not region $A$ is forested (?), but we have clear evidence that region $B$ is forested (T). The truth values of the following statements can be determined from the truth tables in Figure 9.10:

| | |
|---|---|
| Both regions $A$ and $B$ are forested | ? |
| Either region $A$ or $B$, or both, are forested | T |
| If region $A$ is forested, then region $B$ is forested | T |

many-valued logic

degree theory

Kleene logic is one example of a three-valued logic; the bibliographic notes at the end of this chapter provide references to other examples. Three-valued logic is itself a type of *many-valued logic*. Some many-valued logics allow an infinite number of intermediate values to be assigned to statements that lie between absolute truth and absolute falsity. Such logical systems are often termed *degree theories*, because values may be interpreted as "degrees of truth." Fuzzy set theory, discussed in the following section, is an important example of a degree theory.

### 9.3.5  Fuzzy sets

A common approach to locational uncertainty in GIS research is the application of *fuzzy set theory*. In the case of crisp sets (normal sets), an element either belongs to the set or not. With fuzzy sets, a membership function grades the levels of belief in whether an element belongs to the set or not. Although fuzzy set theory appears quantitative, often using real numbers between 0 and 1 as measures of degree of membership, it mostly makes use only of the ordinal properties of these numbers. In this account we will concentrate on the case where membership values are in the real interval $[0, 1]$. In general, membership values may be taken from an algebraic structure called a lattice.

Formally, let $X$ be a universe of discourse. A *fuzzy membership function* is a function $\mu$ from $X$ to the real interval $[0, 1]$, $\mu : X \rightarrow [0, 1]$. A *fuzzy set* $A$ in $X$ is a set of ordered pairs $(u, \mu_A(u))$ for all $x \in X$, where $\mu_A$ is a fuzzy membership function.

fuzzy membership function

fuzzy set

An example of a fuzzy set that describes a region with an uncertain boundary is shown in Figure 9.11. Each pixel is labeled with a fuzzy set value, indicating the strength of belief that each pixel belongs to the region. The degree of belief may arise from evidence about the pixel gained from Earth observation.

| 0 | 0 | 0 | 0 | 0 | 0 | 0 | 0 |
|---|---|---|---|---|---|---|---|
| 0 | 0.1 | 0.2 | 0.2 | 0.2 | 0.1 | 0.1 | 0 |
| 0.1 | 0.2 | 0.4 | 0.4 | 0.6 | 0.2 | 0.1 | 0 |
| 0.2 | 0.3 | 0.4 | 1 | 0.8 | 0.4 | 0.3 | 0.2 |
| 0.1 | 0.3 | 0.8 | 1 | 0.8 | 0.4 | 0.3 | 0.2 |
| 0.1 | 0.2 | 0.6 | 0.6 | 0.6 | 0.3 | 0.3 | 0.1 |
| 0.1 | 0.1 | 0.1 | 0.2 | 0.2 | 0.1 | 0.1 | 0 |
| 0 | 0 | 0 | 0 | 0 | 0 | 0 | 0 |

**Figure 9.11:** Fuzzy region as pixel values

Just like crisp sets, fuzzy sets have properties and operations, some of which are listed below.

- Fuzzy set $A$ is *empty* if $\mu_A(x) = 0$ for all $x \in X$.

- Fuzzy set $A$ is *contained* in $B$ if $\mu_A(x) \leq \mu_B(x)$, for all $x \in X$.

- Fuzzy sets $A$ and $B$ are equal if $\mu_A(x) = \mu_B(x)$, for all $x \in X$.

- The *complement* of fuzzy set $A$ is the set $A'$ with membership function $\mu_{A'}$ such that $\mu_{A'}(x) = 1 - \mu_A(x)$, for all $x \in X$.

- The *union* of fuzzy sets $A$ and $B$ is the set $A \cup B$ with membership function $\max(\mu_A(u), \mu_B(u))$, for all $x \in X$.

- The *intersection* of fuzzy sets $A$ and $B$ is the set $A \cap B$ with membership function $\min(\mu_A(u), \mu_B(u))$, for all $x \in X$.

- The *support* of fuzzy set $A$ is the crisp set containing all elements with non-zero membership of $A$, support$(A) = \{x | \mu_A(x) > 0\}$.

- For $0 \leq \alpha \leq 1$, the $\alpha$-*cut* of fuzzy set $A$ is the crisp set given by $A_\alpha = \{x | \mu_A(x) > \alpha\}$

$\alpha$-cut

What we have given above for union, intersection, and complementation operations are special cases of more general categories. For example, the intersection between two fuzzy sets may be realized, not only by the minimum operation, but by any operation that is a triangular norm (T-norm). A reference to this general theory may be found in the bibliographic notes at the end of the chapter.

The idea of a fuzzy region is rather more general than our example indicates. Define a *fuzzy region* as a fuzzy set whose support is a region. Fuzzy regions have more structure than the more basic fuzzy sets. For example, assuming the regions are based on a square cell grid, as in the example in Figure 9.11, then the cells have many topological and geometrical properties and relations, such as adjacency, area, distance, and bearing. Fusing fuzzy set theory with geometrical or topological structure enables operations such as fuzzy connectivity, convexity, area, and perimeter, to be defined, sometimes termed *fuzzy geometry* or *fuzzy topology*, respectively.

By way of example, let $R$ be a fuzzy region based on a square cell grid with fuzzy membership function $\mu_R$. Then the fuzzy area of $R$, $\mathbf{a}(R)$, may be defined as the sum of the $\mu_R(x)$, for all $x \in X$. In the example of Figure 9.11, the fuzzy area of the region is $14.1$. This notion of fuzzy areas satisfies only some of the basic properties that we would expect an area measure to satisfy. For example, while it is the case that:

$$\mathbf{a}(A \cup B) = \mathbf{a}(A) + \mathbf{a}(B) - \mathbf{a}(A \cap B)$$

it is *not* necessarily the case that:

$$\mathbf{a}(A) = \mathbf{a}(A \cap B') + \mathbf{a}(A \cap B)$$

### 9.3.6   Rough sets

Rough sets are motivated primarily by the need to represent and reason about granularity and indiscernibility in information, as discussed in section 9.2.3. Unlike those techniques presented above, which model our belief in whether a statement is true or not, rough set theory supports reasoning under conditions of granularity and indiscernibility. An important question is how to represent subsets of $X$ at the level of granularity imposed by the indiscernibility relation $\rho$. Rough sets provide a framework in which this question may be answered. First, let set $A$ be a subset of $X$. Then we define the following two constructs:

$$\underline{A} = \{b \in X/\rho \mid b \subseteq A\}$$
$$\overline{A} = \{b \in X/\rho \mid b \cap A \neq \varnothing\}$$

$\overline{A}$ is called the *upper approximation* to set $A$, while $\underline{A}$ is termed the *lower approximation* to set $A$ (with respect to the indiscernibility relation $\rho$ on set $X$). The pair $<\underline{A}, \overline{A}>$ is called the *rough set* (with respect to the indiscernibility relation $\rho$ on set $X$). Note that $\underline{A}$ is always a subset of $\overline{A}$ in $X/\rho$.

fuzzy region

fuzzy geometry

upper approximation

lower approximation

rough set

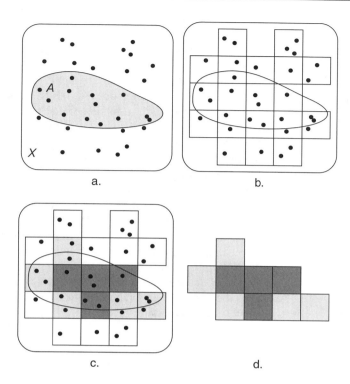

**Figure 9.12:** Rough set construction

Figure 9.12 shows an example of the construction of a rough set. Subset $A$ of $X$ is shown in Figure 9.12a, where the points indicate the elements of $X$. In Figure 9.12b the blocks of the partition induced by the indiscernibility relation are shown. Figure 9.12c shows the construction of $\underline{A}$ and $\overline{A}$, where the blocks of $\underline{A}$ are shown as darker gray and the blocks of $\overline{A} \backslash \underline{A}$ as lighter gray. Finally, Figure 9.12d shows the rough set $<\underline{A}, \overline{A}>$ ($\underline{A}$ is the set of darker blocks, and $\overline{A}$ is the set of all blocks).

Just as for crisp and fuzzy sets, rough sets form an algebra with rough intersection and union defined. Rough set analysis also gives a way of deducing functional dependencies in an information system—an example of a basic data mining technique. References can be found at the end of the chapter that give more details of formal properties of rough set theory, as well as some applications of rough set analysis to GIS.

## 9.4  QUANTITATIVE APPROACHES TO UNCERTAINTY

The previous section has indicated some of the key classes of qualitative approaches to representing and reasoning about uncertainty. In this section, we look in more detail at some important examples of quantitative approaches. As we have already noted, the qualitative-quantitative distinction is itself vague.

### 9.4.1   Probability

Probabilistic methods form the most important and widely used quanti-tative approach to uncertainty. Detailed material on probability may be found in introductory statistics textbooks (see bibliographic notes). This section sketches a rough outline, concentrating on the Bayesian approach. Statistical methods for uncertainty handling for geospatial information are most often used in spatial statistics and spatial analysis. Again, the bibliographic notes provide references to texts on spatial analysis.

*aleatory*    Chance arises when observing the outcome of an *aleatory* (random) experiment, such as the throw of a die. If $X$ denotes the set of possible outcomes, we can specify a chance function $\mathbf{ch} : X \rightarrow [0,1]$. The valuation $\mathbf{ch}(x)$ gives the proportion of times that a particular outcome $x \in X$ might in principle occur, and is determined either from a frequency analysis (throwing the die many times and seeing what happens) or determination from the nature of the experiment (assuming that the die is fair and so 50-50 heads to tails). The function $\mathbf{ch}$ should satisfy the constraint that the sum of chances of all possible outcomes is 1. For a subset $S \subseteq X$, $\mathbf{ch}(S)$ is the chance of an outcome from set $S$.

So, the rules that a chance function must obey are:

$$\mathbf{ch}(\varnothing) = 0 \qquad\qquad (9.1)$$
$$\mathbf{ch}(X) = 1 \qquad\qquad (9.2)$$
$$\text{If } A \cap B = \varnothing, \text{ then } \mathbf{ch}(A \cup B) = \mathbf{ch}(A) + \mathbf{ch}(B) \qquad (9.3)$$

Also, given $n$ independent trials of a single aleatory experiment, the chance of the compound outcome $\mathbf{ch}^n(x_1, ..., x_n)$ is given by:

$$\mathbf{ch}^n(x_1, ..., x_n) = \mathbf{ch}(x_1) * ... * \mathbf{ch}(x_n) \qquad (9.4)$$

Suppose next that an aleatory experiment has been only partly com-pleted and the outcome has been determined to be in the set $V \subseteq X$. If $U \subseteq X$ is the outcome set under consideration, we write the chance of $U$ following the partial determination that the outcome is in $V$ as $\mathbf{ch}(U|V)$

*conditional* (read "the chance of $U$ given $V$" or the *conditional probability* of $U$ given
*probability* $V$). Then:

$$\mathbf{ch}(U|V) = \frac{\mathbf{ch}(U \cap V)}{\mathbf{ch}(V)} \qquad\qquad (9.5)$$

For example, suppose we are looking at the pixels in Figure 9.13 for evidence of a landslide. Assume that the extent of the landslide is small compared with the pixel size, and so the landslide has occurred in exactly one of the pixels. Suppose also, we have evidence from other sources that exactly one landslide has occurred somewhere within the study region. Then, our initial estimate of the landslide being in region $A$ is 4/9 by equation 9.3 above. Now, suppose some new geological evidence becomes available that tells us that the landslide can only be region $B$. Let $U$ be the outcome set given by the statement "the landslide is in region $A$" and $V$ be the outcome set given by the statement "the

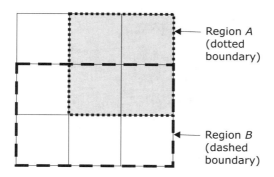

← Region *A*
(dotted
boundary)

**Figure 9.13:**
Conditional
probability
example

← Region *B*
(dashed
boundary)

landslide is in region $B$." Then, by equation 9.5, the chance that the
landslide is in region $A$ now becomes:

$$\mathbf{ch}(U|V) = \frac{\mathbf{ch}(U \cap V)}{\mathbf{ch}(V)} = \frac{(2/9)}{(6/9)} = 1/3$$

### 9.4.2 Bayesian probability

This example leads neatly into a brief word about a theory of belief,
developed by the English clergyman Thomas Bayes (1702–1761). The
notion of chance introduced above is objective, in that it depends not on
our cognitive state but on external conditions. The chance of a fair die
landing a six is $1/6$ independent of our belief. Indeed, the rational way to
set our belief level is based on the laws of chance in this case—that is why
the laws are generally useful. However, if we make "degree of belief" our
prime object of consideration, the laws of chance 9.1–9.3 above, may be
rephrased as follows.

A *degree of belief* with respect to a set $X$ of possibilities (outcomes)  degree of belief
is given by a belief function $\mathbf{Bel} : X \rightarrow [0, 1]$. For a subset $S \subseteq X$,
$\mathbf{Bel}(S)$ is the degree of belief in an outcome from set $S$. Bayesian theory
rests on the following rules that a Bayesian belief function is expected to
obey.

$$\mathbf{Bel}(\varnothing) = 0 \qquad\qquad (9.6)$$
$$\mathbf{Bel}(X) = 1 \qquad\qquad (9.7)$$
If $A \cap B = \varnothing$, then $\mathbf{Bel}(A \cup B) = \mathbf{Bel}(A) + \mathbf{Bel}(B)$ $\quad$ (9.8)

Suppose we begin with a Bayesian belief function $\mathbf{Bel} : X \rightarrow [0, 1]$
and then learn that only a subset of possibilities $V \subseteq X$ is the case. Then,
replace $\mathbf{Bel}$ with a new Bayesian belief function $\mathbf{Bel}_V : X \rightarrow [0, 1]$
given by:

$$\mathbf{Bel}_V(U) = \frac{\mathbf{Bel}(U \cap V)}{\mathbf{Bel}(V)} \qquad\qquad (9.9)$$

or in the language of conditioned belief, and parallel to equation 9.5:

$$\textbf{Bel}(U|V) = \frac{\textbf{Bel}(U \cap V)}{\textbf{Bel}(V)} \tag{9.10}$$

Applying equation 9.10 symmetrically gives:

$$\textbf{Bel}(V|U) = \frac{\textbf{Bel}(U \cap V)}{\textbf{Bel}(U)} \tag{9.11}$$

Eliminating $\textbf{Bel}(U \cap V)$ from 9.10 and 9.11 gives:

$$\textbf{Bel}(U|V) = \textbf{Bel}(U) * \frac{\textbf{Bel}(V|U)}{\textbf{Bel}(V)} \tag{9.12}$$

posterior belief
prior belief

Equation 9.12 is a widely used form of the Bayesian formula. Our *posterior belief* $\textbf{Bel}(U|V)$ is calculated by multiplying our *prior belief* $\textbf{Bel}(U)$ by the likelihood that $V$ will occur if $U$ is the case. $\textbf{Bel}(V)$ acts as a normalizing constant that ensures that $\textbf{Bel}(U|V)$ will lie in the interval $[0, 1]$.

### 9.4.3 Dempster-Shafer theory of evidence

Although at first sight Bayesian theory seems natural, there are some problems. Consider the following example:

$b_1$: there is life in the Sirius system
$b_2$: there is not life in the Sirius system

By equation 9.8, $b_1 + b_2 = 1$, even though there is little evidence either way. So how do we assign belief levels? If we have no evidence either way, the natural thing is to assign equal belief levels, and so $b_1 = b_2 = 0.5$ In this way, high levels of belief are assigned, based on little evidence. Such a model does not appear to accord with our intuition.

The Dempster-Shafer theory of evidence provides a way of pooling the total evidence available. It takes account of evidence both for and against a belief. As with Bayesian theory, a real number between 0 and 1 indicates the degree of support a body of evidence provides for a belief. The Dempster-Shafer theory focuses on the combination of degrees of belief or support from distinct bodies of evidence. Dempster's *rule of combination* provides a method for changing beliefs in the light of new evidence. The example in the previous paragraph pointed to a possible problem when the degrees of independent and exhaustive beliefs are forced to sum to one. To develop this further, take our usual statement:

$p$: "Region $A$ is forested"

and as before, assume that the property "forested" is crisp (so it is definitely the case that either $A$ is forested or not, but not both). Let us consider two cases:

*Case 1 (information scarcity)*:  There are roughly equal small amounts of evidence on both sides.

*Case 2 (information glut)*: There are roughly equal large amounts of evidence on both sides.

Bayesian analysis will give $\mathbf{Bel}(p)$ and $\mathbf{Bel}(\neg p)$ the value roughly 0.5 in both cases, using $\mathbf{Bel}(p) + \mathbf{Bel}(\neg p) = 1$, and so fails to make a distinction between the cases. The way forward is to make a distinction between *credibility* and *plausibility*, or in negative terms, a distinction between *disbelief* and *lack of belief*. The credibility of a state of affairs is the amount of evidence we have in its favor. The plausibility of a state of affairs is the lack of evidence we have against it. In terms of the belief function:

credibility plausibility

$$\mathbf{credibility}(p) = \mathbf{Bel}(p)$$
$$\mathbf{plausibility}(p) = 1 - \mathbf{Bel}(\neg p)$$

Now, the analysis of case 1 is that the credibility of both $p$ and $\neg p$ is small, but the plausibility of both $p$ and $\neg p$ is large. For case 2, the credibility of both $p$ and $\neg p$ is larger, but the plausibility of both $p$ and $\neg p$ is smaller. This is the starting point for the Dempster-Shafer approach. Using Dempster's rule of combination, evidence for and against a state of affairs can be combined. Further details can be found in the bibliographic notes at the end of the chapter.

## 9.5   APPLICATIONS OF UNCERTAINTY IN GIS

This section introduces three specific application domains in which the techniques described earlier in the chapter may be applied. All are based on the notion of an "uncertain region." Classical set theory and logic assume crisp sets and propositions. An element is either definitely in or definitely not in a set; a proposition is either true or false. When applied to the representations of regions, this results in crisp regions, for which each location is either definitely in or definitely not in the region. However, in many practical examples, the situation is not so clear.

### 9.5.1   Uncertain regions

Consider the following definition of a "coastal dune":

> A continuous or nearly continuous mound or ridge of unconsolidated sand landward of, contiguous to, and approximately parallel to the beach, situated so that it may be, but is not necessarily accessible to storm waves and seasonal high waves. (Source: Maui County Code, Hawaii.)

There is clear inherent vagueness in the definition of the entity itself, demonstrated by terms such as "nearly continuous," "approximately parallel," and "[possibly] accessibility to storm waves." Despite this vagueness, it is to be expected that many locations within Maui Country may be classed as definitely "coastal dune" or definitely not "coastal dune." However, there will surely be other locations for which it is unclear whether they form part of the dune or not. Such locations may be part of the boundary to a coastal dune. Such a boundary will likely be more than a line of infinitesimal thickness, and may also be indeterminate.

By far the most common application of fuzzy set theory in the spatial domain is the representation of the locations of uncertain boundaries. Locations definitely in the region are assigned fuzzy membership value 1. Locations outside the region are assigned value 0. Locations within the boundary area are graded between 0 and 1 depending on our level of belief or evidence that the location is in the region, as illustrated previously in Figure 9.11.

The key issue is how to assign the function membership. In the case of an inherently vague region, such as a forest, it might be that the existence and density of various tree species are used as parameters in an assignment formula that computes the fuzzy membership function. For example, Figure 9.14 illustrates the assignment of fuzzy membership values to a "forest" region based on tree density. In our example above of the dune, it might be possible to use elevation of the location with respect to the beach as one such parameter.

**Figure 9.14:**
Assigning fuzzy membership values

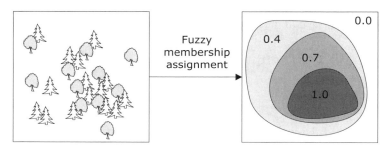

Assuming we are able to assign membership values to a fuzzy region in a principled way, care must still be taken when applying the fuzzy set operators. Suppose we have been able to assign fuzzy values to forest and wetland areas, and now wish to construct the region which is both forest and wetland. Applying the fuzzy intersection operator (defined as the minimum of the values for two membership functions) will result in a new region. However, there is no particular reason to expect this new region to be equivalent to a region derived directly from indicators of "wetland forest" (such as key wetland forest species). These kinds of problems have led to some skepticism as to the application of fuzzy set theory in this domain.

Rough sets provide another means of representing uncertain regions. Let $R$ be such a region. In this case, $\underline{R}$ and $\overline{R}$ provide lower and upper approximations to such a region. Then $\underline{R}$ consists of all those locations that can be said with certainty to be in the region $R$. Similarly, $\overline{R}$ excludes all locations that can be said with certainty not to be in the region $R$. Rough sets give a principled account of indeterminacy arising from change of granularity (for example, in a generalization process). However, in other applications, where assignment of upper and lower approximations depends on level of belief, they are open to the same criticisms as fuzzy sets.

### 9.5.2   Uncertain viewsheds

A *viewshed* is a region of terrain visible from a point or set of points.      viewshed
In the simplest case a viewshed may be represented by a crisp region. In practice, a viewshed depends on many factors that cannot be accurately or precisely predicted. Consequently, a viewshed is more faithfully represented by an uncertain region. The literature distinguishes two types of viewsheds: *probable* and *fuzzy* viewsheds.

*Probable viewshed*: Assuming perfect visibility, crisp terrain, and no vegetation effects, uncertainty in the viewshed will arise through imprecision and inaccuracy in measurements of the elevation of the terrain surface. Based on some model of the uncertainty of elevation, a *probable viewshed* may be calculated. In this case, the boundary of the viewshed will be crisp but its position uncertain due to the measurement errors.

*Fuzzy viewshed*: Assuming perfect measurements, viewshed uncertainty will arise due to atmospheric conditions, light refraction, and seasonal and vegetation effects. In this case, the boundary of the viewshed is not crisp, but broad and graded. Probabilistic techniques are no longer appropriate, and fuzzy regions are often used to model this *fuzzy viewshed*.

### 9.5.3   Regions resulting from vague spatial relations

This example briefly explores the cognitive aspects of uncertainty, using as an example human understanding of the vague concept "nearness" in a university campus. Figure 9.15 shows a sketch map of 22 significant locations on Keele University campus, identified using a preliminary experiment.

In the main experiment, 22 human subjects were divided into two equal groups, the *truth group* and *falsity group*. Each member of each group was then given a series of questionnaires concerning the locations of the significant places with respect to each other. The truth group were asked questions of the form "When is it *true* to say that place $x$ is near place $y$?" The falsity group were asked a similar question: "When is it *false* to say that place $x$ is near place $y$?" Subjects were asked not to refer to campus maps for the duration of the experiment.

By way of illustration, Table 9.2 shows the amalgamated responses to only those questionnaires that concerned nearness to the library (location 16 in Figure 9.15). The table shows that 10 of 11 subjects indicated that it was true that the Chapel was near the Library, whereas none indicated that it was false. Thus, there is strong indication that the Chapel is generally conceptualized by subjects as near to the Library. In the case of the

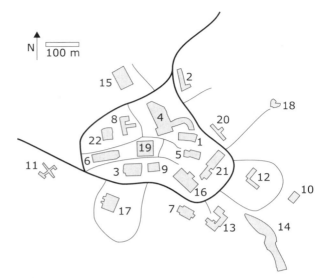

**Figure 9.15:**
Significant
locations on the
Keele
University
Campus

**Table 9.2:**
Tallies of votes
cast for and
against location
being near to
the library
(numbers refer
to Figure 9.15)

| Location | T | F | Location | T | F |
|----------|---|---|----------|---|---|
| 1. Academic Affairs | 5 | 2 | 12. Horwood Hall | 4 | 10 |
| 2. Barnes Hall | 0 | 11 | 13. Keele Hall | 8 | 2 |
| 3. Biological Sciences | 5 | 4 | 14. Lakes | 1 | 11 |
| 4. Chancellor's Building | 4 | 6 | 15. Leisure Center | 0 | 11 |
| 5. Chapel | 10 | 0 | 16. Library | 11 | 0 |
| 6. Chemistry | 4 | 6 | 17. Lindsay Hall | 2 | 8 |
| 7. Clock House | 4 | 6 | 18. Observatory | 0 | 11 |
| 8. Computer Science | 1 | 10 | 19. Physics | 5 | 5 |
| 9. Earth Sciences | 7 | 0 | 20. Reception | 4 | 4 |
| 10. Health Center | 1 | 11 | 21. Student Union | 10 | 0 |
| 11. Holly Cross | 1 | 11 | 22. Visual arts | 1 | 10 |

Leisure Center the opposite view is strongly indicated. In the case of Chemistry, there is no clear weight of evidence either way.

Using a statistical significance test, it is then possible to evaluate the extent to which the pooled responses indicate whether each location is considered near to the other locations. For example, Figure 9.16 shows the pooled results for nearness to the library. The example uses a three-valued logic to represent nearness. Based on the significance test, those places that were significantly near to the library are filled in black in Figure 9.16. Those places that were significantly not near to the library are filled in white. Finally those places that yielded no significant result either way are filled in gray.

Analysis of these results could proceed using a three-valued logic. For example, a three-valued nearness relation $\nu$ might be used to describe the nearness of campus locations to one another. For two places $x$ and $y$, $x\nu y$ will evaluate to $\top$ if $x$ is significantly near to $y$; $\bot$ if $x$ is significantly not near to $y$; and ? if $x\nu y \neq \top$ and $x\nu y \neq \bot$. Further details of this and other experiments on human cognition of spatial relationships are given in the bibliographic notes.

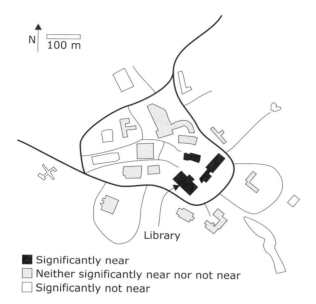

**Figure 9.16:** Nearness relation to Keele library

■ Significantly near
▨ Neither significantly near nor not near
□ Significantly not near

## BIBLIOGRAPHIC NOTES

9.1 The foundation of spatial reasoning is logic. A good introductory book on logic, which includes useful accompanying software, is Barwise and Etchemendy (1999). Hein (2003) also provides a thorough grounding in basic logic. The definitive text on qualitative reasoning is by Kuipers (1994), and a useful survey paper is Cohn and Hazarika (2001). For good introductions to the literature on cognitive aspects of spatial reasoning see Kuipers (2000) and Montello (1992).

9.2.1 The classic work on the channel theory of information flow is that of Shannon and Weaver (1949). Channel theory has been applied to cartographic information, for example, by Clarke and Battersby (2001). A discussion of information theory with particular focus on geographic information can be found in Chapters 2–4 of Duckham et al. (2003).

9.2.2 General material on uncertainty in information systems can be found in Hunter (1997) and Parsons (1996). A good collection of chapters by leading experts on uncertainty management in information systems is contained in Motro and Smets (1996). Fisher (1999) provides an overview of some aspects of uncertainty in spatial information systems.

9.2.3 Vagueness is discussed from a philosophical perspective in Keefe and Smith (1996) and Williamson (1994). Material on granularity, level of detail, and their relationship to geographic scale may be found in Goodchild and Proctor (1997) and Worboys (1998a,b).

9.2.4 The importance of the concepts of semantics and relevance to the value and quality of information are recognized in important books by Dretske (1981) and Sperber and Wilson (1995). A discussion of reasoning under inconsistency may be found in Gabbay and Hunter (1991). Paraconsistent logic is discussed by Hunter (1998). The topic of spatial data quality is discussed in Guptill and Morrison (1995) and Shi et al. (2002).

9.3.1 Introductory-level discussions of possible worlds semantics, modal logic, and many-valued logics may be found in Priest (2001).

9.3.3 Mackinson (1985) discusses belief revision. An abstract set of principles for belief revision was set out by Alchourron et al. (1985). The swan example in section 9.3.3 is from Gärdenfors and Rott (1995).

9.3.5 Fuzzy set theory and fuzzy logic are introduced in Zadeh (1965, 1988). Generalized operators for fuzzy union, intersection, and complementation are discussed in Lowen (1996).

9.3.6 Rough set theory is explained by its originator in Pawlak (1982) and Pawlak (1995). Munakata (1998) gives a concise overview of rough set analysis. Some geospatial applications of rough sets and rough set analysis can be found in Ahlqvist et al. (2000) and Duckham et al. (2001).

9.4 A standard introductory text on probability and statistics is Spiegel et al. (2000). Spatial analysis is covered in Fotheringham et al. (2002) and O'Sullivan and Unwin (2002). Bayesian theory is discussed by Charniak (1991). More information on the Dempster-Shafer theory of evidence can be found in Shafer (1976).

9.5.1 Burrough and Frank (1996) is an edited collection of papers on uncertainty and indeterminacy in spatial regions.

9.5.2 Fisher (1999) introduces the distinction between probabilistic and fuzzy viewsheds.

9.5.3 More detailed discussions of the experimental work and analysis of human understanding of the vague spatial predicate "near to" may be found in Worboys (2001) and Duckham and Worboys (2001). Fisher and Orf (1991) is an earlier example of related work.

# Time

10

**Summary**

A ***spatiotemporal information system*** extends a GIS by storing and managing spatial and temporal information. ***Snapshot*** representations of spatiotemporal phenomena represent the state of the geographic application domain at a particular time. Object **lifelines** extend the snapshot representation by explicitly storing information about changes. More recent advances in spatiotemporal databases have focused on the explicit representation of spatiotemporal entities, such as **events** and **processes**. This chapter examines some of the basic concepts involved in introducing time into a GIS, ranging from the underlying data model to spatiotemporal data structures to facilitate data retrieval.

We have moved from a data-poor to a data-rich information society, and much of this data has both spatial and temporal components. Geographic information systems are now beginning to have some temporal functionality, and a *spatiotemporal information system* manages information that is both (geo)spatially and temporally referenced. While truly spatiotemporal information systems are still primarily a research topic, GISs are beginning to be extended so that they can offer some practical temporal functionality. A wide range of further spatiotemporal functionality is now at the margins of practical application. There is a variety of ways in which such functionality can be integrated. This chapter details some of the issues facing such spatiotemporal systems.

There are many potential application domains for spatiotemporal systems, including environmental change, transportation, socioeconomic and demographic applications, health and epidemiology, multimedia, governance and administration, and defense. In addition to these more traditional spatiotemporal application areas, the increased use of real-time, mobile, and *in situ* sensors is leading to many new potential applications for spatiotemporal data models and systems. Many of these applications

spatiotemporal
information
system

relate to mobile location-aware and pervasive systems (introduced in Chapter 7).

## 10.1   INTRODUCTION: "A BRIEF HISTORY OF TIME"

At a fundamental level, entities in the world may be divided into happenings; objects and their properties; roles and relationships; and a few basic notions like location (in time and space). When we ask a question of an information system, we are generally interested in knowing something about these entities. Traditional databases manage information about objects, along with their properties and relationships, as in, for example, the entity-relationship model (Chapter 2). Adding temporality allows the possibility of managing information about the states of objects at particular times and places, but also the *properties of what happened, when, how, and why*. Spatiotemporal database research is at present primarily focused on answering questions about objects of the form "what–where–when" (for example, "Who owned the land at location $l$ and at time $t$?"). Research into next-generation systems is concerned with answering questions about what happened, where, when, how, and why. To achieve this, functionality for the analysis of relationships between these happenings, such as causality, is needed.

Geographic phenomena have both spatial and temporal components. Figure 10.1 illustrates both natural and artificial spatiotemporal geographic phenomena. Figure 10.1a shows a satellite image of a wildfire in Arizona. Smoke from the burning vegetation can be seen in the center of the image, distinct from atmospheric cloud (bottom left of image). Figure 10.1b shows urban traffic at an intersection in the UK. The images are themselves static, but the underlying phenomena are of course dynamic.

**Figure 10.1:**
Examples of natural and artificial spatiotemporal phenomena

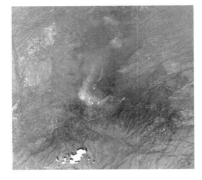

a. Wildfire in Tucson, Arizona, June 2003 (NASA ASTER image)    b. Urban traffic flow in the UK (www.freeimages.co.uk)

Dynamic geographic entities are characterized not only by spatial and attribute components, but also by temporal references. Geographic phenomena, such as events, actions, and processes, are explicitly temporal.

A spatiotemporal information system must manage data about all these time-varying real-world entities. Efforts to incorporate temporal aspects of geographic phenomena into data models and information systems are continuing, but are by no means complete. Consider the following stages in a "brief history of time."

### 10.1.1  Stage zero: Static representations

Traditional spatial and geographic information systems hold only a single state of the "real world." This state is almost always the most recent in time for which the data was captured. Interactions with the system are "timeless," in that only information contained in the single state can be retrieved. Most of the systems and models we have met up to now in this book have been concerned with static representations.

### 10.1.2  Stage one: The snapshot metaphor

Stage one uses the *snapshot* metaphor to show dynamic phenomena as a collection of *timestamped* states. Much research into spatiotemporal information systems has focused on the notion of sequences of snapshots. Indeed, snapshots can be a powerful mechanism for understanding change. For example, Figure 10.2 shows the growth of the University of Maine campus and the surrounding region from 1902 to 1955 as a series of three temporal snapshots.

snapshot

Underlying such stage one representations are models of time. For example, if time is viewed as a linear dimension, shown on a time-line (i.e., timestamps are linearly ordered), then the snapshot collection becomes a sequence. Models for time, including linear, branching, and cyclical structures, as well as distinctions between dense and discrete frameworks, are discussed later in the chapter.

### 10.1.3  Stage two: Object lifelines

The difficulty with stage one is that the static nature of the individual timestamped states dictates the generally static nature of the representation. Thus, it is a problem in stage one representations to identify dynamic phenomena, such as birth, change, and death. Figure 10.3 shows map detail from another part of the region around the University of Maine. The figure shows that an airport was constructed between 1902 and 1946, but only implicitly through comparison of the two states. The snapshot metaphor offers no mechanism for *explicitly* representing information about the time or the occurrence of events such as the construction of an airport.

Stage two begins to address these issues. Object *lifelines* are designed to explicitly represent changes of state in a single object and interactions between different objects. Figure 10.4 shows the consequences of interaction between objects. The events creation, transmission, reappearance, disappearance, transformation, cloning, and deletion are all explicitly

lifeline

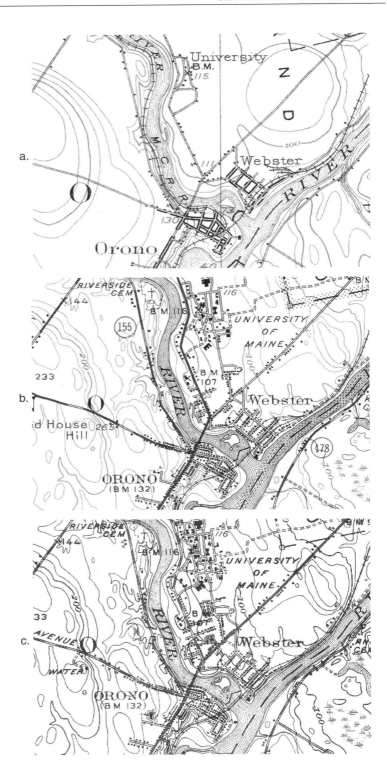

**Figure 10.2:**
Sequence of
temporal
snapshots of
University of
Maine and
Orono region,
Maine, in a.
1902, b. 1946,
and c. 1955
(Source:
USGS)

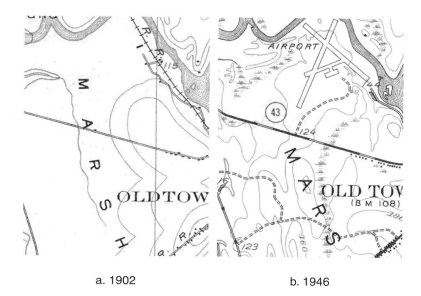

a. 1902                    b. 1946

**Figure 10.3:** Detail of changes between 1902 and 1946 based on topographic maps in Figures 10.2a and 10.2b (Source: USGS)

represented in Figure 10.4. Note the distinction between destruction, where an object is permanently deleted, and disappearance, where an object may reappear.

Figure 10.5 depicts the specific example of the evolution of a plot of land in terms of the changes to the objects located on it. The different types of changes that may occur in an object's lifeline may be summarized as:

*Creation and destruction*:  Creation and destruction occur when an object is first generated and when it is permanently eradicated. For example, the school in Figure 10.5 is created in 1938, while the house on lot two is destroyed in 1958.

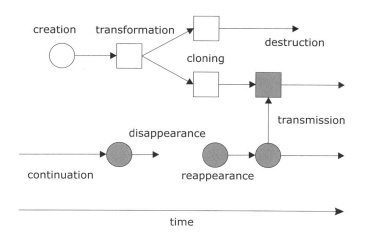

**Figure 10.4:** Object lifelines

*Disappearance and reappearance*: Disappearance and reappearance are distinct from deletion and creation, because changes may not be permanent. For example, administrative changes meant that Rutland, the smallest county in England, disappeared in 1974 (the region was incorporated into a larger county, Leicestershire). Rutland reappeared as a county following further administrative changes in the 1990s.

*Spatial change*: Spatial changes include transformations in the shape, size, and position of an object, such as the transformation of one object from a circle to a square in Figure 10.4.

*Aspatial change*: Aspatial changes involve transformations in the name, color, or other attributes of an object, such as the transformation of the white square into a gray square in Figure 10.4. Aspatial changes also include a change of class or type for an object, for example, the upgrade of a "minor road" object to become a member of the "major road" class.

transmission

*Transmission*: A distinction is often made between a transformation of an object, in which an object's attribute changes independently of any other object, and *transmission*, in which changes are in some way dependent on the attributes of another object (or objects). For example, the transformation of the white square object into a gray square object in Figure 10.4 was the result of the transmission of information about the color attribute from the gray circle object.

*Fission and fusion*: Fission and fusion occur if an object generates or is generated by one or more other objects, for example, cloning in Figure 10.4.

*Mereological change*: Changes to "part of" relationships are termed mereological (see Chapter 2). For example, medieval battles between Scotland and England led to the town of Berwick-upon-Tweed and surrounding region, now part of northern England, to change its nationality (becoming part of Scotland or England) more than a dozen times within a 300-year period.

*Typological change*: Typological change occurs when the underlying classification for a data set changes. Typology changes and schema evolution are mentioned in later sections.

### 10.1.4   Stage three: Events, actions, and processes

The final stage in this evolution is a full-blooded treatment of change, in terms of complexes of events, actions, and processes. Events, actions, and processes are allowed to become explicit entities in stage three representations. From an ontological perspective, in Chapter 4 we made an initial division of those entities that exist in the world into *continuants*,

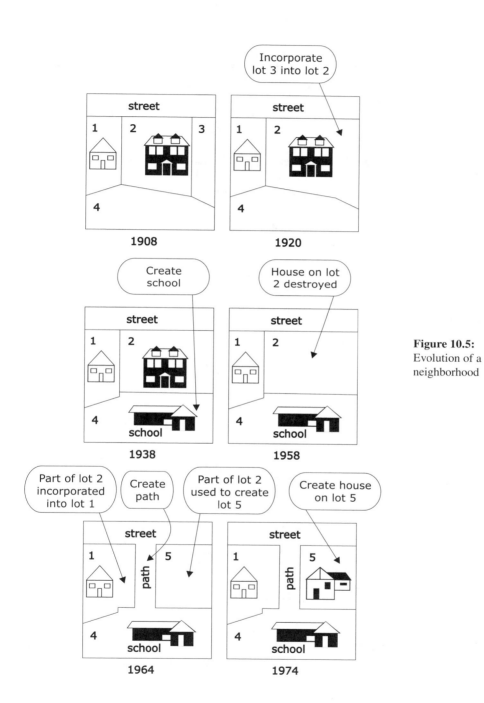

**Figure 10.5:** Evolution of a neighborhood

which endure through time (tables, houses, people) and *occurrents*, which happen or occur (lectures, people's lives, boat races). Continuants can be adequately modeled using the object lifelines of stage two. Modeling occurrents requires stage three.

An initial difficulty arises concerning the meaning of stage three terms. Almost every account uses different definitions for event, process, and action. We present below two distinct classifications of events and processes. It is important to be aware that, while both classifications are valid, they are not interchangeable.

*situation*

*event*

*process*

The first classification, shown in Figure 10.6, is after Mourelatos (1978). The top-level entity in the taxonomy in Figure 10.6 is the general concept *situation*, which then specializes to state and occurrence. Occurrences may be *events* or *processes*, which are point-based or interval-based, respectively.

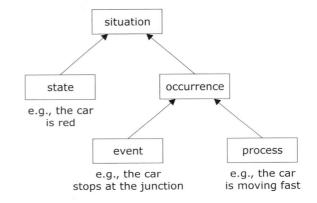

**Figure 10.6:**
A taxonomy of situations after Mourelatos (1978)

**Table 10.1:**
Count and mass nouns; continuants and occurrents

|  | continuant | occurrent |
|---|---|---|
| count noun | thing (e.g., "lake") | event (e.g., "race") |
| mass noun | stuff, (e.g., "water") | process (e.g., "running") |

A second useful classification, summarized in Table 10.1, relates to the distinction between *count nouns* and *mass nouns*. As the name suggests, a count noun refers to something that can be counted, whereas a mass noun refers to something that cannot. In the case of continuants, this distinction manifests itself as a distinction between *things* and *stuff*. For example, you can count lakes (things) but not water (stuff). In the case of occurrents, the same distinction may be used to distinguish events and processes: there may be many athletics races, but "running" cannot be counted.

*count noun*

*mass noun*

Methodologies for event modeling are far less advanced than for object modeling. However, some questions that can be asked of an occurrent include:

- Is it a type or an instance?
- Can it be counted?
- Is it performed by an agent?
- How is it situated in time; what are its boundaries?
- Does it have a purpose?
- Does it have a cause or effect?
- Does it consume resources?
- Does it have attributes, relationships?
- Is it in a partonomic hierarchy or taxonomic hierarchy?
- What is its level of detail?

The third question, about performance by an agent, distinguishes generic events from *actions*. An action is an event that is performed by some agent, human or otherwise. Thus, a "flood" would not usually be classified as an action, but a "murder" would.

action

There are clear analogies between occurrences and objects. Certainly, occurrences may have attributes (e.g., a process happened slowly); belong to a subsumption hierarchy (e.g., moving subsumes walking); have parts ("making toast" is part of "having breakfast"); and have relationships to other events (e.g., a flood event *causing* a landslide). However, the ephemeral nature of events means that identity is more problematic for occurrents.

## 10.2   TEMPORAL INFORMATION SYSTEMS

This section looks at three underlying considerations for temporal systems that are purely temporal in nature. The fusion with spatial aspects is considered in the next section.

### 10.2.1   Valid and transaction time

Two types of temporal references are important to temporal systems: *transaction time* references and *valid time* references (see also "Temporal references" inset, on the following page). Transaction time is the time when the timestamped data was entered in the database. Valid time is the time when the event relating to the capturing of the data actually occurred in the world. These two "temporal dimensions" are orthogonal. Which time(s) need to be represented in the information depends on the application domain. If the key item of information is the situation of real-world objects in real-world time, then valid time is required; if the issue of importance is when the data was entered into the system, then transaction time is required. If both times are needed, then we require a so-called *bitemporal* information system. Bitemporal systems have the disadvantage of needing more complex data structures and query languages.

transaction time

valid time

bitemporal

**Temporal references** *It is conceivable that certain specialized applications might require other types of temporal references in addition to valid and transaction time. For example, it is possible to make a distinction between* valid time, *the time an event actually occurred in the application domain, and* observation time, *the time the effects of that event were observed. Such a distinction might be required by an archaeological application (see illustration below). The time an archaeologist observed a particular artifact (observation time), as well as the time that artifact was made (valid time), and the time the information was entered into the database (transaction time) might all be important in an archaeological information system. However, valid and transaction time are usually regarded as sufficient for most applications.*

Artifact made circa AD 600    Artifact discovered July 2003    Information stored
(valid time)    (observation time)    25 August 2003 at 11:07
(transaction time)

Valid and transaction times have different properties and need to be handled in different ways. A fundamental distinction is that transaction time monotonically increases with the life of the information system, whereas valid time may not. So, if transaction $T_1$ comes before transaction $T_2$, then the transaction time associated with the data entered by $T_1$ must be before the transaction time associated with the data entered by $T_2$. This is not the case for valid time. It is perfectly possible that the real-world event to which $T_1$ refers is *later* than that described by the data in $T_2$. It follows from this that transaction times cannot be changed, whereas valid times can be changed retroactively.

For example, referring back to Figure 10.5 transaction $T_1$ might involve storing the state of the neighborhood in 1974. Subsequently, transaction $T_2$ might involve storing the state of the neighborhood in 1964. The transaction time for $T_1$ is *before* that for $T_2$, but the valid time for $T_1$ is *after* $T_2$. If we subsequently discover that valid time for transaction $T_2$ is actually 1963, not 1964, then a third transaction $T_3$ might be needed to alter the valid time for $T_2$. The transaction times for $T_1$, $T_2$, and $T_3$ can never be altered.

rollback

version

version
management

A transaction time information system maintains the history of system activity, and maintains the capability to *rollback* to previous states of the system. Each state of the system is called a *version*, and the business of organizing this activity is called *version management*. At any time, a transaction time information system has access to the current state and all previous versions.

A valid time information system maintains the history of the real-world activity, and allows the capability to query current, past, and possibly future states of the real-world objects in its database. This history is as best known at the time of access to the system.

### 10.2.2 Temporal extensions to relational database systems

A *temporal information system* can be constructed by extending basic relational database systems. There are two standard ways to add temporality to a relation: *tuple timestamping* and *attribute timestamping*.

temporal information system

#### Tuple timestamping

A bitemporal timestamp is a sequence of ordered pairs $(x, y)$, where $x$ is a transaction time interval and $y$ is a valid time interval. In this way, a *tuple timestamp* indicates that during time period $x$ the database recorded the information in a particular tuple as being valid during time period $y$. Figure 10.7 depicts a house ownership scenario. House $H_1$ is owned by person $P_1$ from (valid) time 1 to 6. At valid time 7, house $H_1$ is bought by person $P_2$, while person $P_1$ purchases house $H_2$ at the same time. A database tracking these changes is updated at (transaction) times 5 and 10. Table 10.2 shows the house ownership scenario in Figure 10.7 using a tuple *bitemporal timestamp*. The valid time for $P_1$ owning $H_1$ is incorrectly recorded in the first update at transaction time interval $[1, 4]$. Consequently, the second update at transaction time interval $[10, 14]$ corrects the original valid time (from $[1, 4]$ to $[1, 6]$).

bitemporal timestamp<br><br>tuple timestamp

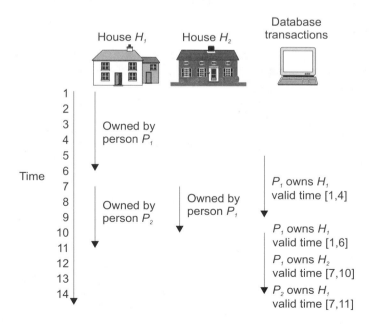

**Figure 10.7:** Temporal house ownership scenario

**Table 10.2:**
Tuple
bitemporal
timestamps for
scenario in
Figure 10.7

| HouseID | PersonID | Time |
|---------|----------|------|
| $H_1$ | $P_1$ | $([5,9],[1,4])$ $([10,15],[1,6])$ |
| $H_2$ | $P_1$ | $([10,15],[7,10])$ |
| $H_1$ | $P_2$ | $([10,15],[7,11])$ |

*Attribute timestamping*

attribute
timestamp

An alternative way of representing temporal information is with attribute timestamping, in which temporal information is associated directly with the attribute values to which it refers. Attribute timestamping relies on complex data types that encapsulate both data values and timestamps for those values. In Table 10.3 information about the same house ownership scenario as above is represented using attribute timestamping.

**Table 10.3:**
Attribute
bitemporal
timestamps for
scenario in
Figure 10.7

| HouseID | PersonID |
|---------|----------|
| $H_1([5,9],[1,4]),([10,14],[1,11]),$ | $P_1([5,9],[1,4]),([10,14],[1,6])$ $P_2([10,14],[7,11])$ |
| $H_2([10,14],[7,10])$ | $P_1([10,14],[7,10])$ |

Note that relations used in timestamping may violate normal form, because timestamps and attributes may not be atomic. Tuple timestamping is conceptually simpler than attribute timestamping, and is more in keeping with the relational model.

*Timestamping in object-oriented databases*   Attribute timestamping can be naturally extended to timestamping in object-oriented databases. For example, Figure 10.8 shows the house $H_1$ of Figure 10.7 with several kinds of temporal reference. In this case the boundary is a spatiotemporal object; information about the owners is represented as a temporally timestamped textual type; the address is atemporal (we assume the address never changes); and registration is a dynamic behavior that updates the registration with a new owner.

### 10.2.3   Schema evolution

schema evolution

The discussion so far has only considered changes to data, but assumes the schema for the data (conceptual model, data types, database structure, and so forth) is fixed. Work on temporal and spatiotemporal information systems normally assumes a fixed schema. However, there may be a range of situations in which the schema changes or evolves over time, termed *schema evolution*. Schema evolution is the subject of an independent area of research not covered in this book (but see bibliographic notes).

Moving object databases  *Spatiotemporal database research can be divided into two broad categories: research dealing with change (e.g., administrative boundary evolution, environmental change) and research dealing with moving objects (vehicles, ships, people). There is a core of similar techniques, but also much that is different in the treatment of these topics. The work of Wolfson et al. (1998) identifies three key issues in moving object databases (MODs) for which conventional spatiotemporal database capabilities are inadequate. First, individual updates in an MOD will often only make small changes to the stored information (an object may only have moved a few meters), but may happen very frequently (for example, a car's location recorded every 30 seconds using a GPS). Second, many basic queries in a MOD rely on extrapolation from and interpolation between the known locations of an object. Such functions must be able to allow for the inherent uncertainty arising from imperfect knowledge of location. Third, conventional static spatial indexes are unable to deal efficiently with moving objects, because every movement would require a change in the spatial index. Efficient MODs are clearly a vital technology for the location-based services discussed in Chapter 7.*

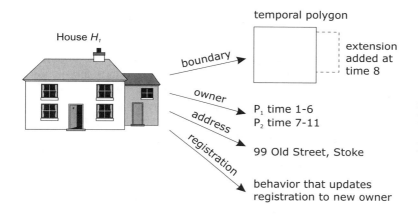

**Figure 10.8:** A spatially and temporally referenced house object, $H_1$, from Figure 10.7

## 10.3  SPATIOTEMPORAL INFORMATION SYSTEMS

Spatiotemporal database technology consists of at least the following components (Figure 10.9):

- standard database technology, whether relational, object-oriented, or another paradigm;

- spatial database technology, including GIS; and

- temporal database technology.

It is clear that efficient spatiotemporal databases will need to integrate the spatial and temporal components into a unified whole. What is not immediately clear is whether spatiotemporal database technology requires any more than simply the union of these three technologies (i.e., whether there is anything in the set labeled '?' in Figure 10.9). The issue here is whether there is more to spatiotemporal than just spatial plus temporal. One area that appears to require more than just the union of spatial and

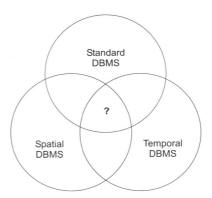

**Figure 10.9:**
Components of
a spatio-
temporal
DBMS

temporal capability is applications involving *movement* rather than simply change (see "Moving object databases" inset, on the previous page).

### 10.3.1   Time, space, and the timeline

Time and space are tantalizingly similar in some basic respects. Time may at one level be thought of as a one-dimensional variant of space, modeled, for example, using the number line as a measuring stick for timestamps. Despite this similarity, entities exist in time very differently from the ways that they exist in space. For example, while an object may move in many directions in space, it is constrained to "move forwards," and not "move backwards," in time. Indeed the very notion of an entity "moving" in time is a metaphor derived from spatial experience (see "Temporal metaphors" inset, on the facing page).

However, there are many ways in which time may usefully be treated as a timeline, analogously to a one-dimensional space. Temporal literals, used as the basis of timestamping, may be either time *instants* (points of the timeline) or time *intervals* (intervals on the timeline). The timeline itself may be modeled as isomorphic to the real numbers (*continuous time*), rational numbers (*dense time*), or the integers (*discrete time*). Most computational approaches to time assume a discrete time model, in accordance with the discrete nature of computation. The discrete timeline then approximates to a timestamp measurement, in the same way that a discretization of the plane approximates to spatial references (see Chapter 4).

discrete time

The basic unit of temporal measurement in a discrete framework is often referred to as a *chronon* or *tick*. A chronon is atomic in the sense that no sub-chronon division is possible. Thus the chronon plays for time the role of the pixel for space. *Duration* is modeled as a whole number of chronons, but not anchored to a particular place on the timeline.

chronon

duration

The discussion above is the simplest, most common, and by far the most useful model of time, so-called *linear time*. Other more elaborate models include *branching time*, in which past and future possibilities may be represented, and *cyclic time*, in which periodic phenomena such as

linear time
branching time
cyclic time

**Temporal metaphors** *Metaphors for time and temporal relationships are often spatial. Future events may be viewed metaphorically as coming toward us, as in the traditional Christmas song "Christmas is coming / The goose is getting fat / Please put a penny in the old man's hat." Possible future events may be viewed metaphorically as having a spatial relationship to us, as in "Love is just around the corner," or "Harry's career can take one of two possible paths from here." The last example shows a movement through time. There are many examples of this, such as "We have just left the 20th century," or "John is approaching retirement." Basic notions of time, like timelines, branching time, and cyclic time (discussed below) also have their roots in physical space. Cognition of events, processes, and actions also has a rich metaphorical structure, with a spatial component. Events may be locations, as in "My father was in the war" (event as container) or "The country is going to war" (event as place). Events may be objects, as in "I have a headache" (event as something to be possessed). Finally, events may be forces, as in "The war changed the boundary of our country" (event having impact on an object). This spatialization is therefore likely to be an important metaphor for HCI in temporal information systems, although spatiotemporal system interfaces using this metaphor would need to avoid possible confusion between geographic and temporal space.*

seasons or times of the day may be represented. However, analysis using these complex variants can often be reduced to the linear timeline. Figure 10.10 illustrates the different types of time.

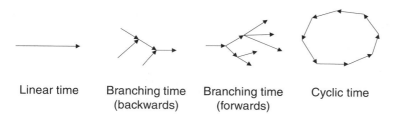

Linear time      Branching time      Branching time      Cyclic time
                 (backwards)         (forwards)

**Figure 10.10:** Linear, branching, and cyclic time

### 10.3.2 Bitemporal spatial models

The bitemporal model described in the previous section can be extended for use within a bitemporal spatial information system. Figure 10.11 illustrates a bitemporal spatial representation of the construction of a bypass around a town. The three frames show three states of the database. In Figure 10.11a, data was entered into the database in 1993 (transaction time) indicating that the bypass was planned to be constructed in 1994 (valid time), along the route $abc$. In Figure 10.11b, data entered into the database in 1994 (transaction time) indicated that the bypass had been built in 1994 (valid time) following the revised route $adefc$. Finally, the data entered into the database in 1995 (transaction time) corrected the previous transaction, indicating that part of the bypass was not actually built until 1995 (valid time).

The bitemporal information associated with an entity may be pictured more clearly using an array. Figure 10.12 illustrates the existence of

Figure 10.11: Bitemporal spatial representation of the construction of a bypass

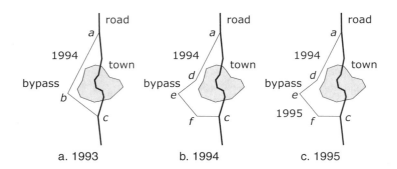

a. 1993            b. 1994            c. 1995

line segment *ef* as a bitemporal array, shaded cells indicating bitemporal timestamps associated with the road segment. The collection of bitemporal timestamps for the entire bypass is given in Figure 10.13.

Figure 10.12: Bitemporal array for bypass line segment *ef* from Figure 10.11

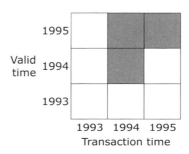

Figure 10.13: Bitemporal references for bypass segments from Figure 10.11

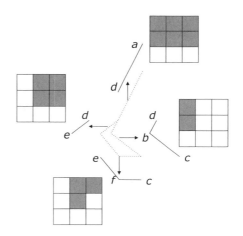

## 10.4  INDEXES AND QUERIES

This section discusses the issue of efficient access to spatiotemporal data in a database using appropriate indexes and queries. As for purely spatial queries, the important indicators for an access method are:

- the space required for the data;

- the amount of processing overhead required on database update; and

- the time taken to retrieve data for a query.

With temporal data, the most naive approach is to physically store each temporal snapshot. However, except in the smallest examples, we would quickly run out of space with this approach. Therefore solutions concentrate on storage of changes (forward and backward), and access methods must take account of this. Throughout this section, we assume a discrete linear time dimension.

### 10.4.1   Temporal data structures: Transaction time

A transaction time database must store current and past states of the database and provide a structure for creation, deletion, and modification transactions. The index for a transaction time database then needs to allow efficient access to objects in both past and current states. There have been several solutions to the problem of indexing data in transaction time databases.

A first solution to indexing transactions would be to simply create a new B-tree index (introduced in Chapter 6) for every change in the database. Figure 10.14 illustrates the idea (cf. Figure 6.5), in which after each update a complete new version of the B-tree is stored (in this case the insertion of the value '10'). Changes (as opposed to insertions or deletions) may be achieved using compound deletion and insertion operations (e.g., changing the value '9' to '10' in Figure 10.14 can be accomplished by deleting '9' and inserting '10' as a single transaction). While this solution does enable efficient query processing for certain types of query, it should be clear that using multiple B-trees for transaction time databases is prohibitively inefficient in terms of storage space.

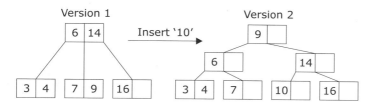

**Figure 10.14:** Multiple B-trees used to index a transaction time database

One simple solution that extends this naive idea is the *overlapping B-tree*. Instead of creating a completely new tree for each update, the overlapping B-tree duplicates only those nodes where some value has changed. Different transactions share any nodes that are unchanged. Figure 10.15 shows the same insert operation as that in Figure 10.14, but now using an overlapping B-tree structure.

overlapping B-tree

In this way, the overlapping B-tree is able to store changes without large amounts of redundant duplicated data. Note that the overlapping B-

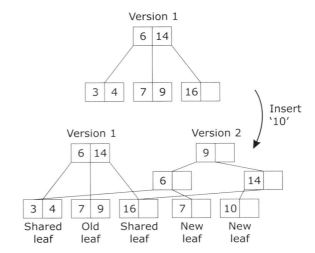

**Figure 10.15:** Overlapping B-tree representation of insert operation in Figure 10.14

tree only stores transaction times at the roots. Overlapping B-trees require a further index for the root nodes themselves.

Overlapping B-trees have the advantage that they are simple to understand and implement. Other more complex temporal indexes provide more advanced capabilities. The *multiversion B-tree*, for example, explicitly represents the insertion and deletion of objects into the database. Deletion time is only implicitly stored in the overlapping B-tree. Attribute changes are modeled as the deletion and creation of a new object. Each object, when it is created, is timestamped with the temporal reference interval $[t_C, \text{NOW}]$, where $t_C$ is the time the creation transaction is made, and NOW is a variable holding the current transaction time. If the object is later deleted, its timestamp is modified to $[t_C, t_D]$, where $t_D$ is the time the deletion transaction is made. Figure 10.16 shows a schematic of the evolution of a transaction time database, in which object creation and deletion are explicitly represented.

*multiversion B-tree*

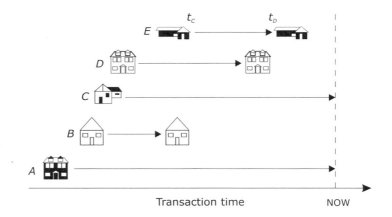

**Figure 10.16:** Evolution in a transaction time database (the states of the database are shown on the temporal axis)

By explicitly storing the creation and deletion time for objects, the multiversion B-tree is able to cluster data within the tree according to whether it is *alive* (referenced by a timestamp of the form $[t_x, \text{NOW}]$) or *dead* (referenced by a timestamp of the form $[t_x, t_y]$), so improving query efficiency. For example, in Figure 10.16 objects $B$, $D$, and $E$ might be clustered together on one branch of the multiversion B-tree, because they are dead and can never change. Objects $A$ and $C$ can change (they may be deleted), and so might be clustered on a separate branch. More information on overlapping B-trees, multiversion B-trees, and many other transaction time indexes, such as *snapshot indexes*, *time-split B-trees*, and *temporal hashing*, may be found in references cited in the bibliographic notes.

### 10.4.2 Temporal indexes: Valid time

A valid time, object-based database holds a dynamic collection of objects, each attribute of which is a collection of interval timestamped values (i.e., attribute histories). In this case the database evolution is not stored, so an update has a permanent effect in that the past database state is lost. A valid-time database needs to store only a single current state, detailing the evolution of its constituent real-world objects. Figure 10.17 illustrates the state of a valid-time database, where in contrast to Figure 10.16 the current database state is represented by the whole figure.

Current database state

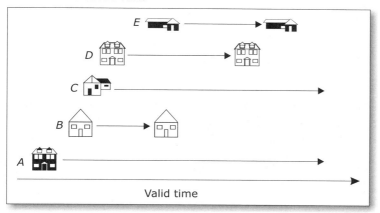

**Figure 10.17:** Current state of a valid-time database

A *segment tree* is one index useful for valid-time databases. A segment tree structures a collection of linear (time) intervals so that it becomes efficient to retrieve all intervals enclosing a given point. For example, a segment tree allows efficient queries of the type:

Given a time $t$, retrieve all objects that were valid at time $t$.

The segment tree structure can be described with reference to an example. Suppose that we have a finite set of time intervals. The first

segment tree

step is to sequence the boundary points in ascending order. For example, if the intervals are:

$$A[1, 10], B[3, 15], C[4, 7], D[4, 9], E[6, 15], F[8, 12],$$
$$G[10, 14], H[16, 22], I[18, 20], J[10, 19]$$

then the sequence is

$$1, 3, 4, 6, 7, 8, 9, 10, 12, 14, 15, 16, 18, 19, 20, 22$$

These points become labels for the leaves of a binary tree, as shown in Figure 10.18. Each segment is now split into maximal chunks and distributed as labels of nodes in the tree. For example, segment $A$ labels a single high-level node $[1, 10]$, whereas $B$ labels leaf-nodes $[3]$ and $[15]$ and non-leaf nodes $[4, 6]$, $[7, 10]$, and $[12, 14]$. Each node therefore represents an interval (or point) and is labeled by zero or more intervals.

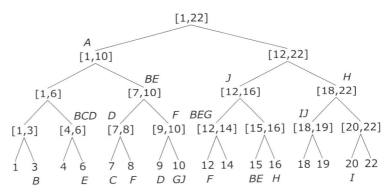

**Figure 10.18:**
Example of a
segment tree

Given a time $t$, the process of searching the tree for intervals containing that point is straightforward. We begin at the root. If the root is labeled with any intervals, insert them into the result list (assuming the root interval contains the search point). Next, examine the immediate descendants of the root. If $t$ does not belong to the interval represented by either of the nodes, then halt the procedure. Otherwise go to the node representing the interval containing $t$, add any intervals labeling the node to the result list, and continue this procedure until the leaf is reached or the procedure halts. The result list will hold all the intervals containing $t$. For example, to search for intervals containing the point 5, start at the root. Proceed to the node $[1, 10]$ and add interval $A$ to the result list. Proceed to node $[1, 6]$, then to node $[4, 6]$ and add intervals $B$, $C$, and $D$ to the result list. Then halt the procedure. The intervals found are $A$, $B$, $C$, and $D$.

The segment tree is a static structure, because the set of interval end-points is given in advance. However, new intervals whose end-points are in the given set may be dynamically inserted into the tree structure by labeling appropriate notes. Also, intervals may be deleted from the structure.

### 10.4.3 Bitemporal indexes

A straightforward solution to providing bitemporal indexes is to treat valid and transaction time as independent dimensions of a MBB (Figure 10.19). Standard spatial indexes, such as the R-tree, can then be used to index each bitemporal MBB. This approach has several advantages, not least simplicity. However, it suffers from considerable overlapping between different bitemporal MBBs. This leads to inefficient queries, because a particular time coordinate (comprising a transaction time $t_t$ and a valid time $t_v$) may lie within several different bitemporal MBBs (see Figure 10.19).

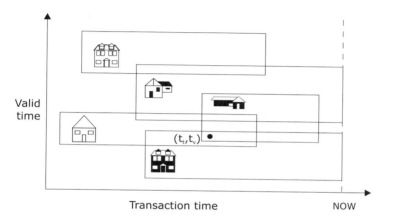

**Figure 10.19:** Bitemporal minimum bounding boxes

One extension to the bitemporal R-tree approach is to use two R-tree indexes, one to store objects that are alive (from a transaction time perspective) and another to store those that are dead (again from a transaction time perspective). Using two R-tree indexes in this way tends to reduce the amount of overlapping, because it is only necessary to store the first element of the transaction timestamp in the "alive" R-tree. By definition, the second element will be NOW for all objects objects in the alive R-tree. The bibliographic notes contain references to more details on bitemporal indexes.

### 10.4.4 Spatiotemporal indexes and queries

The appropriate structure for a spatiotemporal index will depend on the temporal (valid, transaction, bitemporal) and spatial nature of the data stored and access required. Most spatiotemporal indexes are based on merging spatial and temporal indexes, by treating time as essentially another spatial dimension. For example, for a single temporal dimension and two spatial dimensions spatiotemporal data can be represented and stored as three-dimensional spatiotemporal objects. An R-tree might be then used to index the *minimum bounding cuboid* for the spatiotemporal data. This approach is simple, but may suffer from similar overlap problems as with the bitemporal R-tree discussed above.

historical R-tree

An alternative approach is the *historical R-tree* (HR-tree), which uses an approach very similar to that of the overlapping B-tree described above. Spatial data is indexed as a conventional R-tree (rather than the B-tree in Figure 10.15). New versions introduce new root nodes that share unchanged leaves. The *historical $R^+$-tree* goes one step further and allows the explicit representation of both insertion times and deletion times for spatiotemporal objects, in much the same way as the multiversion B-tree extends the capabilities of the overlapping B-tree. The bibliographic notes contain references to further information on spatiotemporal indexes.

historical $R^+$-tree

Queries to a spatiotemporal database, whether valid or transaction time, may be seen in their simplest form as answering the following questions:

- What was the state of the world or database at a particular location and given time?

- Where was the world or database in a particular state at a given time?

- When was the world or database in a particular state at a given location?

As discussed in Chapter 6, the two basic categories of query are point and range queries. In a spatiotemporal database both point and range queries may take values from spatial, temporal, or attribute domains. In the case of point queries, we might specify a single value of an attribute, a point location, or an instant in time. For example, "Where was John at 3.00 pm, July 20, 2002?" is a temporal point query. Range queries allow the specification of a value of a collection type attribute, extended location, or temporal period. For example, "When was John within 100 miles of Boston?" is a spatial range query.

The most complex and demanding queries involve a combination of spatial, temporal, and attribute points and ranges. For example, "Retrieve the locations of all salespersons within 100 miles of Boston Headquarters in April" involves both a spatial and temporal range query. Performance for such queries will depend heavily on the actual index used.

temporal SQL

SQL, introduced in Chapter 2, is recognized as the standard for database query languages. Potentially, temporal and spatiotemporal queries can be formulated within SQL, but such queries rapidly become highly complex. An emerging standard for temporal databases is *temporal SQL* (TSQL2), which is an extension of SQL-92. Temporal SQL allows time-stamping tuples with both transaction and valid timestamps, using the tuple-based extension to the relational model, described in the previous section. Spatial functions will also be required to extend TSQL2 to a spatiotemporal query language.

## BIBLIOGRAPHIC NOTES

10.1 Langran (1992) published a groundbreaking book on the topic of spatiotemporal information. More recently, some of the conceptual background to this work is discussed in Peuquet (2001). Temporal reasoning is reviewed in Shoham (1988) and Allen (1983, 1984). Hornsby and Egenhofer (2000) discuss object-based models of spatiotemporal change. An excellent treatment of reasoning about change is provided by Galton (2000).

10.1.3 The land ownership example is a modified version of one given in Al Taha (1992).

10.1.4 Ontological and philosophical issues related to time, events, actions, and processes are discussed in Pianesi and Varzi (1996) and in Higginbottom et al. (2000). The first chapter of Higginbottom et al. is particularly recommended as an overview of the issues.

10.2.1 For a discussion of transaction time, valid time, and bitemporal databases, in the context of spatiotemporal information systems, see Worboys (1994b).

10.2.2 Attribute and tuple timestamping are considered in the context of temporal query languages by Tansel and Tin (1997).

10.2.3 For a survey of schema evolution issues see, for example, Andany et al. (1991) and Roddick (1996).

10.3 A useful survey of spatiotemporal database issues, with a leaning to constraint database models is provided by Chomicki and Revesz (1997). Another excellent survey is that of Abraham and Roddick (1999). For a survey of database technology for managing moving objects, see Güting et al. (2000). A European project focused on movement was the Chorochronos project (1999).

10.3.1 Metaphors for time and events are discussed in Lakoff and Johnson (1999).

10.3.2 The bitemporal road bypass example is taken from Worboys (1994b).

10.4 An excellent treatment of temporal and spatiotemporal indexing is contained in Manolopoulos et al. (1999).

10.4.1 An overview of temporal hashing is contained within Kollios and Tsotras (2002). Other transaction time indexes are summarized in Manolopoulos et al. (1999). The multiversion B-tree is introduced by Becker et al. (1996).

10.4.2 Blankenagel and Güting (1994) have extended the scope of the segment tree in a non-trivial way to the *external segment tree* that provides a dynamic balanced tree structure with efficient storage properties and update algorithms.

10.4.3 An example of an approach to bitemporal indexing can be found in Kumar et al. (1998).

10.4.4 Chomicki (1994) provides a useful survey of query languages for temporal databases. The temporal query language TSQL2 is discussed in Snodgrass (1995).

# Appendix A

## CINEMA RELATIONAL DATABASE EXAMPLE

A Microsoft Access database containing all the tables in this section may be downloaded from the book website, at http://worboys.duckham.org. The Access database also contains the queries relating to Tables 2.3, 2.4, and 2.6.

### Cinema database schema

CINEMA (<u>CIN ID</u>, NAME, MANAGER, TELNO, TOWN, GRID_REF)

SCREEN (<u>CINEMA ID</u>, <u>SCREEN NO</u>, CAPACITY)

FILM (<u>TITLE</u>, DIRECTOR, CNTRY, YEAR, LNGTH)

SHOW (<u>CINEMA ID</u>, <u>SCREEN NO</u>, <u>FILM NAME</u>, STANDARD, LUXURY)

STAR (<u>NAME</u>, BIRTH_YEAR, GENDER, NTY)

CAST (<u>FILM STAR</u>, <u>FILM TITLE</u>, ROLE)

(See page 46.)

### SCREEN relation

| CINEMA_ID | SCREEN_NO | CAPACITY |
|---|---|---|
| 1 | 1 | 800 |
| 1 | 2 | 750 |
| 1 | 3 | 250 |
| 2 | 1 | 800 |
| 2 | 2 | 800 |
| 3 | 1 | 500 |

**SHOW relation**

| CINEMA_ID | SCREEN_NO | FILM_NAME | STANDARD | LUXURY |
|---|---|---|---|---|
| 1 | 1 | X2 | 5.50 | 7.00 |
| 1 | 2 | American Beauty | 5.50 | 6.50 |
| 1 | 3 | The Hours | 5.00 | |
| 2 | 1 | Training Day | 5.00 | 6.00 |
| 2 | 2 | Traffic | 4.50 | 6.00 |
| 3 | 1 | X2 | 6.00 | |

**CAST relation**

| FILM_STAR | FILM_TITLE | ROLE |
|---|---|---|
| Benicio Del Toro | The Hunted | Aaron Hallam |
| Benicio Del Toro | The Usual Suspects | Fred Fenster |
| Benicio Del Toro | Traffic | Javier Rodriguez Rodriguez |
| Denzel Washington | Malcolm X | Malcolm X |
| Denzel Washington | Philadelphia | Joe Miller |
| Denzel Washington | Training Day | Alonzo Harris |
| Halle Berry | Die Another Day | Jinx |
| Halle Berry | Monster's Ball | Leticia Musgrove |
| Halle Berry | X2 | Storm |
| Kevin Spacey | A Bug's Life | Hopper |
| Kevin Spacey | American Beauty | Lester Burnham |
| Kevin Spacey | Midnight in the Garden of Good and Evil | James Williams |
| Kevin Spacey | The Usual Suspects | Roger "Verbal" Kint |
| Nicole Kidman | Eyes Wide Shut | Alice Harford |
| Nicole Kidman | Moulin Rouge | Satine |
| Nicole Kidman | The Hours | Virginia Woolf |

**CINEMA relation**

| CIN_ID | NAME | MANAGER | TELNO | TOWN | GRID_REF |
|---|---|---|---|---|---|
| 1 | Majestic | Julie Jones | 1782764127 | Stoke | SJ 878450 |
| 2 | Regal | Harry Emms | 1782749253 | Hanley | SJ 880471 |
| 3 | Regal | Mary Carr | 1782530152 | Newcastle | SJ 852462 |

## FILM relation

| TITLE | DIRECTOR | CNTRY | YEAR | LNGTH |
|---|---|---|---|---|
| A Bug's Life | Lasseter | USA | 1998 | 96 |
| Traffic | Soderbergh | USA | 2000 | 147 |
| Die Another Day | Tamahori | UK | 2002 | 132 |
| Malcolm X | Lee | USA | 1992 | 194 |
| American Beauty | Mendes | USA | 1999 | 122 |
| Eyes Wide Shut | Kubrick | USA | 1999 | 159 |
| Moulin Rouge | Luhrmann | USA | 2001 | 127 |
| The Usual Suspects | Singer | USA | 1995 | 106 |
| The Hours | Daldry | USA | 2002 | 114 |
| Monster's Ball | Forster | USA | 2001 | 111 |
| Midnight in the Garden of Good and Evil | Eastwood | USA | 1997 | 155 |
| The Hunted | Friedkin | USA | 2003 | 94 |
| X2 | Singer | USA | 2003 | 133 |
| Philadelphia | Demme | USA | 1993 | 125 |
| Training Day | Fuqua | USA | 2001 | 120 |

## STAR relation

| NAME | BIRTH_YEAR | GENDER | NTY |
|---|---|---|---|
| Benicio Del Toro | 1967 | M | Puerto Rico |
| Denzel Washington | 1954 | M | USA |
| Halle Berry | 1968 | F | USA |
| Kevin Spacey | 1959 | M | USA |
| Nicole Kidman | 1967 | F | USA |

# Appendix B

## ACRONYMS AND ABBREVIATIONS

*1NF*: first normal form

*ALU*: arithmetic/logic unit

*ASCII*: American standard code for information interchange

*ASDI*: Australian spatial data infrastructure

*ATM*: automated teller machine

*BSP*: binary space partitioning

*CAD*: computer-aided design

*CGDI*: Canadian geospatial data infrastructure

*CISC*: complex instruction set computer

*CORBA*: common object request broker architecture

*CPU*: central processing unit

*CRC*: cyclic redundancy check

*CRM*: customer relationship management

*CSCW*: computer-supported cooperative work

*DAG*: directed acyclic graph

*DBMS*: database management system

*DCEL*: doubly connected edge list

*DCOM*: distributed component object model

*DDBMS*: distributed database management system

*DDL*: data definition language

*DEM*: digital elevation model

*DSL*: digital subscriber line

*DTD*: document type definition

*DTM*: digital terrain model

*E911*: enhanced 911

*EEPROM*: electrically erasable programmable read-only memory

*EER*: extended entity-relationship

*EPROM*: erasable programmable read-only memory

*E-R*: entity-relationship

*ESRI*: Environmental Systems Research Institute

*FCS*: frame check sequence

*FIFO*: first-in-first-out

*FTP*: file transfer protocol

*GIS*: geographic information system

*GML*: geography markup language

*GOMS*: goals, operations, methods, selection

*GPS*: global positioning system

*HCI*: human-computer interaction

*HSV*: hue, saturation, value

*HTA*: hierarchical task analysis

*HTTP*: hypertext transfer protocol

*IBIS*: issue-based information system

*IEEE*: Institute of Electrical and Electronic Engineers

*IIOP*: Internet inter-orb protocol

*IO*: input/output

*IVR*: interactive voice response

*LAN*: local area network

*LBS*: location-based services

*LIFO*: last-in-first-out

*LOBS*: location/orientation-based services

*MAN*: metropolitan area network

*MAT*: medial axis transformation

*MBB*: minimum bounding box

*MOD*: moving object database

*NAA*: node-arc-area

*NAN*: neighborhood area network

*NGDI*: national geospatial data infrastructure

*NSDI*: national spatial data infrastructure

*NTF*: neutral tranfer format

*OOA*: object-oriented analysis

*OOD*: object-oriented design

*OODBMS*: object-oriented database management system

*OOPL*: object-oriented programming language

*ORB*: object request broker

*ORDBMS*: object-relational database management system

*p2p*: peer-to-peer

*PAN*: personal area network

*PC*: personal computer

*PDA*: personal digital assistant

*PGIS*: participatory geographic information system

*PROM*: programmable read-only memory

*QTM*: quaternary triangular mesh

*RAM*: random access memory

*RCC*: region connection calculus

*RDBMS*: relational database management system

*RISC*: reduced instruction set computer

*RLE*: run length encoding

*RMI*: remote method invocation

*RMSE*: root mean square error

*ROM*: read-only memory

*SAIF*: spatial archive interchange format

*SDTS*: spatial data transfer standard

*SEQUEL*: structured English query language

*SIS*: spatial information system

*SOAP*: simple object access protocol

*SQL*: structured query language

*TACIS*: tactile acoustic computer interaction system

*TIN*: triangulated irregular network

*TTFF*: time to first fix

*TTS*: text-to-speech

*UDDI*: universal description, discovery, and integration (protocol)

*UML*: unified modeling language

*USGS*: United States Geological Survey

*UTM*: universal transverse Mercator

*VDU*: visual display unit

*VR*: (immersive) virtual reality

*W3C*: World Wide Web consortium

*WAN*: wide area network

*Wi-Fi*: wireless fidelity

*WIMP*: windows, icons, menus, pointers

*WSDL*: web services description language

*WWW*: World Wide Web

*XML*: extensible markup language

*XSLT*: extensible stylesheet language translation

# Bibliography

Abadi, M. and Cardelli, L. (1996). *A Theory of Objects*. Berlin: Springer.

Abdelmoty, A., Williams, M., and Paton, N. W. (1993). Deduction and deductive databases for geographic data handling. In Abel, D. and Ooi, D. C. (Eds), *Advances in Spatial Databases*, pp. 443–464. Berlin: Springer.

Abel, D. J. and Mark, D. M. (1990). A comparative analysis of some two-dimensional orderings. *International Journal of Geographical Information Systems*, 4(1):21–31.

Abiteboul, S., Hull, R., and Vianu, V. (1995). *Foundations of Databases*. Reading, MA: Addison-Wesley.

Abowd, G., Mynatt, E. D., and Rodden, T. (2002). The human experience. *Pervasive Computing*, 1(1):48–57.

Abraham, T. and Roddick, J. F. (1999). Survey of spatiotemporal databases. *GeoInformatica*, 3:61–69.

ACM (2003). Association for computing machinery classic of the month website. http://www.acm.org/classics. Accessed 9 August 2003.

Ahlqvist, O., Keukelaar, J., and Oukbir, K. (2000). Rough classification and accuracy assessment. *International Journal of Geographic Information Science*, 14:475–496.

Aho, A. V., Hopcroft, J. E., and Ullman, J. D. (1974). *The Design and Analysis of Computer Algorithms*. Reading, MA: Addison-Wesley.

Akima, H. (1978). A method of bivariate interpolation and smooth surface fitting for irregularly distributed data points. *ACM Transactions on Mathematical Software*, 4(2):148–159.

Al Taha, K. K. (1992). *Temporal Reasoning in Cadastral Systems*. PhD thesis, University of Maine, Orono, ME.

Alchourron, C., Gärdenfors, P., and Makinson, D. (1985). On the logic of theory change: Partial meet functions for contraction and revision. *Journal of Symbolic Logic*, 50:510–530.

Allen, J. F. (1983). Maintaining knowledge about temporal intervals. *Communications of ACM*, 26:832–843.

Allen, J. F. (1984). Towards a general theory of action and time. *Artificial Intelligence*, 23:123–154.

Andany, J., Leonard, M., and Palisser, C. (1991). Management of schema evolution in databases. In *Proc. 17th International Conference on Very Large Data Bases (VLDB)*.

Armstrong, M. A. (1979). *Basic Topology*. Maidenhead, England: McGraw-Hill.

Barwise, J. and Etchemendy, J. (1999). *Language, Proof, and Logic*. New York: Seven Bridges.

Baumgart, B. (1975). A polyhedron representation for computer vision. In *Proc. AFIPS National Conference 44*, pp. 589–596.

Becker, B., Gschwind, S., Ohler, T., Seeger, B., and Widmayer, P. (1996). An asymptotically optimal multiversion B-tree. *VLDB Journal*, 5(4):264–275.

Bentley, J. L. and Friedman, J. H. (1979). Data structures for range searching. *ACM Computing Surveys*, 11(4):397–409.

Beresford, A. and Stajano, F. (2003). Location privacy in pervasive computing. *IEEE Pervasive Computing*, 2(1):46–55.

Bertin, J. (1967). *Sémiologie Graphique*. Paris: Mouton.

Bertolotto, M., De Floriani, L., and Marzano, P. (1994). An efficient representation for pyramidal terrain models. In Pissinou, N. and Makki, K. (Eds), *Proc. Second ACM Workshop on Advances in Geographic Information Systems*, pp. 129–136. Gaithersburg, MD: National Institute for Standards and Technology.

Bézier, P. (1972). *Numerical Control—Mathematics and Applications*. London: Wiley. Translated from *Emploi des Machines á Commande Numérique* (1970), Masson et Cie: Paris, by Forrest, A. R. and Pankhurst, A. F.

Blankenagel, G. and Güting, R. H. (1994). External segment trees. *Algorithmica*, 12:498–532.

Booch, G. (1993). *Object-Oriented Analysis and Design with Applications*. Reading, MA: Addison-Wesley, 2nd edition.

Booch, G., Rumbaugh, J., and Jacobson, I. (1998). *The Unified Modeling Language User Guide*. Reading, MA: Addison-Wesley.

Bowyer, A. (1981). Computing Dirichlet tessellations. *Computer Journal*, 24:162–166.

Brodie, M., Mylopoulos, J., and Schmidt, J. (Eds) (1984). *On Conceptual Modeling*. Berlin: Springer.

Brookshear, J. G. (1989). *Theory of Computation: Formal Languages, Automata, and Complexity*. Redwood City, CA: Benjamin/Cummings.

Bryant, R. and Singerman, D. (1985). Foundations of the theory of maps on surfaces with boundaries. *Quarterly Journal of Mathematics*, 36(2):17–41.

Bryant, V. (1985). *Metric Spaces: Iteration and Application*. Cambridge: Cambridge University Press.

Burrough, P. (1981). Fractal dimension of landscapes and other environmental data. *Nature*, 294:240–242.

Burrough, P. and Frank, A. (1996). *Geographic Objects with Indeterminate Boundaries*. London: Taylor & Francis.

Burrough, P. A. and McDonnell, R. A. (1998). *Principles of Geographical Information Systems*. Oxford: Oxford University Press, 2nd edition.

Buttenfield, B. P. and McMaster, R. B. (Eds) (1991). *Map Generalization: Making Rules for Knowledge Representation*. London: Longman.

Camara, A. S. and Raper, J. (Eds) (1999). *Spatial Multimedia and Virtual Reality*. London: Taylor & Francis.

Cardelli, L. and Wegner, P. (1985). On understanding types, data abstraction, and polymorphism. *ACM Computing Surveys*, 17(4):480–521.

Chamberlin, D. and Boyce, R. (1974). SEQUEL: A structured English query language. In *Proc. ACM SIGMOD Conference*, pp. 249–264. New York: ACM Press.

Charniak, E. (1991). Bayesian networks without tears. *AI Magazine*, 91:50–63.

Chazelle, B. (1991). Triangulating a simple polygon in linear time. *Discrete Computational Geometry*, 6:485–524.

Chen, P. P.-S. (1976). The entity-relationship model—toward a unified view of data. *ACM Transactions on Database Systems*, 1(1):9–36.

Chess, D., Grosof, B., Harrison, C., Levine, D., Parris, C., and Tsudik, G. (1995). Itinerant agents for mobile computing. *IEEE Personal Communications*, 2(5):34–49.

*Chicago Tribune* (2001). Rental firm uses GPS in speeding fine. July 2nd, p9. Chicago, IL: Associated Press.

Chomicki, J. (1994). Temporal query languages: A survey. In *Proc. 1st International Conference on Temporal Logic (ICTL'94)*.

Chomicki, J. and Revesz, P. (1997). Constraint-based interoperability of spatiotemporal databases. In *Advances in Spatial Databases*, volume 1262 of *Lecture Notes in Computer Science*, pp. 142–161. Berlin: Springer.

Chorochronos Participants (1999). Chorochronos: A research network for spatiotemporal database systems. *SIGMOD Record*, 28(3):12–20.

Chrisman, N. (1975). Topological information systems for geographic representation. In *Proc. AUTOCARTO 2*, pp. 346–351. Falls Church, VA: ASPRS/ACSM.

Chrisman, N. (1978). Concepts of space as a guide to cartographic data structures. In Dutton, G. (Ed), *Proc. First International Advanced Study Symposium on Topological Data Structures for Geographic Information Systems*, pp. 1–19. Cambridge, MA: Harvard Laboratory for Computer Graphics and Spatial Analysis.

Chrisman, N. (1997). *Exploring Geographic Information Systems*. New York: John Wiley.

Chrisman, N. (1998). Rethinking levels of measurement in cartography. *Cartography and GIS*, 25(4):231–242.

Clarke, B. L. (1981). A calculus of individuals based on connection. *Notre Dame Journal of Formal Logic*, 22(3).

Clarke, B. L. (1985). Individuals and points. *Notre Dame Journal of Formal Logic*, 26(1).

Clarke, K. C. (2002). *Getting Started with GIS*. Upper Saddle River, NJ: Prentice-Hall, 4th edition.

Clarke, K. C. and Battersby, S. E. (2001). The coordinate digit density function and map information content analysis. In *Proc. ACSM Annual Meeting*, Las Vegas, NV.

Clarke, K. C., Parks, B. O., and Crane, M. P. (Eds) (2001). *Geographic Information Systems and Environmental Modeling*. Upper Saddle River, NJ: Prentice-Hall.

Clements, P. (1996). From subroutine to subsystems: Component-based software development. In Brown, A. W. (Ed), *Component Based Software Engineering*, pp. 3–6. Los Alamitos, CA: IEEE Computer Society Press.

Cliff, A. D. and Ord, J. K. (1981). *Spatial Processes, Models, and Applications*. London: Pion.

Coad, P. and Yourdon, E. (1991a). *Object-Oriented Analysis*. Upper Saddle River, NJ: Prentice-Hall/Yourdon Press.

Coad, P. and Yourdon, E. (1991b). *Object-Oriented Design*. Upper Saddle River, NJ: Prentice-Hall/Yourdon Press.

Codd, E. (1970). A relational model for large shared data banks. *Communications of the ACM*, 13(6):377–387.

Codd, E. (1979). Extending the relational database model to capture more meaning. *ACM Transactions on Database Systems*, 4(4):397–434.

Cohn, A. G., Bennett, B., Gooday, J., and Gotts, N. M. (1997). Qualitative spatial representation and reasoning with the region connection calculus. *GeoInformatica*, 1(3):275–316.

Cohn, A. G. and Hazarika, S. M. (2001). Qualitative spatial reasoning: An overview. *Fundamenta Informaticae*, 46(1–2).

Coleman, D. J. (1999). GIS in networked environments. In Longley, P., Goodchild, M. F., Maguire, D. J., and Rhind, D. W. (Eds), *Geographical Information Systems*, chapter 22. New York: John Wiley & Sons, 2nd edition.

Comer, D. (1979). The ubiquitous B-tree. *ACM Computing Surveys*, 11(2):121–137.

Connolly, T. and Begg, C. (1999). *Database Systems: A Practical Approach to Design, Implementation, and Management*. Reading, MA: Addison-Wesley.

Coppock, J. T. and Rhind, D. W. (1991). The history of GIS. In Maguire, D. J., Goodchild, M. F., and Rhind, D. W. (Eds), *Geographical Information Systems*, volume 1, pp. 21–43. Essex, England: Longman.

Couclelis, H. (1992). Beyond the raster-vector debate in GIS. In Frank, A. U., Campari, I., and Formentini, U. (Eds), *Theories of Spatio-Temporal Reasoning in Geographic Space*, volume 639 of *Lecture Notes in Computer Science*, pp. 65–77. Berlin: Springer.

Coxeter, H. S. M. (1961). *Introduction to Geometry*. New York: Wiley.

Cui, Z., Cohn, A. G., and Randell, D. A. (1993). Qualitative and topological relationships in spatial databases. In Abel, D. and Ooi, B. C. (Eds), *Advances in Spatial Databases, Proc. SSD'93*, volume 692 of *Lecture Notes in Computer Science*, pp. 296–315. Berlin: Springer.

Date, C. J. (2003). *An Introduction to Database Systems*. Reading, MA: Addison-Wesley, 8th edition.

David, B., Raynal, L., Schorter, G., and Mansart, V. (1993). GeO$_2$: Why objects in a geographical DBMS? In Abel, D. and Ooi, B. C. (Eds), *Advances in Spatial Databases*, volume 692 of *Lecture Notes in Computer Science*, pp. 264–276. Berlin: Springer.

Davies, N., Cheverst, K., Mitchell, K., and Efrat, A. (2001). Using and determining location in a context-sensitive tour guide. *IEEE Computer*, 34(8):35–41.

de Berg, M., van Kreveld, M., Overmars, M., and Schwarzkopf, O. (2000). *Computational Geometry: Algorithms and Applications*. Berlin: Springer, 2nd edition.

De Floriani, L. (1989). A pyramidal data structure for triangle-based surface description. *IEEE Computer Graphics and Applications*, 9(2):67–78.

De Floriani, L., Marzano, P., and Puppo, E. (1993). Spatial queries and data models. In Frank, A. U. and Campari, I. U. (Eds), *Spatial Information Theory*, volume 716 of *Lecture Notes in Computer Science*, pp. 113–138. Berlin: Springer.

De Floriani, L., Marzano, P., and Puppo, E. (1994). Hierarchical terrain models: Survey and formalization. In *Proc. SAC'94*, pp. 323–327, Phoenix, AR.

De Floriani, L. and Puppo, E. (1992). A hierarchical triangle-based model for terrain description. In Frank, A. U., Campari, I., and Formentini, U. (Eds), *Theories of Spatio-Temporal Reasoning in Geographic Space*, volume 639 of *Lecture Notes in Computer Science*, pp. 236–251. Berlin: Springer.

Di Biase, D., MacEachren, A. M., Krygier, J., and Reeves, C. (1992). Animation and the role of map design in scientific visualization. *Cartography and Geographic Information Systems*, 23(4):345–370.

Dijkstra, E. W. (1959). A note on two problems in connexion with graphs. *Numerische Mathematik*, (1):269–271.

Dirichlet, G. L. (1850). Über die Reduction der positeven quadratischen Formen mit drei unbestimmten ganzen Zahlen. *Journal für die Reine und Angewandte Mathematik*, 40:209–227.

Dix, A., Finlay, J., Abowd, G., and Beale, R. (1998). *Human-Computer Interaction*. London: Prentice-Hall, 2nd edition.

Dodge, M. and Kitchin, R. (2002). *Atlas of Cyberspace*. Reading, MA: Addison-Wesley.

Douglas, D. (1990). It makes me so CROSS. In Peuquet, D. J. and Marble, D. F. (Eds), *Introductory Readings in Geographic Information Systems*. London: Taylor & Francis. Reprinted from an unpublished manuscript, Harvard Laboratory for Computer Graphics and Spatial Analysis, 1974.

Dretske, F. (1981). *Knowledge and the Flow of Information*. Cambridge, MA: MIT Press.

Duckham, M., Drummond, J. E., and Forrest, D. (2000). Assessment of error in digital vector data using fractal geometry. *International Journal of Geographical Information Science*, 14(1):67–84.

Duckham, M., Goodchild, M. F., and Worboys, M. F. (Eds) (2003). *Foundations of Geographic Information Science*. London: Taylor & Francis.

Duckham, M., Mason, K., Stell, J. G., and Worboys, M. F. (2001). A formal approach to imperfection in geographic information. *Computer, Environment, and Urban Systems*, 25:89–103.

Duckham, M. and Worboys, M. F. (2001). Computational structure in three-valued nearness relations. In Montello, D. R. (Ed), *Spatial Information Theory: Foundations of Geographic Information Science*, volume 2205 of *Lecture Notes in Computer Science*, pp. 76–91. Berlin: Springer.

Dutton, G. H. (1984). Geodesic modeling of planetary relief. *Cartographica*, 21(2 & 3):188–207.

Dutton, G. H. (1989). Planetary modeling via hierarchical tessellation. In *Proc. AUTOCARTO 9*, pp. 462–471. Baltimore, MD: ASPRS/ACSM.

Dutton, G. H. (1990). Locational properties of quaternary triangular meshes. In Brassel, K. and Kishimoto, H. (Eds), *Proc. 4th International Symposium on Spatial Data Handling*, pp. 901–910, Zurich.

Dutton, G. H. (1999). *A Hierarchical Coordinate System for Geoprocessing and Cartography*, volume 79 of *Lecture Notes in Earth Sciences*. Berlin: Springer.

Dwyer, R. A. (1987). A fast divide-and-conquer algorithm for constructing Delaunay triangulations. *Algorithmica*, 2:137–151.

Dykes, J. A. (1997). Exploring spatial data representations with dynamic graphics. *Computers and Geosciences*, 23(4):345–370.

Egenhofer, M. and Kuhn, W. (1999). Interacting with geographic information. In Longley, P., Goodchild, M. F., Maguire, D. J., and Rhind,

D. W. (Eds), *Geographical Information Systems*, chapter 28, pp. 401–412. New York: John Wiley & Sons, 2nd edition.

Egenhofer, M. J. (1989). A formal definition of binary topological relationships. In Litwin, W. and Schek, H.-J. (Eds), *Proc. Third International Conference on Foundations of Data Organization and Algorithms (FODO)*, volume 367 of *Lecture Notes in Computer Science*, pp. 457–472. Berlin: Springer.

Egenhofer, M. J. (1991). Reasoning about binary topological relations. In Günther, O. and Schek, H.-J. (Eds), *Advances in Spatial Databases, Proc. SSD'91*, volume 525 of *Lecture Notes in Computer Science*, pp. 143–160. Berlin: Springer.

Egenhofer, M. J. and Frank, A. (1992). Object-oriented modeling for GIS. *URISA Journal*, 4:3–19.

Egenhofer, M. J., Frank, A. U., and Jackson, J. (1989). A topological data model for spatial databases. In Günther, O. and Smith, T. (Eds), *Design and Implementation of Large Spatial Databases, Proc. SSD'89*, volume 409 of *Lecture Notes in Computer Science*, pp. 189–211. Berlin: Springer.

Egenhofer, M. J. and Franzosa, R. D. (1991). Point-set topological spatial relations. *International Journal of Geographical Information Systems*, 5(2):161–176.

Egenhofer, M. J. and Franzosa, R. D. (1995). On the equivalence of topological relations. *International Journal of Geographical Information Systems*, 9(2):133–152.

Egenhofer, M. J. and Herring, J. R. (1990). A mathematical framework for the definition of topological relationships. In Brassel, K. and Kishimoto, H. (Eds), *Proc. Fourth International Symposium on Spatial Data Handling*, pp. 803–813, Zurich.

Egenhofer, M. J. and Herring, J. R. (1991). High-level spatial data structures for GIS. In Maguire, D. J., Goodchild, M. F., and Rhind, D. W. (Eds), *Geographical Information Systems*, volume 1, pp. 227–237. Essex, England: Longman.

Egenhofer, M. J. and Kuhn, W. (1998). Beyond desktop GIS. In *Proceedings GIS PlaNET*, Lisbon.

Elmasri, R. E. and Navathe, S. B. (2003). *Fundamentals of Database Systems*. Reading, MA: Addison-Wesley, 3rd edition.

Fabrikant, S. I. and Buttenfield, B. P. (2001). Formalizing spaces for information access. *Annals of the Association of American Geographers*, 91:263–280.

Faloutsos, C., Sellis, T., and Roussopoulos, N. (1987). Analysis of object-oriented spatial access methods. In *Proc. 16th ACM SIGMOD Conference*, pp. 426–439. New York: ACM Press.

Fekete, G. and Davis, L. S. (1984). Property spheres: A new representation for 3-D object recognition. In *Proc. Workshop on Computer Vision: Representation and Control*, pp. 176–186, Annapolis, MD.

Fernández, R. N. and Rusinkiewicz, M. (1993). A conceptual design of a soil database for a geographical information system. *International Journal of Geographical Information Systems*, 7(6):525–539.

Finkel, R. A. and Bentley, J. L. (1974). Quad trees: A data structure for retrieval on composite keys. *Acta Informatica*, 4(1):1–9.

Fisher, P. F. (1994). Hearing the reliability in classified remotely sensed images. *Cartography and Geographical Information Systems*, 21(1):31–36.

Fisher, P. F. (1999). Models of uncertainty in spatial data. In Longley, P., Goodchild, M. F., Maguire, D. J., and Rhind, D. W. (Eds), *Geographical Information Systems*, chapter 13. New York: John Wiley & Sons, 2nd edition.

Fisher, P. F. and Orf, T. (1991). Investigation of the meaning of near and close on a university campus. *Computers, Environment, and Urban Systems*, 15:23–35.

Flanagan, N., Jennings, C., and Flanagan, C. (1994). Automatic GIS data capture and conversion. In Worboys, M. F. (Ed), *Innovations in GIS*, pp. 25–38. London: Taylor & Francis.

Foley, J. D., van Dam, A., Feiner, S. K., and Hughes, J. F. (1995). *Computer Graphics: Principles and Practice*. Reading, MA: Addison-Wesley, 2nd edition.

Foresman, T. (Ed) (1998). *The History of Geographic Information Systems*. Upper Saddle River, NJ: Prentice-Hall.

Forman, G. H. and Zahorjan, J. (1994). The challenges of mobile computing. *IEEE Computer*, 27(4):38–47.

Fortune, S. (1987). A sweepline algorithm for voronoi diagrams. *Algorithmica*, 2:153–174.

Fotheringham, A. S., Brunsdon, C., and Charlton, M. (2002). *Geographically Weighted Regression: The Analysis of Spatially Varying Relationships*. New York: John Wiley.

Frank, A. and Barrera, R. (1989). The field-tree: A data structure for geographic information systems. In Günther, O. and Smith, T. (Eds), *Design and Implementation of Large Spatial Databases, Proc. SSD'89*, volume 409 of *Lecture Notes in Computer Science*, pp. 29–44. Berlin: Springer.

Freeman, J. (1975). The modeling of spatial relations. *Computer Graphics and Image Processing*, 4:156–171.

Freeston, M. (1989). A well-behaved file structure for the storage of spatial objects. In Günther, O. and Smith, T. (Eds), *Design and Implementation of Large Spatial Databases, Proc. SSD'89*, volume 409 of *Lecture Notes in Computer Science*, pp. 287–300. Berlin: Springer.

Freeston, M. (1993). Begriffsverzeichnis: A concept index. In Worboys, M. F. and Grundy, A. F. (Eds), *Advances in Databases, Proc. 11th British National Conference on Databases, BNCOD11*, volume 696 of *Lecture Notes in Computer Science*, pp. 1–22. Berlin: Springer.

Freundschuh, S. and Egenhofer, M. (1997). Human conceptions of spaces: Implications for GIS. *Transactions in GIS*, 2(4):361–375.

Fuchs, H., Abram, G. D., and Grant, E. D. (1983). Near real-time shaded display of rigid objects. *Computer Graphics*, 17(3):65–72.

Fuchs, H., Kedem, Z. M., and Naylor, B. F. (1980). On visible surface generation by *a priori* tree structures. *Computer Graphics*, 14(3):124–133.

Gabbay, D. and Hunter, A. (1991). Making inconsistencies respectable: Part 1—A logical framework for inconsistency in reasoning. In Jorrand, P. and Keleman, J. (Eds), *Foundations of Artificial Intelligence Research*, volume 535 of *Lecture Notes in Computer Science*, pp. 19–32. Berlin: Springer.

Galton, A. (2000). *Qualitative Spatial Change*. Oxford: Oxford University Press.

Galton, A. (2003). On the ontological status of geographical boundaries. In Duckham, M., Goodchild, M. F., and Worboys, M. F. (Eds), *Foundations of Geographic Information Science*, pp. 151–171. London: Taylor & Francis.

Gärdenfors, P. and Rott, H. (1995). Belief revision. In Gabbay, D. M., Hogger, C. J., and Robinson, J. A. (Eds), *Handbook of Logic in Artificial Intelligence and Logic Programming*, volume 4, pp. 35–132. Oxford: Oxford University Press.

Garey, M. R., Johnson, D. S., Preparata, F. P., and Tarjan, R. E. (1978). Triangulating a simple polygon. *Information Processing Letters*, 7:175–179.

Garlan, D. and Shaw, M. (1993). An introduction to software architecture. In Ambriola, V. and Tortora, G. (Eds), *Advances in Software Engineering and Knowledge Engineering*, volume 1. River Edge, NJ: World Scientific Publishing.

Gatrell, A. C. (1991). Concepts of space and geographical data. In Maguire, D. J., Goodchild, M. F., and Rhind, D. W. (Eds), *Geographical Information Systems*, volume 1, pp. 119–134. Essex, England: Longman.

Gaver, W. W. (1986). Auditory icons: Using sound in computer interfaces. *Human-Computer Interaction*, 2:167–177.

Giblin, P. J. (1977). *Graphs, Surfaces, and Homology: An Introduction to Algebraic Topology*. London: Chapman & Hall.

Gold, C. (1994). Three approaches to automated topology, and how computational geometry helps. In Waugh, T. and Healey, R. G. (Eds), *Advances in GIS Research, Proc. SDH'94*, pp. 145–158.

Gold, C. and Cormack, S. (1987). Spatially ordered networks and topographic reconstructions. *International Journal of Geographical Information Systems*, 1:137–148.

Goodchild, M. F. (1988). Lakes on fractal surfaces: A null hypothesis for lake rich landscapes. *Mathematical Geology*, 20(6):615–629.

Goodchild, M. F. (1989). Tiling large geographical databases. In Günther, O. and Smith, T. (Eds), *Design and Implementation of Large Spatial Databases, Proc. SSD'89*, volume 409 of *Lecture Notes in Computer Science*, pp. 137–146. Berlin: Springer.

Goodchild, M. F. (1991). Geographical data modeling. *Computers and Geosciences*, 18:400–408.

Goodchild, M. F. (1997). Towards a geography of geographic information in a digital world. *Computers, Environment and Urban Systems*, 21(6):377–391.

Goodchild, M. F. and Proctor, J. (1997). Scale in a digital geographic world. *Geographical and Environmental Modelling*, 1(1):5–23.

Goodchild, M. F. and Shiren, Y. (1992). A hierarchical data structure for global geographical information systems. *Graphical Models and Image Processing*, 54(1):31–44.

Greene, D. and Yao, F. (1986). Finite resolution computational geometry. In *Proc. 27th IEEE Symposium on the Foundations of Computer Science*, pp. 143–152.

Guibas, L. and Stolfi, J. (1985). Primitives for the manipulation of general subdivisions and the computation of Voronoi diagrams. *ACM Transactions on Graphics*, 4(2):74–123.

Günther, O. (1988). *Efficient Structures for Geometric Data Management*, volume 337 of *Lecture Notes in Computer Science*. Berlin: Springer.

Guptill, S. C. and Morrison, J. L. (Eds) (1995). *Elements of Spatial Data Quality*. Oxford: Pergamon Press.

Güting, R. H., Böhlen, M. H., Erwig, M., Jensen, C. S., Lorentzos, N. A., Schneider, M., and Vazirgiannis, M. (2000). A foundation for representing and querying moving objects. *ACM Transactions on Database Systems*, 25(1):1–42.

Güting, R. H. and Schneider, M. (1995). Realm-based spatial data types: The ROSE algebra. *VLDB Journal*, 4(2):243–286.

Guttman, A. (1984). R-trees: A dynamic index structure for spatial searching. In *Proc. 13th ACM SIGMOD Conference*, pp. 47–57. New York: ACM Press.

Hammer, M. M. and McLeod, D. J. (1981). Database description with SDM. *ACM Transactions on Database Systems*, 6(3):351–386.

Harary, F. (1969). *Graph Theory*. Reading, MA: Addison-Wesley.

Hein, J. L. (2003). *Discrete Mathematics*. Sudbury, MA: Jones & Bartlett, 2nd edition.

Henle, M. (1979). *A Combinatorial Introduction to Topology*. San Francisco, CA: Freeman.

Herring, J. (1991). The mathematical modeling of spatial and non-spatial information in geographic information systems. In Mark, D. M. and Frank, A. U. (Eds), *Cognitive and Linguistic Aspects of Geographic Space*, pp. 313–350. Dordrecht, the Netherlands: Kluwer Academic.

Higginbottom, J., Pianesi, F., and Varzi, A. (Eds) (2000). *Speaking of Events*. Oxford: Oxford University Press.

Hightower, J. and Boriello, G. (2001). Location systems for ubiquitous computing. *IEEE Computer*, 34(8):57–66.

Hoffmann, C. M. (1989). *Geometric and Solid Modeling: An Introduction*. San Mateo, CA: Morgan Kaufmann.

Hornsby, K. and Egenhofer, M. (2000). Identity-based change: A foundation for spatio-temporal knowledge representation. *International Journal of Geographical Information Science*, 14(3):207–224.

Hull, R. and King, R. (1987). Semantic data modeling: Survey, applications, and research issues. *ACM Computer Surveys*, 19:201–260.

Hunter, A. (1997). *Uncertainty in Information Systems: An Introduction to Techniques and Applications*. New York: McGraw-Hill.

Hunter, A. (1998). Paraconsistent logics. In Besnard, P. and Hunter, A. (Eds), *Handbook of Defeasible Reasoning and Uncertainty Management Systems*, volume 2, pp. 11–36. Dordrecht, the Netherlands: Kluwer.

Jankowski, P. and Nyerges, T. (2001). *Geographic Information Systems for Group Decision Making*. London: Taylor & Francis.

Jayawardena, D. P. W. and Worboys, M. F. (1995). The role of triangulation in spatial data handling. In Fisher, P. F. (Ed), *Innovations in GIS*, pp. 7–17. London: Taylor & Francis.

João, E. (1998). *Causes and Consequences of Map Generalisation*. London: Taylor & Francis.

Johnson, M. (1987). *The Body in the Mind*. Chicago: University of Chicago Press.

Jones, C. B. (1997). *Geographical Information Systems and Computer Cartography*. Reading, MA: Addison-Wesley.

Keefe, R. and Smith, P. (1996). *Vagueness: A Reader*. Cambridge, MA: MIT Press.

Kemp, K. K. (1992). Environmental modeling with GIS: A strategy for dealing with spatial continuity. In *Proc. GIS/LIS Annual Conference, ASPRS and ACSM*, pp. 397–406, Bethseda.

Knuth, D. (1973). *The Art of Computer Programming: Sorting and Searching*, volume 3. Reading, MA: Addison-Wesley.

Kollios, G. and Tsotras, V. J. (2002). Hashing methods for temporal data. *IEEE Transactions on Data and Knowledge Engineering*, 14(4):902–919.

Kriegel, H. P., Brinkhoff, T., and Schneider, R. (1993). Efficient spatial query processing in geographic database systems. *IEEE Data Engineering Bulletin*, 6(3):10–15.

Krygier, J. (1994). Sound and geographic visualization. In MacEachren, A. M. and Taylor, D. R. F. (Eds), *Visualization in Modern Cartography*, pp. 1–12. Oxford: Pergamon.

Kuhn, W. and Frank, A. U. (1991). A formalization of metaphors and image-schemas in user interfaces. In Mark, D. M. and Frank, A. U. (Eds), *Cognitive and Linguistic Aspects of Geographic Space*, pp. 419–434. Dordrecht, the Netherlands: Kluwer Academic.

Kuipers, B. (1978). Modeling spatial knowledge. *Cognitive Science*, 2:129–153.

Kuipers, B. (1994). *Qualitative Reasoning: Modeling and Simulation with Incomplete Knowledge*. Cambridge, MA: MIT Press.

Kuipers, B. (2000). The spatial semantic hierarchy. *Artificial Intelligence*, 119:191–233.

Kumar, A., Tsotras, V. J., and Faloutsos, C. (1998). Designing access methods for bitemporal databases. *IEEE Transactions on Data and Knowledge Engineering*, 10(1):1–20.

Kurose, J. F. and Ross, K. W. (2002). *Computer Networking: A Top-Down Approach Featuring the Internet*. Reading, MA: Addison-Wesley, 2nd edition.

Lakoff, G. and Johnson, M. (1999). *Philosophy in the Flesh: The Embodied Mind and Its Challenge to Western Thought*. New York: Basic Books.

Lam, N. S.-N. and De Cola, L. (Eds) (1993). *Fractals in Geography*. Upper Saddle River, NJ: Prentice-Hall.

Langran, G. (1992). *Time in Geographic Information Systems*. London: Taylor & Francis.

Lee, D.-T. and Schachter, B. J. (1980). Two algorithms for constructing the Delaunay triangulation. *International Journal of Computer and Information Sciences*, 9(3):219–242.

Lee, J. (1991). Analyses of visibility sites on topographic surfaces. *International Journal of Geographical Information Systems*, 1:413–429.

Lennes, N. J. (1911). Theorems on the simple finite polygon and polyhedron. *American Journal of Mathematics*, 33:37–62.

Lipschutz, S. (1997). *Schaum's Outline of Discrete Mathematics*. New York: McGraw-Hill, 2nd edition.

Longley, P., Goodchild, M. F., Maguire, D. J., and Rhind, D. W. (Eds) (1999). *Geographical Information Systems*. New York: John Wiley & Sons, 2nd edition.

Longley, P. A., Goodchild, M. F., Maguire, D. J., and Rhind, D. W. (2001). *Geographic Information Systems and Science*. New York: John Wiley & Sons.

Lowen, R. (1996). *Fuzzy Set Theory: Basic Concepts, Techniques, and Bibliography*. Dordrecht, the Netherlands: Kluwer Academic Publishers.

Luger, G. F. and Stubblefield, W. A. (1998). *Artificial Intelligence: Structures and Strategies for Complex Problem Solving*. Reading, MA: Addison-Wesley, 3rd edition.

MacEachren, A. (1994a). Time as a cartographic variable. In Hearnshaw, H. M. and Unwin, D. J. (Eds), *Visualization in Geographical Information Systems*, chapter 13, pp. 115–130. New York: John Wiley.

MacEachren, A. M. (1994b). Visualization in modern cartography: Setting the agenda. In MacEachren, A. M. and Taylor, D. R. F. (Eds), *Visualization in Modern Cartography*, pp. 1–12. Oxford: Pergamon.

MacEachren, A. M. (1995). *How Maps Work*. New York: Guilford Press.

MacEachren, A. M., Howard, D., von Wyss, M., Askov, D., and Taormino, T. (1993). Visualizing the health of Chesapeake Bay: An uncertain endeavor. In *Proc. GIS/LIS*, pp. 449–458. http://www.geovista.psu.edu/publications/RVIS/index.html, accessed 24 July 2003.

Mackinson, D. (1985). How to give it up: A survey of some formal aspects of the logic of theory change. *Synthese*, 62:347–363.

Madsen, K. H. (1994). A guide to metaphorical design. *Communications of the ACM*, 37(12):57–62.

Maguire, D. J., Goodchild, M. F., and Rhind, D. W. (Eds) (1991). *Geographical Information Systems*. Chicago: Longman, 1st edition. Some chapters available online from http://www.wiley.com/gis.

Mandelbrot, B. B. (1982). *The Fractal Geometry of Nature*. San Francisco, CA: Freeman.

Manolopoulos, Y., Theodoridis, Y., and Tsotras, V. J. (1999). *Advanced Database Indexing*. Dordrecht, the Netherlands: Kluwer Academic Publishers.

Mark, D. M. (1979). Phenomenon-based data structuring and digital terrain modeling. *Geo-Processing*, 1:27–36.

Mark, D. M. (1989). Cognitive image-schemata for geographic information: Relations to users' views and GIS interfaces. In *Proc. GIS/LIS*, volume 2, pp. 551–560. Falls Church, VA: ASPRS/ACSM.

Mark, D. M. and Frank, A. U. (1996). Experiential and formal models of geographic space. *Environment and Planning B*, 23:3–24.

Miller, C. L. and Laflamme, R. A. (1958). The digital terrain model—theory and application. *Photogrammetric Engineering*, 24(3):433–442.

Miller, H. J. and Shaw, S.-L. (2002). *Geographic Information Systems for Transportation: Principles and Applications*. Oxford: Oxford University Press.

Moellering, H. (Ed) (1991). *Spatial Database Transfer Standards: Current International Status*. London: Elsevier.

Moellering, H. (Ed) (1997). *Spatial Database Transfer Standards 2: Characteristics for Assessing Standards and Full Descriptions of the National and International Standards in the World*. London: Elsevier.

Monmonier, M. (1996). *How to Lie with Maps*. Chicago: University of Chicago Press, 2nd edition.

Monmonier, M. (2002). *Spying with Maps: Surveillance Technologies and the Future of Privacy*. Chicago: University of Chicago Press.

Montello, D. R. (1992). The geometry of environmental knowledge. In Frank, A. U., Campari, I., and Formentini, U. (Eds), *Theories and Methods of Spatio-Temporal Reasoning in Geographic Space*, volume 639 of *Lecture Notes in Computer Science*, pp. 136–152. Berlin: Springer.

Montello, D. R. (1993). Scale and multiple psychologies of space. In Frank, A. U. and Campari, I. (Eds), *Spatial Information Theory: A Theoretical Basis for GIS*, volume 716 of *Lecture Notes in Computer Science*, pp. 312–321. Berlin: Springer.

Montello, D. R. (2003). Regions in geography. In Duckham, M., Goodchild, M. F., and Worboys, M. F. (Eds), *Foundations of Geographic Information Science*, pp. 173–189. London: Taylor & Francis.

Morehouse, S. (1990). The role of semantics in geographic data modeling. In Brassel, K. and Kishimoto, H. (Eds), *Proc. Fourth International Symposium on Spatial Data Handling*, pp. 689–700, Zurich.

Motro, A. and Smets, P. (1996). *Uncertainty Management in Information Systems: From Needs to Solutions*. Boston: Kluwer.

Mourelatos, A. P. D. (1978). Events, processes, and states. *Linguistics and Philosophy*, 2:415–434.

Munakata, T. (1998). *Fundamentals of the New Artificial Intelligence: Beyond Traditional Paradigms*. Berlin: Springer.

Musavi, M. T., Shirvaikar, M. V., Ramanathan, E., and Nekorei, A. R. (1988). A vision-based method to automate map processing. *International Journal of Pattern Recognition*, 21(4):319–326.

Nelson, R. C. and Samet, H. (1986). A consistent hierarchical representation for vector data. *Computer Graphics*, 20(4):197–206.

Nievergelt, J., Hinterberger, H., and Sevcik, K. C. (1984). The grid file: An adaptable, symmetric, multikey file structure. *ACM Transactions on Database Systems*, 9(1):38–71.

Norman, D. A. (1988). *The Design of Everyday Things*. New York: Doubleday.

Ohya, T., Iri, M., and Murota, K. (1984a). A fast Voronoi-diagram algorithm with quaternary tree bucketing. *Information Processing Letters*, 18(4):227–231.

Ohya, T., Iri, M., and Murota, K. (1984b). Improvements of the incremental method for the Voronoi diagram with computational comparisons of various algorithms. *Journal of the Operations Research Society of Japan*, 27:306–336.

Okabe, A., Boots, B., and Sugihara, K. (1992). *Spatial tessellations: Concepts and Applications of Voronoi Diagrams*. Chichester, England: Wiley.

O'Leary, T. J. and O'Leary, L. I. (2003). *Computing Essentials*. New York: McGraw-Hill.

Onsrud, H. J., Johnson, J., and Lopez, X. (1994). Protecting personal privacy in using geographic information systems. *Photogrammetric Engineering and Remote Sensing*, 60(9):1083–1095.

Openshaw, S. (1991). Developing appropriate spatial analysis methods for GIS. In Maguire, D. J., Goodchild, M. F., and Rhind, D. W. (Eds), *Geographical Information Systems*, volume 1, pp. 389–402. Essex, England: Longman.

O'Rourke, J. (1998). *Computational Geometry in C*. Cambridge, UK: Cambridge University Press, 2nd edition.

O'Sullivan, D. and Unwin, D. (2002). *Geographic Information Analysis*. New York: John Wiley & Sons.

Oszu, M. T. and Valduriez, P. (1999). *Principles of Distributed Database Systems*. Upper Saddle River, NJ: Prentice-Hall, 2nd edition.

Otoo, E. J. and Zhu, H. (1993). Indexing on spherical surfaces using semi-quadcodes. In Abel, D. and Ooi, B. C. (Eds), *Advances in Spatial Databases, Proc. SSD'93*, volume 692 of *Lecture Notes in Computer Science*, pp. 510–529. Berlin: Springer.

Parsons, S. (1996). Current approaches to handling imperfect information in data and knowledge bases. *IEEE Transactions in Knowledge and Data Engineering*, 8(3):353–372.

Paton, N., Abdelmoty, A., and Williams, M. (1996). Programming spatial databases: A deductive object-oriented approach. In Parker, D. (Ed), *Innovations in GIS 3*, pp. 69–78. London: Taylor & Francis.

Pawlak, Z. (1982). Rough sets. *International Journal of Computer and Information Sciences*, 11:341–356.

Pawlak, Z. (1995). Vagueness and uncertainty: A rough set perspective. *Computational Intelligence*, 11:227–232.

Peitgen, H.-O., Jurgens, H., and Saupe, D. (1992). *Chaos and Fractals: New Frontiers of Science*. Berlin: Springer.

Peterson, M. P. (1994). Spatial visualization through cartographic animation: Theory and practice. In *Proc. GIS/LIS*, pp. 619–628. http://wwwsgi.ursus.maine.edu/gisweb/spatdb/gis-lis/gi94078.html, accessed 24 July 2003.

Peucker, T. K. (1978). Data structures for digital terrain models: Discussion and comparison. In *Harvard Papers on Geographic Information Systems*, number 5. Cambridge, MA: Harvard University Press.

Peucker, T. K., Fowler, R. J., Little, J. J., and Mark, D. M. (1978). The triangulated irregular network. In *Proc. ASP Digital Terrain Models Symposium*, pp. 516–540, Falls Church, VA.

Peuquet, D. J. (1984). A conceptual framework and comparison of spatial data models. *Cartographica*, 21(4):66–113. Reproduced in Peuquet and Marble (1990).

Peuquet, D. J. (2001). Making space for time: Issues in space-time data representation. *GeoInformatica*, 5:11–32.

Peuquet, D. J. and Marble, D. M. (Eds) (1990). *Introductory Readings in Geographic Information Systems*. London: Taylor & Francis.

Peuquet, D. J. and Zhan, C.-X. (1987). An algorithm to determine the directional relationship between arbitrarily-shaped polygons in the plane. *Pattern Recognition*, 20:65–74.

Phillips, A. D. M. (1993). *The Potteries: Continuity and Change in the Staffordshire Conurbation*. Thrupp, UK: Sutton Publishing.

Pianesi, F. and Varzi, A. C. (1996). Events, topology, and temporal relations. *The Monist*, 78:89–116.

Polidori, L., Chorowicz, J., and Guillande, R. (1991). Description of terrain as a fractal surface and application to digital elevation model quality assessment. *Photogrammetric Engineering and Remote Sensing*, 57(10):1329–1332.

Pradhan, S., Brignone, C., Cui, J.-H., McReynolds, A., and Smith, M. T. (2001). Websigns: Hyperlinking physical locations to the web. *IEEE Computer*, 34(8):42–48.

Preece, J., Rogers, Y., and Sharp, H. (2002). *Interaction Design: Beyond Human-Computer Interaction*. New York: John Wiley & Sons.

Preparata, F. P. and Shamos, M. I. (1985). *Computational Geometry: An Introduction*. Berlin: Springer.

Priest, G. (2001). *An Introduction to Non-Classical Logic*. Cambridge, UK: Cambridge University Press.

Pullar, D. and Egenhofer, M. J. (1988). Towards formal definitions of spatial relationships among spatial objects. In *Proc. Third International Symposium on Spatial Data Handling*, pp. 225–242. Columbus, OH: International Geographical Union.

Ramakrishnan, R. and Gehrke, J. (2000). *Database Management Systems*, volume 2nd. New York: McGraw-Hill.

Randell, D. A., Cui, Z., and Cohn, A. G. (1992). A spatial logic based on regions and connection. In Nebel, B., Rich, C., and Swartout, W. (Eds), *Principles of Knowledge Representation and Reasoning: Proc. Third International Conference KR'92*, pp. 165–176. San Mateo, CA: Morgan Kaufmann.

Rigaux, P., Scholl, M. O., and Voisard, A. (2001). *Spatial Databases: With Application to GIS*. San Mateo, CA: Morgan Kaufmann.

Ritter, G. X., Wilson, J., and Davidson, J. (1990). Image algebra: An overview. *Computer Vision, Graphics, and Image Processing*, 49:297–331.

Robinson, A. H., Morrison, J. L., Muehrcke, P. C., Kimerling, A. J., and Guptill, S. C. (1995). *Elements of Cartography*. New York: John Wiley & Sons, 6th edition.

Roddick, J. (1996). A survey of schema versioning issues for database systems. *Information and Software Technology*, 37(7):383–393.

Rothermel, K. and Schwehm, M. (1999). Mobile agents. In Kent, A. and Williams, J. G. (Eds), *Encyclopedia for Computer Science and Technology*, volume 40, pp. 155–176. New York: Marcel Dekker.

Rumbaugh, J., Blaha, M., Premerlani, W., Eddy, F., and Lorenson, W. (1990). *Object-Oriented Modeling and Design*. Upper Saddle River, NJ: Prentice-Hall.

Russell, S. J. and Norvig, P. (2002). *Artificial Intelligence: A Modern Approach*. Upper Saddle River, NJ: Prentice-Hall, 2nd edition.

Samet, H. (1990a). *Applications of Spatial Data Structures: Computer Graphics, Image Processing, and GIS*. Reading, MA: Addison-Wesley.

Samet, H. (1990b). *The Design and Analysis of Spatial Data Structures*. Reading, MA: Addison-Wesley.

Samet, H. (1995). Spatial data structures. In Kim, W. (Ed), *Modern Database Systems*, chapter 18, pp. 361–385. New York: ACM Press.

Samet, H. and Aref, W. G. (1995). Spatial data models and query processing. In Kim, W. (Ed), *Modern Database Systems*, chapter 17, pp. 338–360. New York: ACM Press.

Samet, H. and Webber, R. E. (1985). Storing a collection of polygons using quadtrees. *ACM Transactions on Graphics*, 4(3):182–222.

Shafer, G. (1976). *A Mathematical Theory of Evidence*. Princeton, NJ: Princeton University Press.

Shamos, M. I. (1975). Geometric complexity. In *Proc. Seventh Annual ACM Symposium on Theory of Computing*, pp. 224–233.

Shamos, M. I. and Hoey, D. (1975). Closest point problems. In *Proc. 16th Annual IEEE Symposium on Foundations of Computer Science*, pp. 151–162.

Shannon, C. and Weaver, W. (1949). *Mathematical Theory of Communication*. Urbana, IL: University of Illinois Press. Republished 1963.

Shekhar, S. and Chawla, S. (2002). *Spatial Databases: A Tour*. Upper Saddle River, NJ: Prentice-Hall.

Sheth, A. and Larson, J. L. (1990). Federated databases: Architectures and integration. *ACM Computing Surveys*, 22(3):183–236.

Shi, W., Goodchild, M. F., and Fisher, P. F. (Eds) (2002). *Spatial Data Quality*. London: Taylor & Francis.

Shneiderman, B. (1997). *Designing the User Interface*. Reading, MA: Addison-Wesley, 3rd edition.

Shoham, Y. (1988). *Reasoning about Change*. Cambridge, MA: MIT Press.

Smith, B. (1995). On drawing lines on a map. In Frank, A. U. and Kuhn, W. (Eds), *Spatial Information Theory: A Theoretical Basis for GIS*, volume 988 of *Lecture Notes in Computer Science*. Berlin: Springer.

Smith, B. (1998). The basic tools of formal ontology. In Guarino, N. (Ed), *Formal Ontology in Information Systems*, pp. 19–28. Amsterdam: IOS Press.

Smith, B. and Varzi, A. C. (2000). Fiat and bona fide boundaries. *Philosophy and Phenomenological Research*, 60(2):401–420.

Smith, J. M. and Smith, D. C. P. (1977). Database abstractions: Aggregation and generalization. *ACM Transactions on Database Systems*, 2(2):105–133.

Snodgrass, R. (Ed) (1995). *The TSQL2 Temporal Query Language*. Dordrecht, the Netherlands: Kluwer Academic Publishers.

Sondheim, M., Gardels, K., and Buehler, K. (1999). GIS interoperability. In Longley, P., Goodchild, M. F., Maguire, D. J., and Rhind, D. W. (Eds), *Geographical Information Systems*, chapter 24. New York: John Wiley & Sons, 2nd edition.

Sperber, D. and Wilson, D. (1995). *Relevance: Communication and Cognition*. Oxford: Blackwell, 2nd edition.

Spiegel, M. R., Schiller, J. J., and Srinivasan, R. A. (2000). *Schaum's Outline of Probability and Statistics*. New York: McGraw-Hill, 2nd edition.

Spyns, P., Meersman, R., and Jarrar, M. (2002). Data modelling versus ontology engineering. *SIGMOD Record*, 31(4):12–17.

Stallings, W. (1999). *Data and Computer Communications*. Upper Saddle River, NJ: Prentice-Hall, 6th edition.

Stevens, S. S. (1946). On the theory of scales and measurement. *Science*, 103:677–680.

Stonebraker, M., Sellis, T., and Hanson, E. (1986). An analysis of rule indexing implementations in data base systems. In *Proc. First International Conference on Expert Database Systems*, pp. 353–364, Charleston, SC.

Sutherland, W. A. (1975). *Introduction to Metric and Topological Spaces*. Oxford: Clarendon Press.

Szyperski, C. (2002). *Component Software*. Reading, MA: Addison-Wesley, 2nd edition.

Takayama, M. and Couclelis, H. (1997). Map dynamics integrating cellular automata and GIS through geo-algebra. *International Journal of Geographical Information Science*, 11(1):73–91.

Tannenbaum, A. S. (2002). *Computer Networks*. Upper Saddle River, NJ: Prentice-Hall, 4th edition.

Tansel, A. U. and Tin, E. (1997). The expressing power of temporal relational query languages. *IEEE Transactions of Data and Knowledge Engineering*, 9(1):120–134.

Tarjan, R. E. and Wyk, C. J. V. (1988). An $O(n \log \log n)$-time algorithm for triangulating a simple polygon. *SIAM Journal of Computing*, 17:143–178.

Teorey, T. J., Yang, D., and Fry, J. P. (1986). Logical design methodology for relational databases. *ACM Computing Surveys*, 18(2):197–222.

Theodoridis, Y. (2003). Ten benchmark database queries for location-based services. *The Computer Journal*, 46(5).

Thompson, R. B. and Thompson, B. F. (2002). *PC Hardware in a Nutshell*. Sebastopol, CA: O'Reilly, 2nd edition.

Timpf, S. (2003). Geographic activity models. In Duckham, M., Goodchild, M. F., and Worboys, M. F. (Eds), *Foundations of Geographic Information Science*, pp. 241–254. London: Taylor & Francis.

Tobler, W. (1970). A computer movie simulating urban growth in the Detroit region. *Economic Geography*, 46(2):234–240.

Tobler, W. (1993). Non-isotropic geographic modeling. Technical Report 93-1, National Center for Geographic Information and Analysis, Santa Barbara, CA. In Three Presentations on Geographical Analysis and Modeling.

Tomlin, C. D. (1983). A map algebra. In *Proc. Harvard Computer Graphics Conference*, Cambridge, MA.

Tomlin, C. D. (1990). *Geographic Information Systems and Cartographic Modeling*. Upper Saddle River, NJ: Prentice-Hall.

Tomlin, C. D. (1991). Cartographic modeling. In Maguire, D. J., Goodchild, M. F., and Rhind, D. W. (Eds), *Geographical Information Systems*, volume 1, pp. 361–374. Essex, England: Longman.

Tsai, V. J. D. (1993). Delaunay triangulations in TIN creation: An overview and a linear-time algorithm. *International Journal of Geographical Information Systems*, 7(6):501–524.

Unwin, D. J. (1981). *Introductory Spatial Analysis*. London: Methuen.

van Oosterom, P. J. M. (1993). *Reactive Data Structures for Geographic Information Systems*. Oxford: Oxford University Press.

van Oosterom, P. J. M. (1999). Spatial access methods. In Longley, P., Goodchild, M. F., Maguire, D. J., and Rhind, D. W. (Eds), *Geographical Information Systems*, chapter 27. New York: John Wiley & Sons, 2nd edition.

Vckovski, A. (1998). *Interoperable and Distributed Processing in GIS*. London: Taylor & Francis.

Voronoi, G. (1908). Nouvelles applications des paramètres continus à la théorie des formes quadratiques, deuxième memoire, recherches sur les parallelloèdres primitifs. *Journal für die Reine und Angewandte Mathematik*, 134:198–287.

Want, R., Hopper, A., Falcao, V., and Gibbons, J. (1992). The Active Badge location system. *ACM Transactions on Information Systems*, 10(1):91–102.

Want, R. and Pering, T. (2003). New horizons for mobile computing. In *Proc. First International Conference on Pervasive Computing and Communications (PerCom '03)*, pp. 3–8.

Ware, C. (2000). *Information Visualization: Perception for Design*. San Francisco: Morgan Kaufmann.

Warwick, K. (2000). Cyborg 1.0. *Wired*, 8(2).

Waugh, T. C. and Healey, R. (1987). The GEOVIEW design: A relational database approach to geographic data handling. *International Journal of Geographical Information Systems*, 1:101–118.

Wegner, P. (1996). Interoperability. *ACM Computing Surveys*, 28(1):285–287.

Weibel, R. and Heller, M. (1991). Digital terrain modeling. In Maguire, D. J., Goodchild, M. F., and Rhind, D. W. (Eds), *Geographical Information Systems*, volume 1, pp. 269–297. Essex, England: Longman.

Weibel, R. and Jones, C. B. (1998). Computational perspectives on map generalization. *Geoinformatica*, 2(4):301–314.

Weiler, K. (1985). Edge-based data structures for solid modeling in curved-surface environments. *Computer Graphics Applications*, 5(1):21–40.

Weiser, M. (1991). The computer for the twenty-first century. *Scientific American*, pp. 94–104.

Werbach, K. (2000). Location-based computing: Wherever you go, there you are. *Release 1.0*, 18(6):1–26.

Wiederhold, G. (1992). Mediators in the architecture of future information systems. *IEEE Computer*, 25(3):38–49.

Williams, B. K., Sawyer, S. C., and Hutchinson-Clifford, S. (2001). *Using Information Technology*. New York: McGraw-Hill, 4th edition.

Williamson, T. (1994). *Vagueness*. London: Routledge.

Wolfson, O., Xu, B., Chamberlain, S., and Jiang, L. (1998). Moving objects databases: Issues and solutions. In *Proc. 10th International Conference on Scientific and Statistical Database Management, SSDBM98*, pp. 111–122, Capri, Italy.

Wood, J. (2002). *Java Programming for Spatial Scientists*. London: Taylor & Francis.

Worboys, M. F. (1992). A generic model for planar geographic objects. *International Journal of Geographical Information Systems*, 6:353–372.

Worboys, M. F. (1994a). Object-oriented approaches to geo-referenced information. *International Journal of Geographical Information Systems*, 8(4):385–399.

Worboys, M. F. (1994b). A unified model of spatial and temporal information. *Computer Journal*, 37(1):26–34.

Worboys, M. F. (1998a). Computation with imprecise geospatial data. *Computers, Environment, and Urban Systems*, 22(2):85–106.

Worboys, M. F. (1998b). Imprecision in finite resolution spatial data. *Geoinformatica*, 2(3):257–280.

Worboys, M. F. (2001). Nearness relations in environmental space. *International Journal of Geographical Information Science*, 15(7):633–651.

Worboys, M. F. and Duckham, M. (2004). *GIS: A Computing Perspective.* Boca Raton, FL: CRC Press, 2nd edition.

Worboys, M. F., Hearnshaw, H. M., and Maguire, D. J. (1990). Object-oriented data modeling for spatial databases. *International Journal of Geographical Information Systems*, 4:369–383.

Xia, Z.-G. and Clarke, K. C. (1997). Approaches to scaling of spatial data in the geosciences. In Quattrochi, D. A. and Goodchild, M. F. (Eds), *Scaling in Remote Sensing and GIS*, pp. 309–360. Boca Raton, FL: CRC Lewis.

Zadeh, L. A. (1965). Fuzzy sets. *Information and Control*, 8:338–353.

Zadeh, L. A. (1988). Fuzzy logic. *IEEE Computer*, 21:83–93.

Zhang, T. Y. and Suen, C. Y. (1984). A fast parallel algorithm for thinning digital patterns. *Communications of the ACM*, 27(3):236–239.

Zubin, D. (1989). Natural language understanding and reference frames. In Mark, D., Frank, A., Egenhofer, M., Freundschuh, S., McGranaghan, M., and White, R. M. (Eds), *Languages of Spatial Relations: Initiative 2 Specialist Meeting Report*, pp. 13–16. Santa Barbara, CA: National Center for Geographic Information and Analysis.

# Index

Bold folios refer to marginal terms. Italic folios refer to terms within inset boxes. The preface contains further information on index formatting.